Chemical Process
Equipment Design

Chemical Process Equipment Design

Richard Turton

Joseph A. Shaeiwitz

PRENTICE
HALL

Boston · Columbus · Indianapolis · New York · San Francisco · Amsterdam · Cape Town
Dubai · London · Madrid · Milan · Munich · Paris · Montreal · Toronto · Delhi · Mexico City
São Paulo · Sydney · Hong Kong · Seoul · Singapore · Taipei · Tokyo

For information about buying this title in bulk quantities, or for special sales opportunities (which may include electronic versions; custom cover designs; and content particular to your business, training goals, marketing focus, or branding interests), please contact our corporate sales department at corpsales@pearsoned.com or (800) 382-3419.

For government sales inquiries, please contact governmentsales@pearsoned.com.

For questions about sales outside the U.S., please contact intlcs@pearson.com.

Visit us on the Web: informit.com/ph

Library of Congress Control Number: 2016957093

ISBN-13: 978-0-13-380447-8
ISBN-10: 0-13-380447-X

1   17

We would like to dedicate this book to our long-suffering wives, Becky and Terry, without whose love and support this work would never have been completed.

Contents

Contents

Preface

Why write a book entitled *Chemical Process Equipment Design* when there are many good books on the market that cover this topic, often in much greater detail than in this text? The purpose of this book is not to repeat what others have written but rather to present material to undergraduate students in such a way that they can easily use this information to make preliminary designs of important equipment found in the chemical process industry. Rather than give a compendium of design equations for a vast array of process equipment, this text aims to provide a set of algorithms and methods that the undergraduate chemical engineer can use to design the majority of process equipment found in a typical chemical plant that processes gases and/or liquids. Little time is devoted to deriving formulae or detailing the theoretical background for the equations, which can be found in other textbooks. The approach used here is to state what equation applies to a given situation and then to show how to use the equations to design and evaluate the performance of equipment.

To this end, approximately 80 worked examples are provided in this text covering most of the fundamental concepts found in the undergraduate curriculum for fluid mechanics, heat transfer, separation processes, heterogeneous reactions/reactors, and basic vapor-liquid separators. The text is meant as a supplemental resource for these courses and provides information often omitted in the standard undergraduate texts in the area of equipment design. Moreover, the authors have paid particular attention to addressing how the performance of existing equipment can be estimated when operating conditions differ from the design case, which is how most equipment operates throughout its lifetime. Coverage of the performance or rating problem is often omitted in the chemical engineering curriculum, but its understanding is essential for both the neophyte and experienced engineer. For example, how is the change in exit temperature from a heat exchanger estimated when the process flowrate is increased by 20%, or what is the maximum liquid flow possible through a pumped circulation loop, or by how much can the flowrate to an exothermic reactor be changed without having a temperature runaway occur in the reactor? These are important practical questions that are often not addressed in standard chemical engineering courses.

The book is organized into five chapters—fluid mechanics, heat transfer, separations, reactors, and phase separators and steam ejector systems. Chapter 1, "Process Fluid Mechanics," could be the basis of a practical undergraduate fluid mechanics class, and Chapter 2, "Process Heat Transfer," could be the basis of a practical undergraduate class in heat transfer. Chapter 3, "Separation Equipment," and Chapter 4, "Reactors," are meant to be supplements to the popular textbooks used for these classes. Finally, Chapter 5, "Other Equipment," provides information pertinent to the senior design course. Alternatively, this text could be used as a recommended book in curricula that have a separate course for the design of equipment. It could also be used as a supplement to the senior design course, especially when that course is where the majority of process equipment design is covered. If the design of process equipment is distributed throughout the curriculum, then this text would be a good resource for several courses typically taught in the junior year (fluids, heat transfer, reactors, and separations).

Register your copy of *Chemical Process Equipment Design* at informit.com for convenient access to downloads, updates, and corrections as they become available. To start the registration process, go to informit.com/register and log in or create an account. Enter the product ISBN (9780133804478) and click Submit. Once the process is complete, you will find any available bonus content under "Registered Products."

Acknowledgments

We are grateful to the teams at Prentice Hall for helping to coordinate and enabling us to publish this text.

We are also grateful for the detailed technical reviews of the book provided by Drs. John Hwalek and Susan Montgomery. We are certain that their reviews were very time consuming and greatly appreciate their willingness to spend the time to do this. Thanks also goes to Dr. Troy Vogel for his input.

About the Authors

Richard Turton is the WVU Bolton Professor and has taught in the Chemical and Biomedical Engineering Program at West Virginia University for the past 30 years. He received his B.Sc. from the University of Nottingham, UK, and his M.S. and Ph.D. from Oregon State University with all degrees in chemical engineering. He also worked in industry for Pullman-Kellogg, Fluor E&C, and Shell Oil. His research interests include particle technology, design education, and simulation and control of advanced power plants. He is a registered professional engineer in the state of West Virginia.

Joseph A. Shaeiwitz is an emeritus professor of chemical engineering at West Virginia University and is currently a visiting professor in chemical engineering at Auburn University, where he teaches design and other unit operations classes. He received his B.S. from the University of Delaware and his M.S. and Ph.D. from Carnegie Mellon University. He is a fellow of both ASEE and AIChE. His professional interests are in engineering education, accreditation, assessment, and design.

CHAPTER

1

Process
Fluid Mechanics

WHAT YOU WILL LEARN

- The basic relationships for fluid flow—mass, energy, and force balances
- The primary types of fluid flow equipment—pipes, pumps, compressors, valves
- How to design a system for incompressible and compressible frictional flow of fluid in pipes
- How to design a system for frictional flow of fluid with submerged objects
- Methods for flow measurement
- How to analyze existing fluid flow equipment
- How to use the concept of net positive suction head (NPSH) to ensure safe and appropriate pump operation
- The analysis of pump and system curves
- How to use compressor curves and when to use compressor staging

---

1.0   INTRODUCTION

The purpose of this chapter is to introduce the concepts needed to design piping systems, including pumps, compressors, turbines, valves, and other components, and to evaluate the performance of these systems once designed and implemented. The scope is limited to steady-state situations. Derivations are minimized, and the emphasis is on providing a set of useful, working equations that can be used to design and evaluate the performance of piping systems.

1.1   BASIC RELATIONSHIPS IN FLUID MECHANICS

In expressing the basic relationships for fluid flow, a general control volume is used, as illustrated in Figure 1.1. This control volume can be the fluid inside the pipes and equipment connected by the pipes, with the possibility of multiple inputs and multiple outputs. For the simple case of one input and one output, the subscript 1 refers to the upstream side and the subscript 2 refers to the downstream side.

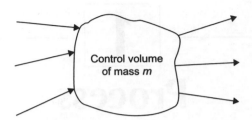

Figure 1.1   General control volume

1.1.1   Mass Balance

At steady state, mass is conserved, so the total mass flowrate ($\dot{m}$, mass/time) in must equal the total mass flowrate out. For a device with m inputs and n outputs, the appropriate relationship is given by Equation (1.1). For a single input and single output, Equation (1.2) is used.

$$\sum_{i=1}^{m} \dot{m}_{i,\,in} = \sum_{i=1}^{n} \dot{m}_{i,\,out} \tag{1.1}$$

$$\dot{m}_1 = \dot{m}_2 \tag{1.2}$$

In describing fluid flow, it is necessary to write the mass flowrate in terms of both volumetric flowrate ($\dot{v}$, volume/time) and velocity (u, length/time). These relationships are

$$\dot{m} = \rho\dot{v} = \rho A u \tag{1.3}$$

where ρ is the density (mass/volume) and A is the cross-sectional area for flow (length^2). From Equation (1.3), it can be seen that, for an incompressible fluid (constant density) at steady state, the volumetric flowrate is constant, and the velocity is constant for a constant cross-sectional area for flow. However, for a compressible fluid flowing with constant cross-sectional area, if the density changes, the volumetric flowrate and velocity both change in the opposite direction, since the mass flowrate is constant. Accordingly, if the density decreases, the volumetric flowrate and velocity both increase. For problems involving compressible flow, it is useful to define the superficial mass velocity, G (mass/area/time), as

$$G = \frac{\dot{m}}{A} = \rho u \tag{1.4}$$

The advantage of defining a superficial mass velocity is that it is constant for steady-state flow in a constant cross-sectional area, unlike density and velocity, and it shows that the product of density and velocity remains constant.

For a system with multiple inputs and/or multiple outputs at steady state, as is illustrated in Figure 1.2, the total mass flowrate into the system must equal the total mass flowrate out,

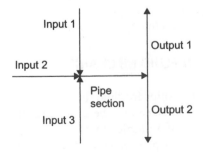

Figure 1.2   System with multiple inputs and outputs

Equation (1.1). However, the output mass flowrate in each section differs depending on the size, length, and elevation of the piping involved. These problems are discussed later.

Example 1.1

Two streams of crude oil (specific gravity of 0.887) mix as shown in Figure E1.1. The volumetric flowrate of Stream 1 is 0.006 m³/s, and its pipe diameter is 0.078 m. The volumetric flowrate of Stream 2 is 0.009 m³/s, and its pipe diameter is 0.10 m.

   a. Determine the volumetric and mass flowrates of Stream 3.

   b. Determine the velocities in Streams 1 and 2.

   c. If the velocity is not to exceed 1 m/s in Stream 3, determine the minimum possible pipe diameter.

   d. Determine the superficial mass velocity Stream 3 using the pipe diameter calculated in Part (c).

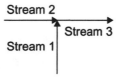

Figure E1.1   Physical situation in Example 1.1

Solution

   a. Since the density is constant, the volumetric flowrate of Stream 3 is the sum of the volumetric flowrates of Streams 1 and 2, 0.015 m³/s. To obtain the mass flowrate, $\dot{m} = \rho \dot{v}_3$, so $\dot{m}_3 = (887\ \text{kg/m}^3)\,(0.015\ \text{m}^3/\text{s}) = 13.3\ \text{kg/s}$. Alternatively, the mass flowrate of Streams 1 and 2 could be calculated and added to get the same result.

   b. From Equation (1.3), at constant density $u = \dot{v}/A$. Therefore,

$$u_1 = \frac{\dot{v}_1}{A_1} = \frac{4\dot{v}_1}{\pi D_1^2} = \frac{4(0.006\ \text{m}^3/\text{s})}{\pi(0.078\ \text{m})^2} = 1.26\ \text{m/s} \tag{E1.1a}$$

$$u_2 = \frac{\dot{v}_2}{A_2} = \frac{4\dot{v}_2}{\pi D_2^2} = \frac{4(0.009\ \text{m}^3/\text{s})}{\pi(0.1\ \text{m})^2} = 1.15\ \text{m/s} \tag{E1.1b}$$

   c. The diameter at which $u_3 = 1$ m/s can be calculated from Equation (1.3) at constant density.

$$\dot{v}_3 = u_3 A_3 \Rightarrow 0.015\ \text{m}^3/\text{s} = (1\ \text{m/s})\left(\frac{\pi D^2}{4}\right) \qquad \therefore D = 0.138\ \text{m} \tag{E1.1c}$$

If the diameter were smaller, the cross-sectional area would be smaller, and from Equation (1.3), the velocity would be larger. Hence, the result in Equation (E1.1c) is the minimum possible diameter. As shown later, actual pipes are available only in discrete sizes, so it is necessary to use the next higher pipe diameter.

   d. From Equation (1.4), using the rounded values,

$$G_1 = \frac{\dot{m}_3}{A_3} = \frac{4\dot{m}_3}{\pi D_1^2} = \frac{4(13.3\ \text{kg/s})}{\pi(0.138\ \text{m})^2} = 889.2\ \text{kg/m}^2/\text{s} \tag{E1.1d}$$

1.1.2   Mechanical Energy Balance

The mechanical energy balance represents the conversion between different forms of energy in piping systems. With the exception of temperature changes for a gas undergoing compression or expansion with no phase change, temperature is assumed to be constant. The mechanical energy balance is

$$\int_1^2 \frac{dP}{\rho} + \frac{1}{2}\Delta\left(\frac{<u^3>}{<u>}\right) + g\Delta z + e_f - W_s = 0 \tag{1.5}$$

In Equation (1.5) and throughout this chapter, the difference, Δ, represents the value at Point 2 minus the value at Point 1, that is, out − in. The units in Equation (1.5) are energy/mass or length^2/time^2. In SI units, since 1 J = 1 kg m^2/s^2, it is clear that 1 J/kg = 1 m^2/s^2. In American Engineering units, since 1 lb$_f$ = 32.2 ft lb$_m$/sec^2, this conversion factor (often called g_c) must be used to reconcile the units. The notation < > represents the appropriate average quantity.

The first term in Equation (1.5) is the enthalpy of the system. On the basis of the constant temperature assumption, only pressure is involved. For incompressible fluids, such as liquids, density is constant, and the term reduces to

$$\int_1^2 \frac{dP}{\rho} = \frac{\Delta P}{\rho} \tag{1.6}$$

For compressible fluids, the integral must be evaluated using an equation of state.

The second term in Equation (1.5) is the kinetic energy term. For turbulent flow, a reasonable assumption is that

$$\frac{<u^3>}{<u>} \approx <u>^2 \tag{1.7}$$

For laminar flow

$$\frac{<u^3>}{<u>} \approx 2<u>^2 \tag{1.8}$$

For simplicity, $<u^2>$ is hereafter represented as $<u>^2$, which is shortened to u^2.

The third term in Equation (1.5) is the potential energy term. Based on the general control volume, Δz is positive if Point 2 is at a higher elevation than Point 1.

The fourth term in Equation (1.5) is often called the energy "loss" due to friction. Of course, energy is not lost—it is just expended to overcome friction, and it manifests as a change in temperature. The procedures for calculating frictional losses are discussed later.

The last term in Equation (1.5) represents the shaft work, that is, the work done on the system (fluid) by a pump or compressor or the work done by the system on a turbine. These devices are not 100% efficient. For example, more work must be applied to the pump than is transferred to the fluid, and less work is generated by the turbine than is expended by the fluid. In this book, work is defined as positive if done on the system (pump, compressor) and negative if done by the fluid (turbine). This convention is consistent with the flow of energy in or out of the system; however, many textbooks use the reverse sign convention. Equipment such as pumps, compressors, and turbines are described in terms of their power, where power is the rate of doing work. Therefore, a device power (W_s, energy/time) is defined as the product of the mass flowrate (mass/time) and the shaft work (energy/mass):

$$\dot{W}_s = \dot{m}W_s \tag{1.9}$$

When efficiencies are included, the last term in Equation (1.5) becomes

$$\eta_p W_s = \frac{\eta_p \dot{W}_s}{\dot{m}} \quad \text{pump/compressor} \tag{1.10}$$

$$\frac{W_s}{\eta_t} = \frac{\dot{W}_s}{\eta_t \dot{m}} \quad \text{turbine} \tag{1.11}$$

Example 1.2

Water in an open (source or supply) tank is pumped to a second (destination) tank at a rate of 5 lb/sec with the water level in the destination tank 25 ft above the water level in the source tank, and it is assumed that the water level does not change with the flow of water. The destination tank is under a constant 30 psig pressure. The pump efficiency is 75%. Neglect friction.

a. Determine the required horsepower of the pump.

b. Determine the pressure increase provided by the pump assuming the suction and discharge lines have the same diameter.

Solution

a. Turbulent flow in the pipes is assumed. The mechanical energy balance is

$$\frac{\Delta P}{\rho} + \frac{1}{2}\Delta u^2 + g\Delta z + e_f - \frac{\eta_p \dot{W}_s}{\dot{m}} = 0 \tag{E1.2a}$$

Figure E1.2 is an illustration of the system.

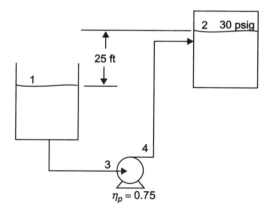

Figure E1.2 Physical situation for Example 1.2

The control volume is the water in the tanks, the pipes, and the pump, and the locations of Points 1 and 2 are illustrated. The integral in the first term of Equation (1.2) is simplified to the first term in Equation (E1.2a), since the density of water is a constant. In general, the fluid velocity in tanks is assumed to be zero because tank diameters are much larger than pipe diameters, so the kinetic energy term for the liquid surface in the tank is essentially zero. Any fluid in contact with the atmosphere is at atmospheric pressure, so $P_1 = 1$ atm = 0 psig. The friction term is assumed to be zero in this problem, as stated. So, Equation (E1.2a) reduces to

$$\frac{P_2 - P_1}{\rho} + g(z_2 - z_1) - \frac{\eta_p \dot{W}_s}{\dot{m}} = 0 \qquad\qquad \text{(E1.2b)}$$

and

$$\frac{(30-0)\ \text{lb}_f/\text{in}^2 (12\ \text{in}/\text{ft})^2}{62.4\ \text{lb}/\text{ft}^3} + \frac{32.2\ \text{ft}/\text{sec}^2}{32.2\ \text{ft}\,\text{lb}/\text{lb}_f/\text{sec}^2}(25-0)\ \text{ft} - \frac{0.75\dot{W}_s}{5\ \text{lb}/\text{sec}} = 0 \qquad\qquad \text{(E1.2c)}$$

so, $\dot{W}_s = 628.2\ \text{ft}\ \text{lb}_f/\text{sec}$.

Converting to horsepower yields

$$\dot{W}_s = \frac{628.2\ \text{ft}\ \text{lb}_f/\text{sec}}{550\ \text{ft}\ \text{lb}_f/\text{hp}/\text{sec}} = 1.14\ \text{hp} \qquad\qquad \text{(E1.2d)}$$

b. To determine the pressure rise in the pump, the control volume is now taken as the fluid in the pump. So, the mechanical energy balance is written between points 3 and 4. The mechanical energy balance reduces to

$$\frac{\Delta P}{\rho} - \frac{\eta_p \dot{W}_s}{\dot{m}} = 0 \qquad\qquad \text{(E1.2e)}$$

The kinetic energy term is zero because the suction and discharge pipes have the same diameter. Frictional losses are assumed to be zero in this example. The potential energy term is also assumed to be zero across the pump; however, since the discharge line of a pump may be higher than the suction line, in a more detailed analysis, that potential energy difference might be included. Solving

$$\frac{P(\text{lb}_f/\text{in}^2)(12\ \text{in}/\text{ft})^2}{62.4\ \text{lb}/\text{ft}^3} - \frac{0.75(628.2\ \text{ft}\ \text{lb}_f/\text{sec})}{5\ \text{lb}/\text{sec}} = 0 \qquad\qquad \text{(E1.2f)}$$

so, $\Delta P = 40.8\ \text{lb}_f/\text{in}^2$.

Example 1.3

A nozzle is a device that converts pressure into kinetic energy by forcing a fluid through a small-diameter opening. Turbines work in this way because the fluid (usually a gas) with a high kinetic energy impinges on turbine blades, causing spinning, and allowing the energy to be converted to electric power.

Consider a nozzle that forces 2 gal/min of water at 50 psia in a tube of 1-in inside diameter through a 0.1-in nozzle from which it discharges to atmosphere. Calculate the discharge velocity.

Solution

The system is illustrated in Figure E1.3. It is assumed that the velocity at a small distance from the end of the nozzle is identical to the velocity in the nozzle, but the contact with the atmosphere makes the pressure atmospheric.

Figure E1.3   Illustration of nozzle for Example 1.3

For the case when frictional losses may be neglected, the mechanical energy balance reduces to

$$\frac{P_2 - P_1}{\rho} + \frac{u_2^2 - u_1^2}{2} = 0 \qquad (E1.3a)$$

which yields

$$\frac{(14.7 - 50)\,lb_f/in^2(12\,in/ft)^2}{62.4\,lb/ft^3} + \frac{u_2^2 - \left[\dfrac{4(2\,gal/min)(ft^3/7.48\,gal)(min/60\,sec)}{\pi(1/12\,ft)^2}\right]^2}{2(32.2\,ft\,lb/lb_f/sec^2)} = 0 \qquad (E1.3b)$$

so $u_2 = 72.4$ ft/sec. For a real system, there would be some frictional losses and the actual discharge velocity would be lower than calculated here.

This problem was solved under the assumption of turbulent flow. The criterion for turbulent flow is introduced later; however, for this system, the Reynolds number is about 2×10^5, which is well into the turbulent flow region.

1.1.3   Force Balance

The force balance is essentially a statement of Newton's law. A common form for flow in pipes is

$$\Delta(\dot{m}\underline{u}) = \sum \underline{F} \qquad (1.12)$$

where F are the forces on the system. The underlined parameters indicate vectors, since there are three spatial components of a force balance. For steady-state flow and the typical forces involved in fluid flow, Equation (1.12) reduces to

$$\dot{m}\Delta(\underline{u}) = \underline{F}_p + \underline{F}_d + \underline{F}_g + \underline{R} \qquad (1.13)$$

where $\underline{F}_p$ is the pressure force on the system, $\underline{F}_d$ is the drag force on the system, $\underline{F}_g$ is the gravitational force on the system, and $\underline{R}$ is the restoring force on the system, that is, the force necessary to keep the system stationary. The term on the left side of Equation (1.12) is acceleration, confirming that Equation (1.12) is a statement of Newton's law. The most common application of Equation (1.13) is to determine the restoring forces on an elbow. These problems are not discussed here.

1.2   FLUID FLOW EQUIPMENT

The basic characteristics of fluid flow equipment are introduced in this section. The performance of pumps and compressors is dictated by their characteristic curves and, for pumps, the net positive suction head curve. The performance of these pieces of equipment is discussed in Section 1.5.

1.2.1   Pipes

Pipes and their associated fittings that are used to transport fluid through a chemical plant are usually made of metal. For noncorrosive fluids under conditions that are not of special concern, carbon steel is typical. For more extreme conditions, such as higher pressure, higher temperature, or corrosive fluids, stainless steel or other alloy steels may be needed. It may even be necessary, for very-high-temperature service such as for the flow of molten metals, to use refractory-lined pipes.

Pipes are sized using a nominal diameter and a schedule number. The higher the schedule number, the thicker the pipe walls, making pipes with a higher schedule number more suitable for higher-pressure operations. The nominal diameter is a number such as 2 in; however, there is no dimension of the pipe that is actually 2 in until the diameter reaches 14 in. For pipes with a diameter of 14 in or larger, the nominal diameter is the outside diameter. Pipes typically have integer nominal diameters; however, for smaller diameters, they can be in increments of 0.25 in. At larger diameters, the nominal diameters may only be even integer values. Table 1.1 shows the dimensions of schedule pipe.

Table 1.1   Dimensions of Standard Steel Pipe

Nominal Size (in)	Outside Diameter		Schedule Number	Wall Thickness		Inside Diameter		Inside Cross-Sectional Area	
	in	mm		in	mm	in	mm	10^2ft^2	10^4m^2
1/8	0.405	10.29	40	0.068	1.73	0.269	6.83	0.040	0.3664
			80	0.095	2.41	0.215	5.46	0.025	0.2341
1/4	0.540	13.72	40	0.088	2.24	0.364	9.25	0.072	0.6720
			80	0.119	3.02	0.302	7.67	0.050	0.4620
3/8	0.675	17.15	40	0.091	2.31	0.493	12.52	0.133	1.231
			80	0.126	3.20	0.423	10.74	0.098	0.9059
1/2	0.840	21.34	40	0.109	2.77	0.622	15.80	0.211	1.961
			80	0.147	3.73	0.546	13.87	0.163	1.511
3/4	1.050	26.67	40	0.113	2.87	0.824	20.93	0.371	3.441
			80	0.154	3.91	0.742	18.85	0.300	2.791
1	1.315	33.40	40	0.133	3.38	1.049	26.64	0.600	5.574
			80	0.179	4.45	0.957	24.31	0.499	4.641
1 1/4	1.660	42.16	40	0.140	3.56	1.380	35.05	1.040	9.648
			80	0.191	4.85	1.278	32.46	0.891	8.275
1 1/2	1.900	48.26	40	0.145	3.68	1.610	40.89	1.414	13.13
			80	0.200	5.08	1.500	38.10	1.225	11.40
2	2.375	60.33	40	0.154	3.91	2.067	52.50	2.330	21.65
			80	0.218	5.54	1.939	49.25	2.050	19.05
2 1/2	2.875	73.03	40	0.203	5.16	2.469	62.71	3.322	30.89
			80	0.276	7.01	2.323	59.00	2.942	27.30

(continued)

Table 1.1   Dimensions of Standard Steel Pipe   (*Continued*)

Nominal Size (in)	Outside Diameter		Schedule Number	Wall Thickness		Inside Diameter		Inside Cross-Sectional Area	
	in	mm		in	mm	in	mm	$10^2 ft^2$	$10^4 m^2$
3	3.500	88.90	40	0.216	5.59	3.068	77.92	5.130	47.69
			80	0.300	7.62	2.900	73.66	4.587	42.61
3 1/2	4.000	101.6	40	0.226	5.74	3.548	90.12	6.870	63.79
			80	0.318	8.08	3.364	85.45	6.170	57.35
4	4.500	114.3	40	0.237	6.02	4.026	102.3	8.840	82.19
			80	0.337	8.56	3.826	97.18	7.986	74.17
5	5.563	141.3	40	0.258	6.55	5.047	128.2	13.90	129.1
			80	0.375	9.53	4.813	122.3	12.63	117.5
6	6.625	168.3	40	0.280	7.11	6.065	154.1	20.06	186.5
			80	0.432	10.97	5.761	146.3	18.10	168.1
8	8.625	219.1	40	0.322	8.18	7.981	202.7	34.74	322.7
			80	0.500	12.70	7.625	193.7	31.71	294.7
10	10.75	273.1	40	0.365	9.27	10.02	254.5	54.75	508.6
			80	0.594	15.09	9.562	242.8	49.87	463.3
12	12.75	304.8	40	0.406	10.31	11.94	303.3	77.73	722.1
			80	0.688	17.48	11.37	288.8	70.56	655.5
14	14	355.6	40	0.438	11.13	13.12	333.2	93.97	873.0
			80	0.750	19.05	12.50	317.5	85.22	791.7
16	16	406.4	40	0.500	12.70	15.00	381.0	122.7	1140
			80	0.844	21.44	14.31	363.5	111.7	1038
18	18	457.2	40	0.562	14.27	16.88	428.8	155.3	1443
			80	0.938	23.83	16.12	409.4	141.8	1317
20	20	508.0	40	0.597	15.16	18.81	477.8	193.0	1793
			80	1.031	26.19	17.94	455.7	175.5	1630
24	24	635.0	40	0.688	17.48	22.62	574.5	279.2	2594
			80	1.219	30.96	21.56	547.6	253.6	2356

Source: Adapted from Geankoplis (2003) and Perry and Green (1984).

Tubing is commonly used in heat exchangers. The dimensions and use of tubing are discussed in Chapter 2.

Pipes are typically connected by screw threads, flanges, or welds. Welds and flanges are more suitable for larger diameters and higher-pressure operation. Proper welds are stronger and do not leak, whereas screwed or flanged connections can leak, especially at higher pressures. Changes in direction are usually accomplished by elbows or tees, and those changes in direction are usually 90°.

1.2.2   Valves

Valves are found in piping systems. Valves are about the only way to regulate anything in a chemical process. Valves serve several functions. They are used to regulate flowrate, reduce pressure by adding resistance, or to isolate (turn flow on/off) equipment.

Two common types of valves are gate valves and globe valves. Figure 1.3 shows illustrations of several common types of valves.

Gate valves are used for on/off control of fluid flow. The flow path through a gate valve is roughly straight, so when the valve is fully open, the pressure drop is very small. However, gate valves are not suitable for flowrate regulation because the flowrate does not change much until the "gate" is almost closed. There are also ball valves, in which a quarter turn opens a flow channel, and they can also be used for on/off regulation.

Globe valves are more suitable than gate valves for flowrate and pressure regulation. Because the flow path is not straight, globe valves have a higher pressure drop even when wide open. Globe valves are well suited for flowrate regulation because the flowrate is responsive to valve position. In a control system, the valve stem is raised or lowered pneumatically (by instrument air) or via an electric motor in response to a measured parameter, such as a flowrate. Pneumatic systems can be designed for the valve to fail open or closed, the choice depending on the service. Failure is defined as loss of instrument air pressure. For example, for a valve controlling the flowrate of a fluid removing heat from a reactor with a highly exothermic reaction, the valve would be designed to fail open so that the reactor cooling is not lost.

Check valves, such as the swing check valve, are used to ensure unidirectional flow. In Figure 1.3(c), if the flow is left to right, the swing is opened and flow proceeds. If the flow is right to left, the swing closes, and there is no flow in that direction. Such valves are often placed on the discharge side of pumps to ensure that there is no flow reversal through the pump.

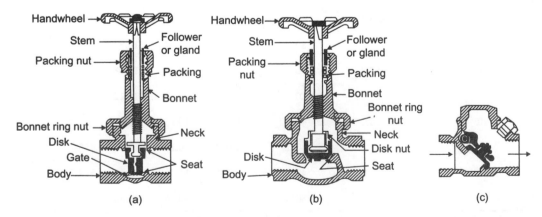

Figure 1.3   Common types of valves: (a) gate, (b) globe, (c) swing check (Reproduced by permission from Couper et al. [2012])

1.2.3   Pumps

Pumps are used to transport liquids, and pumps can be damaged by the presence of vapor, a phenomenon discussed in Section 1.5.2. The two major classifications for pumps are *positive displacement* and *centrifugal*. For a more detailed summary of all types of pumps, see Couper et al. (2012) or Green and Perry (2008).

Positive-displacement pumps are often called *constant-volume pumps* because a fixed amount of liquid is taken into a chamber at a low pressure and pushed out of the chamber at a high pressure. The chamber has a fixed volume, hence the name. An example of a positive-displacement pump is a reciprocating pump, illustrated in Figure 1.4(a). Specifically, this is an example of a piston pump in which the piston moves in one direction to pull liquid into the chamber and then moves in the opposite direction to discharge liquid out of the chamber at a higher pressure. There are other variations of positive-displacement pumps, such as rotary pumps in which the chamber moves between the inlet and discharge points. In general, positive-displacement pumps can increase pressure more than centrifugal pumps and run at higher pressures overall. These characteristics define their applicability. Efficiencies tend to be between 50% and 80%. Positive-displacement pumps are preferred for higher pressures, higher viscosities, and anticipated viscosity variations.

In centrifugal pumps, which are a common workhorse in the chemical industry, the pressure is increased by the centrifugal action of an impeller. An impeller is a rotating shaft with blades, and it might be tempting to call it a propeller because an impeller resembles a propeller. (While there might be a resemblance, the term *propeller* is reserved for rotating shafts with blades that move an

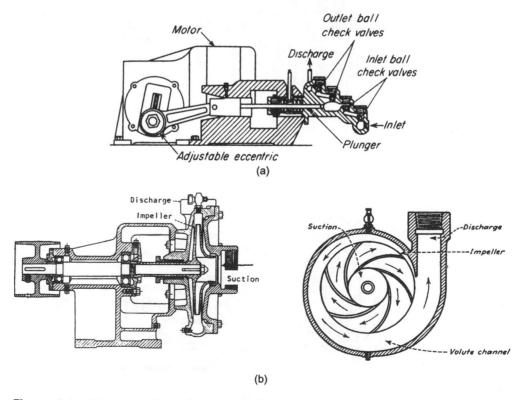

Figure 1.4   (a) Inner workings of positive-displacement pump, (b) inner workings of centrifugal pump (a reproduced by permission from McCabe, Smith, and Harriott [1993]; b reproduced by permission from Couper et al. [2012])

object, such as a boat or airplane.) The blades of an impeller have small openings, known as *vanes*, that increase the kinetic energy of the liquid. The liquid is then discharged through a *volute* in which the kinetic energy is converted into pressure. Figure 1.4(b) shows a centrifugal pump. Centrifugal pumps often come with impellers of different diameters, which enable pumps to be used for different services (different pressure increases). Of course, shutdown is required to change the impeller. Although standard centrifugal pump impellers only spin at a constant rate, variable-speed centrifugal pumps also are available.

Centrifugal pumps can handle a wide range of capacities and pressures, and depending on the exact type of pump, the efficiencies can range from 20% to 90%.

1.2.4   Compressors

Devices that increase the pressure of gases fall into three categories: fans, blowers, and compressors. Figure 1.5 illustrates some of this equipment. For a more detailed summary of all types of pumps, see Couper et al. (2012) or Green and Perry (2008).

Fans provide very low-pressure increases (<1 psi [7 kPa]) for low volumes and are typically used to move air. Blowers are essentially mini-compressors, providing a maximum pressure of about 30 psi (200 kPa). Blowers can be either positive displacement or centrifugal, and while their

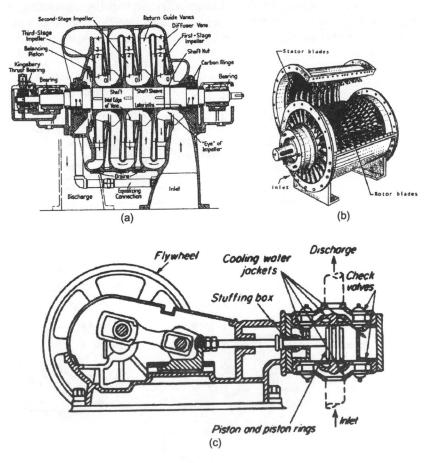

Figure 1.5 Inner working of compressors: (a) centrifugal, (b) axial, (c) positive displacement (a and b reproduced by permission from Couper et al. [2012]; c reproduced by permission from McCabe, Smith, and Harriott [1993])

general construction is similar to pumps, there are many internal differences. Compressors, which can also be either positive displacement or centrifugal, can provide outlet pressures of 1500 psi (10 MPa) and sometimes even 10 times that much.

In a centrifugal compressor, the impeller may spin at tens of thousands of revolutions per minute. If liquid droplets or solid particles are present in the gas, they hit the impeller blades at such high relative velocity that the impeller blades will erode rapidly and may cause bearings to become damaged, leading to mechanical failure. The compressor casing also may crack. Therefore, it is important to ensure that the gas in a centrifugal compressor does not contain solids and liquids. A filter can be used to keep particles out of a compressor, and a packed-bed adsorbent can also be used, for example, to remove water vapor from inlet air. Knockout drums are often provided between compressor stages with intercooling to allow the disengagement of any condensed drops of liquid and are covered in more detail in Chapter 5, Section 5.2. The seals on compressors are temperature sensitive, so a maximum temperature in one stage of a compressor is generally not exceeded, which is another reason for staged, intercooled compressor systems. It should also be noted that compressors are often large and expensive pieces of equipment that often have a large number of auxiliary systems associated with them. The coverage given in this text is very simplified but allows the estimate of the power required.

Positive-displacement compressors typically handle lower flowrates but can produce higher pressures compared to centrifugal compressors. Efficiencies for both types of compressor tend to be high, above 75%.

1.3   FRICTIONAL PIPE FLOW

1.3.1   Calculating Frictional Losses

The fourth term in Equation (1.5) must be evaluated to include friction in the mechanical energy balance. There are different expressions for this term depending on the type of flow and the system involved. In general, the friction term is

$$e_f = \frac{2fLu^2}{D} = \frac{32fL\dot{v}^2}{\pi^2 D^5} \tag{1.14}$$

where L is the pipe length, D is the pipe diameter, and f is the Fanning friction factor. (The Fanning friction factor is typically used by chemical engineers. There is also the Moody friction factor, which is four times the Fanning friction factor. Care must be used when obtaining friction factor values from different sources. It is even more confusing, since the plot of friction factor versus Reynolds number is called a *Moody plot* for both friction factors.) The friction factor is a function of the Reynolds number (Re = $Dv\rho/\mu$, where μ is the fluid viscosity), and its form depends on the flow regime (laminar or turbulent), and for turbulent flow, f is also a function of the pipe roughness factor (e, a length that represents small asperities on the pipe wall; values are given at the top of Figure 1.6), which is a tabulated value. Historically, the friction factor was measured and the data were plotted in graphical form. Figure 1.6 is such a plot. A key observation from Figure 1.6 is that, with the exception of smooth pipes, the friction factor asymptotically approaches a constant value above a Reynolds number of approximately 10^5. This is called fully developed turbulent flow, and the friction factor becomes constant and can be used to simplify certain calculations, examples of which are presented later. Typical values for the pipe roughness for some common materials are shown at the top of Figure 1.6.

The friction factor for laminar flow is a theoretical result derivable from the Hagen-Poiseuille equation (Bird, Stewart, and Lightfoot, 2006) and is valid for Re < 2100.

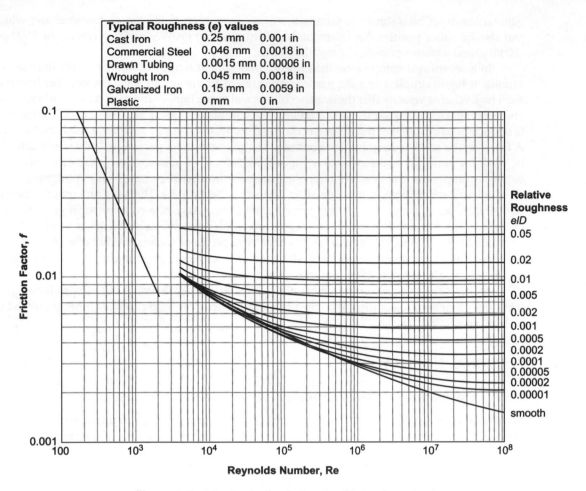

Figure 1.6   Moody plot for the Fanning friction factor in pipes

$$f = \frac{16}{\mathrm{Re}} = \frac{16\mu}{Du\rho} \tag{1.15}$$

For turbulent flow, the data have been fit to equations. One such equation is the Pavlov equation (Pavlov, Romankov, and Noskov, 1981 [cited in Levenspiel, 2014]):

$$\frac{1}{f^{0.5}} = -4\log_{10}\left[\frac{e}{3.7D} + \left(\frac{6.81}{\mathrm{Re}}\right)^{0.9}\right] \tag{1.16}$$

The Pavlov equation provides results within a few percent of the measured data. There are more accurate equations; however, they are not explicit in the friction factor. Any of these curve fits provides significantly more accuracy than reading a graph.

For flow in pipes containing valves, elbows, and other pipe fittings, there are two common methods for including the additional frictional losses created by this equipment. One is the *equivalent length* method, whereby additional pipe length is added to the value of L in Equation (1.14). The other method is the *velocity head* method, in which a value (K_i) is assigned to each valve, fitting, and so on, and an additional frictional loss term is added to the frictional loss term in Equation (1.14). These terms are of the form

$$\sum_i \frac{K_i u_i^2}{2} \tag{1.17}$$

where the index i indicates a sum over all valves, elbows, and similar components in the system. If there are different pipe diameters within the system, the velocity in Equation (1.17) is specific to each section of pipe, and a term for each section of pipe must be included. It should be noted that the equivalent K_i value for straight pipe (K_{pipe}) is given by

$$K_{pipe} = \frac{4\,fL}{D} \tag{1.18}$$

Tables 1.2 and 1.3 show equivalent lengths and K_i values for some common items found in pipe networks, for turbulent flow and for laminar flow, respectively. The values are different for laminar and turbulent flow. Darby (2001) presents analytical expressions for the K values that can be used for more exact calculations.

Table 1.2   Frictional Losses for Turbulent Flow

Type of Fitting or Valve	Frictional Loss, Number of Velocity Heads, K_f	Frictional Loss, Equivalent Length of Straight Pipe, in Pipe Diameters, L_{eq}/D
45° Elbow	0.35	17
90° Elbow	0.75	35
Tee	1	50
Return bend	1.5	75
Coupling	0.04	2
Union	0.04	2
Gate valve, wide open	0.17	9
Gate valve, half open	4.5	225
Globe valve, wide open	6.0	300
Globe valve, half open	9.5	475
Angle valve, wide open	2.0	100
Check valve, ball	70.0	3500
Check valve, swing	2.0	100
Contraction	$0.55(1 - A_2/A_1)$	$27.5(1 - A_2/A_1)$
Contraction $A_2 \ll A_1$	0.55	27.5
Expansion	$(1 - A_1/A_2)^2$	$50(1 - A_1/A_2)^2$
Expansion $A_1 \ll A_2$	1	50

Source: From Geankoplis (2003) citing Perry and Green (1984).

Table 1.3   Frictional Loss for Laminar Flow

	Frictional Loss, Number of Velocity Heads, K_f					
Reynolds Number	50	100	200	400	1000	Turbulent
90° Elbow	17	7	2.5	1.2	0.85	0.75
Tee	9	4.8	3.0	2.0	1.4	1.0
Globe Valve	28	22	17	14	10	6.0
Check Valve, Swing	55	17	9	5.8	3.2	2.0

Source: From Geankoplis (2003) citing Kittredge and Rowley (1957).

Another common situation involves frictional loss in a packed bed, that is, a vessel packed with solids. One application is if the solids are catalysts, making the packed bed a reactor. The frictional loss term for packed beds is obtained from the Ergun equation, which yields a friction term for a packed bed as

$$e_f = \frac{Lu_s^2(1-\varepsilon)}{\varepsilon^3 D_p}\left[\frac{150(1-\varepsilon)\mu}{D_p u_s \rho}+1.75\right] \tag{1.19}$$

where u_s is the superficial velocity (based on pipe diameter, not particle diameter), D_p is the particle diameter (assumed spherical here; corrections are available for nonspherical shape), and ε is the packing void fraction, which is the volume fraction in the packed bed not occupied by solids. When Equation (1.19) is used in the mechanical energy balance, one unknown parameter, such as velocity, pressure drop, or particle diameter, can be obtained.

For incompressible flow in packed beds, the Ergun equation, Equation (1.19), is used for the friction term in the mechanical energy balance.

For the expansion and contraction losses, A_i is the cross-sectional area of the pipe, subscript 1 is the upstream area, and subscript 2 is the downstream area.

1.3.2   Incompressible Flow

1.3.2.1   Single-Pipe Systems
Incompressible flow problems fall into three categories:

- Any parameter unknown in the mechanical energy balance other than velocity (flowrate) or diameter
- Unknown velocity (flowrate)
- Unknown diameter

For turbulent flow problems with any unknown other than velocity (or flowrate) or diameter, in the mechanical energy balance, Equation (1.5), there is a second unknown: the friction factor. The friction factor can be calculated from Equation (1.15). The solution method can use a sequential calculation, solving Equation (1.5) for the unknown once the friction factor is calculated. If there are valves, elbows, and so on, the length term in Equation (1.15) can be adjusted appropriately or Equation (1.17) can be used. Alternatively, Equations (1.14) and (1.16) can be solved simultaneously to yield all the unknowns. Example 1.5 shows both of these calculation methods. For laminar flow problems, Equation (1.15) can be combined with Equation (1.14) in the mechanical energy balance to solve any problem analytically.

For turbulent flow, if the velocity is unknown, Equations (1.5) and (1.15) must be solved simultaneously for the velocity or flowrate and the friction factor. When solving for a velocity directly, if the pump work term must be included, it is necessary to express the mass flowrate in terms of velocity. If solving for the volumetric flowrate, the second equality in Equation (1.13) must be used, and if a kinetic energy term is required in the mechanical energy balance, the velocities must be expressed in terms of volumetric flowrate. In the friction factor equation, the Reynolds number also needs to be expressed in terms of the volumetric flowrate as follows:

$$\text{Re} = \frac{Du\rho}{\mu} = \frac{D\rho}{\mu}\frac{\dot{v}}{A} = \frac{D\rho}{\mu}\frac{4\dot{v}}{\pi D^2} = \frac{4\dot{v}\rho}{\pi D\mu} \tag{1.20}$$

For laminar flow, an analytical solution is possible simply by using Equation (1.14) for the friction factor in the mechanical energy balance.

For turbulent flow, if the diameter is unknown, Equations (1.5) and (1.13) (second equality involving flowrate and diameter to the fifth power) must be solved simultaneously, using Equation (1.20) for the Reynolds number. For laminar flow, an analytical solution may once again be possible by using Equation (1.12) for the friction factor in the mechanical energy balance. If kinetic energy terms are involved, an unknown diameter will appear when expressing velocity in terms of flowrate. If minor losses are involved, the equivalent length will include a diameter term, and the K-value method will include a diameter in the conversion between flowrate and velocity.

Examples 1.4 and 1.5 illustrate the methods for solving these types of problems.

Example 1.4

Consider a physical situation similar to that in Example 1.2. The flowrate between tanks is 10 lb/sec. The source-tank level is 10 ft off of the ground, and the discharge-tank level is 50 ft off of the ground. For this example, both tanks are open to the atmosphere. The suction-side pipe is 2-in, schedule-40, commercial steel, and the discharge-side pipe is 1.5-in, schedule-40, commercial steel. The length of the suction line is 25 ft, and the length of the discharge line is 60 ft. The pump efficiency is 75%. Losses due to fittings, expansions, and contractions may be assumed negligible for this problem.

a. Determine the required horsepower of the pump.

b. Determine the pressures before and after the pump.

Solution

a. The physical situation is depicted in Figure E1.4.

For the control volume of the fluid in both tanks, the pipes, and the pump, the mechanical energy balance reduces to

$$g\Delta z + e_{f,suct} + e_{f,disch} - \frac{\eta_p \dot{W}_s}{\dot{m}} = 0 \tag{E1.4a}$$

The pressure term is zero, because both tanks are open to the atmosphere ($P_1 = P_2 = 1$ atm). The kinetic energy term is zero, because the velocities of the fluid at the surfaces of both tanks are assumed to be zero. There are two friction terms, one for the suction side of the pump and one for the discharge side of the pump, because the friction factors are different due to the different pipe diameters.

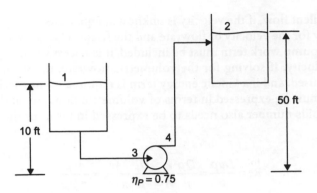

Figure E1.4

To calculate the friction terms, the Reynolds numbers must be calculated first for each section to determine whether the flow is laminar or turbulent. Since a temperature is not provided, the density is assumed to be 62.4 lb/ft³, and the viscosity is assumed to be 1 cP = 6.72 × 10⁻⁴ lb/ft/sec. Using Table 1.1 for the schedule pipe diameter and cross-sectional area, the Reynolds number for the suction side is

$$\text{Re} = \frac{Du\rho}{\mu} = \frac{(2.067/12\ \text{ft})\left(\dfrac{10\ \text{lb/sec}}{(0.0233\ \text{ft}^2)(62.4\ \text{lb/ft}^3)}\right)(62.4\ \text{lb/ft}^3)}{6.72\times10^{-4}\ \text{lb/ft/sec}} = 110{,}000 \qquad \text{(E1.4b)}$$

Similarly, the Reynolds number for the discharge side is 141,200. Therefore, the flow is turbulent in both sections of pipe. The friction factor is now calculated for each section of pipe. For the suction side, with commercial-steel pipe (e = 0.0018 in from the top of Figure 1.6),

$$\frac{1}{f^{0.5}} = -4\log_{10}\left[\frac{0.0018\ \text{in}}{3.7(2.067\ \text{in})} + \left(\frac{6.81}{110{,}010}\right)^{0.9}\right] \qquad \text{(E1.4c)}$$

so f_{suct} = 0.0054. Similarly, f_{disch} = 0.0055. Now, the mechanical energy balance on the entire system is used to solve for the pump power:

$$\frac{(32.2\ \text{ft/sec}^2)(40\ \text{ft})}{32.2\ \text{ft lb/lb}_f/\text{sec}^2} + \frac{2(0.0054)(25\ \text{ft})\left(\dfrac{10\ \text{lb/sec}}{(0.0233\ \text{ft}^2)(62.4\ \text{lb/ft}^3)}\right)^2}{\left(\dfrac{2.067}{12}\ \text{ft}\right)(32.2\ \text{ft lb/lb}_f/\text{sec}^2)}$$

$$+ \frac{2(0.0055)(60\ \text{ft})\left(\dfrac{10\ \text{lb/sec}}{(0.01414\ \text{ft}^2)(62.4\ \text{lb/ft}^3)}\right)^2}{\left(\dfrac{1.61}{12}\ \text{ft}\right)(32.2\ \text{ft lb/lb}_f/\text{sec}^2)} - \frac{(0.75)\dot{W}_s(550\ \text{ft lb}_f/\text{sec/hp})}{(10\ \text{lb/sec})} = 0$$

$$\text{(E1.4d)}$$

Solving Equation (E1.4d) gives $\dot{W}_s$ = 1.5 hp. If the contribution of each term is enumerated, 0.97 hp is to overcome the potential energy, 0.48 hp is to overcome the discharge line friction, with 0.056 hp to overcome the suction line friction. Generally, potential energy differences and pressure differences are more significant than frictional losses.

b. To obtain the pressure on the suction side of the pump, the mechanical energy balance is written on the control volume of the fluid in the tank and pipes before the pump.

$$\frac{P_3-P_1}{\rho}+\frac{u_3^2}{2}+g\Delta z+e_f=0 \tag{E1.4e}$$

$$\frac{[P_3-(14.7)](144)\mathrm{lb}_f/\mathrm{ft}^2}{62.4\,\mathrm{lb/ft}^3}+\frac{\left(\dfrac{10\,\mathrm{lb/sec}}{\left(0.0233\,\mathrm{ft}^2\right)\left(62.4\,\mathrm{lb/ft}^3\right)}\right)^2}{2(32.2\,\mathrm{ft\,lb/lb}_f/\mathrm{sec}^2)}+\frac{(32.2\,\mathrm{ft/sec}^2)(-10\,\mathrm{ft})}{32.2\,\mathrm{ft\,lb/lb}_f/\mathrm{sec}^2}$$
$$+\frac{2(0.0054)(15\,\mathrm{ft})\left(\dfrac{10\,\mathrm{lb/sec}}{\left(0.0233\,\mathrm{ft}^2\right)\left(62.4\,\mathrm{lb/ft}^3\right)}\right)^2}{\left(\dfrac{2.067}{12}\,\mathrm{ft}\right)(32.2\,\mathrm{ft\,lb/lb}_f/\mathrm{sec}^2)}=0 \tag{E1.4f}$$

So, P_3 = 17.7 psi. It is observed that the height change in the potential energy term is negative, since the point at the pump entrance is below the liquid level in the tank, noting that the z-coordinate system is positive in the upward direction.

There are two ways to obtain the discharge-side pressure. One is to solve the mechanical energy balance on the control volume between points 4 and 2. The other method is to write the mechanical energy balance on the fluid in the pump (pressure, kinetic energy, and work terms) to obtain the pressure rise in the pump. Both methods give the same result of P_4 = 28.9 psi.

The discharge line of a pump is at a slightly higher elevation than the suction line, as illustrated. This height difference is small and is neglected in this analysis.

Example 1.5

Determine the required horsepower of the pump in Example 1.4 if the presence of one 90° elbow and one wide-open gate valve in the suction line and one wide-open gate valve, one half-open globe valve, and two 90° elbows is included.

Solution

The solution to this problem starts with Equation (E1.4d). Friction terms must be added for each item in each section of pipe. Using the equivalent length method for the suction line, L_{eq} = 25 ft + (2.067/12 ft) (35 + 9 + 27.5) = 37.3 ft, where the equivalent length terms for the elbow, gate valve, and contraction upon leaving the source tank, respectively, are obtained from Table 1.2. For the discharge line, L_{eq} = 60 ft + (1.61/12 ft) [2(35) + 9 + 475 + 50] = 141.04 ft, where the equivalent length terms are for the two elbows, gate valve, globe valve, and expansion upon entering the destination tank, respectively. In terms of friction, these items add significantly to the frictional losses, especially the half-open globe valve in the discharge line. The result is that $\dot{W}_s$ = 2.71 hp.

It is also possible to use the velocity heads method. For the suction side, once again referring to Table 1.2, ΣK_i = 0.75 + 0.17 + 0.55 = 1.47, so a term of $1.47u_1^2/2/32.2$ is added to the mechanical energy balance. For the discharge side, K = 2(0.75) + 0.17 + 9.5 + 1 = 12.17, so a term of $12.17u_2^2/2/32.2$ is added to the mechanical energy balance. The result is 2.12 hp, which illustrates that the two methods do not give exactly the same results. The difference is because both methods are empirical and are subject to uncertainties. Either method is within the typical tolerance of a design specification. To provide flexibility and since pumps are typically available at fixed values, at least a 3 hp pump would probably be used here, and valves would be used to adjust the flowrate to the desired value.

Example 1.6

A fuel oil ($\mu = 70 \times 10^{-3}$ kg/m/s, SG = 0.9) is pumped through 2.5-in, schedule-40 pipe for 500 m at 3 kg/s. The discharge point is 5 m above the inlet, and the source and destination are both at 101 kPa. If the pump is 80% efficient, what power is required?

Solution

The situation is shown in Figure E1.6.

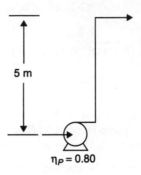

$$\eta_P = 0.80$$

Figure E1.6

The control volume is the fluid in the pipe between the source and destination. The mechanical energy balance contains only the potential energy, friction, and work terms, since there is only one pipe (velocity constant) and since the pressures are identical at the source and destination. The mechanical energy balance is

$$g\Delta z + e_f - \frac{\eta_p \dot{W}_s}{\dot{m}} = 0 \qquad (E1.6a)$$

As in Example 1.4, the Reynolds number should be calculated first:

$$\text{Re} = \frac{Du\rho}{\mu} = \frac{(0.06271\ \text{m})\left(\dfrac{3\ \text{kg/s}}{\left(0.003089\ \text{m}^2\right)\left(900\ \text{kg/m}^3\right)}\right)(900\ \text{kg/m}^3)}{70\times10^{-3}\ \text{kg/m/s}} = 870 \qquad (E1.6b)$$

Therefore, the flow is laminar, and the friction factor $f = 16/\text{Re}$. A hint that the flow might be laminar is that the fluid is 70 times more viscous than water. This emphasizes the need to check the Reynolds number before proceeding.

The mechanical energy balance is then

$$(9.8\ \text{m/s}^2)(5\ \text{m}) + \frac{2\left(\dfrac{16}{870}\right)(500\ \text{m})\left(\dfrac{3\ \text{kg/s}}{(0.003089\ \text{m}^2)(900\ \text{kg/m}^3)}\right)^2}{0.06271\ \text{m}} - \frac{0.8\dot{W}_s}{3\ \text{kg/s}} = 0 \qquad (E1.6c)$$

which gives $\dot{W}_s = 1464$ W.

Example 1.7

Water flows from a constant-level tank at atmospheric pressure through 8 m of 1-in, schedule-40, commercial-steel pipe. It discharges to atmosphere 4 m below the level in the source tank. Calculate the mass and volumetric flowrates, neglecting entrance and exit losses.

Solution

Since the flowrate is unknown, the velocity is unknown, so the Reynolds number cannot be calculated, which means that the friction factor cannot be calculated initially. A simultaneous solution of the friction factor equation and the mechanical energy balance is necessary. Since the fluid is water, turbulent flow will be assumed, but it must be checked once the velocity or flowrate has been calculated.

The control volume is the fluid in the tank and the discharge pipe. In Figure E1.7, Point 1 is the level in the tank, which is at zero velocity, and Point 2 is the pipe discharge to the atmosphere.

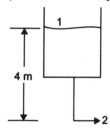

Figure E1.7

The mechanical energy balance reduces to

$$\frac{u_2^2}{2} + g\Delta z + \frac{2fLu_2^2}{D} = 0 = \frac{u_2^2}{2} + (9.8 \text{ m/s}^2)(-4 \text{ m}) + \frac{2f(8 \text{ m})u_2^2}{0.02664 \text{ m}} \quad \text{(E1.7a)}$$

and the friction factor is

$$\frac{1}{f^{0.5}} = -4\log_{10}\left[\frac{4.6\times10^{-5} \text{ m}}{3.7(0.02664 \text{ m})} + \left(\frac{6.81(10^{-3} \text{ kg/m/s})}{(0.02664 \text{ m})(u_2)(1000 \text{ kg/m}^3)}\right)^{0.9}\right] \quad \text{(E1.7b)}$$

Equations (E1.7a) and (E1.7b) are solved simultaneously to give $f = 0.0062$ and $u_2 = 3.04$ m/s. Using the relationships between velocity, volumetric flowrate, and mass flowrate, the results are $\dot{v}_2 = 1.69\times10^{-3}$ m^3/s and $\dot{m} = 1.69$ kg/s. Now, the Reynolds number must be checked using the calculated velocity, and Re = 80,960, so the turbulent flow assumption is valid.

Example 1.8

Number 6 fuel oil ($\mu = 800$ cP, $\rho = 62$ lb/ft^3) flows in a 1.5-in, schedule-40 pipe over a distance of 1000 ft. The discharge point is 20 ft above the inlet, and the source and discharge are both at 1 atm. A 15 hp pump that is 75% efficient is used. What is the flowrate in the pipe?

Solution

The mechanical energy balance reduces to

$$g\Delta z + \frac{32 fL\dot{v}^2}{\pi^2 D^5} - \frac{\eta_p \dot{W}_s}{\rho\dot{v}} = 0 \quad \text{(E1.8a)}$$

The friction expression is in terms of the volumetric flowrate, and in the third term, the mass flow-rate in the denominator is also expressed in terms of the volumetric flowrate. The volumetric flowrate is the unknown variable. Given the high viscosity, laminar flow is assumed. This assumption must be checked once a flowrate is calculated. For laminar flow, since $f = 16/\mathrm{Re}$, Equation (E1.8a) becomes

$$g\Delta z + \frac{32L\dot{v}^2}{\pi^2 D^5}\frac{16\pi D\mu}{4\dot{v}\rho} - \frac{\eta_p \dot{W}_s}{\rho\dot{v}} = 0 \tag{E1.8b}$$

where the fourth equality in Equation (1.20) is used for the Reynolds number. All terms are known other than the volumetric flowrate, so

$$32.2\ \mathrm{ft/sec^2}(20\ \mathrm{ft}) + \frac{128(1000\ \mathrm{ft})(800\ \mathrm{cP})(6.72\times10^{-4}\ \mathrm{lb/ft/sec/cP})\dot{v}}{\pi\left(\dfrac{1.610\ \mathrm{in}}{12\ \mathrm{in/ft}}\right)^4 (62\ \mathrm{lb/ft^3})}$$

(E1.8c)

$$-\frac{0.75(15\ \mathrm{hp})(550\ \mathrm{ft\ lb_f/sec/hp})(32.2\ \mathrm{ft\ lb/lb_f/sec^2})}{\dot{v}(62\ \mathrm{lb/ft^3})} = 0$$

The solution is $\dot{v} = 0.054\ \mathrm{ft^3/sec}$. Checking the Reynolds number,

$$\mathrm{Re} = \frac{4\dot{v}\rho}{\pi D\mu} = \frac{4(0.054\ \mathrm{ft^3/sec})(62\ \mathrm{lb/ft^3})}{\pi\left(\dfrac{1.610\ \mathrm{in}}{12\ \mathrm{in/ft}}\right)(800\ \mathrm{cP})(6.72\times10^{-4}\ \mathrm{lb/ft/sec/cP})} = 59.1 \tag{E1.8d}$$

so the flow is indeed laminar.

1.3.2.2   Multiple-Pipe Systems

For complex, multiple-pipe systems, including branching or mixing pipe systems, as illustrated in Figure 1.7, there are two sets of key relationships.

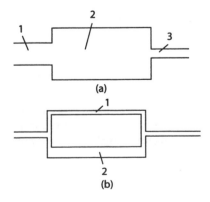

Figure 1.7   Multiple pipe systems: (a) pipes of different diameters in series, (b) pipes of different diameters in parallel

For pipes in series, the mass flowrate is constant and the pressure differences are additive:

$$\dot{m}_1 = \dot{m}_2 = \dot{m}_3 \tag{1.21}$$

$$\Delta P = \Delta P_1 + \Delta P_2 + \Delta P_3 \tag{1.22}$$

For pipes in parallel, the mass flowrates are additive and the pressure differences are equal:

$$\dot{m} = \dot{m}_1 + \dot{m}_2 + \dot{m}_3 \tag{1.23}$$

$$\Delta P_1 = \Delta P_2 = \Delta P_3 \tag{1.24}$$

Equation (1.21) is just the mass balance; the mass flowrate through each section must be constant. Equation (1.22) just means that the pressure drops in all sections in series are additive.

In the case of parallel flow, Equation (1.23) means that the mass flowrates in and out of the parallel section are additive, since mass must be conserved. Equation (1.24) means that the pressure drops in parallel sections are equal. This is because mixing streams must be designed to be at the same pressure, or the flowrates will readjust so the pressures at the mixing point are identical. This concept is discussed in more detail later.

The solution method is to write all of the relevant equations, including the mechanical energy balance, friction factor expression, and mass balance, along with the appropriate constraints from Equations (1.21) through (1.24), and solve the equations simultaneously. It is understood that this method applies to any number of pipes in series or parallel.

Example 1.9

Water flows through a pipe, splits into two parallel pipes, and then the fluids mix into another single pipe, as in Figure 1.7(b). All piping is commercial steel. The equivalent length of Branch 1 is 75 m, and the equivalent length of Branch 2 is 50 m. The elevation at the split point is the same as the elevation at the mixing point. Branch 1 is 2-in, schedule-40 pipe, and Branch 2 is 1.5-in, schedule-40 pipe. The pressure drop across Branch 1 is fixed at 100 kPa. Determine the volumetric flowrate in each branch and the total volumetric flowrate. What information could be obtained if the pressure drop was not provided?

Solution

The mechanical energy balance for both sections reduces to

$$\frac{\Delta P_i}{\rho} + \frac{2 f_i L_i u_i^2}{D_i} = 0 \tag{E1.9a}$$

where the subscript i denotes a parallel section of pipe. The kinetic energy terms are not present in Equation (E1.9a) because the control volume is the parallel pipes not including the feed pipe, the mixing point, the split point, and the discharge pipe. There are four unknowns, the friction factor and velocity in each section. The mechanical energy balance for each section is Equation (E1.9a), and there are two expressions for the friction factor, so the problem can be solved. Because the two branches are in parallel and then mix, the pressure drop in each section is the same, as shown in Equation (1.24), and it is negative, since the downstream pressure is less than the upstream pressure. Initially, turbulent flow will be assumed. The equations are

$$\frac{-100,000 \text{ Pa}}{1000 \text{ kg/m}^3} + \frac{2 f_1 (75 \text{ m}) u_1^2}{0.0525 \text{ m}} = 0 \tag{E1.9b}$$

$$\frac{-100,000\ \text{Pa}}{1000\ \text{kg/m}^3}+\frac{2f_2(50\ \text{m})u_2^2}{0.04089\ \text{m}}=0 \tag{E1.9c}$$

$$\frac{1}{f_1^{0.5}}=-4\log_{10}\left[\frac{4.6\times10^{-5}\ \text{m}}{3.7(0.0525\ \text{m})}+\left(\frac{6.81(10^{-3}\ \text{kg/m/s})}{(0.0525\ \text{m})(u_1)(1000\ \text{kg/m}^3)}\right)^{0.9}\right] \tag{E1.9d}$$

$$\frac{1}{f_2^{0.5}}=-4\log_{10}\left[\frac{4.6\times10^{-5}\ \text{m}}{3.7(0.04089\ \text{m})}+\left(\frac{6.81(10^{-3}\ \text{kg/m/s})}{(0.04089\ \text{m})(u_2)(1000\ \text{kg/m}^3)}\right)^{0.9}\right] \tag{E1.9e}$$

Solving Equations (E1.9b) to (E1.9e) simultaneously gives f_1 = 0.0053, u_1 = 2.57 m/s, f_2 = 0.0056, u_2 = 2.69 m/s. The volumetric flowrates are $\dot{v}_1$ = 0.0056 m^3/s and $\dot{v}_2$ = 0.0035 m^3/s. While Branch 2 is shorter, the smaller diameter has a stronger effect on the friction, as seen by the fifth-power dependence in Equation (1.14), so Branch 2 has a smaller flowrate.

Finally, the Reynolds numbers must be calculated to prove that the flow is turbulent. The results are Re_1 = 134,700, and Re_2 = 110,160, so the flow is indeed turbulent.

When streams mix, the pressure will be the same. If a pipe system is designed such that the pressures at a mixing point are not the same, the flowrates will adjust (as illustrated in Example 1.9) to make the mixing-point pressures identical, and the flowrates will not be as designed. This is important because steady-state process simulators allow streams to be mixed at different pressures, and the lowest pressure is taken as the outlet pressure unless an outlet pressure or a mixing-point pressure drop is specified. Just because steady-state process simulators allow this to be done does not make it physically correct. Valves are used to reduce higher pressures to make the pressures equal at a mixing point. When using simulators, it is the user's responsibility to include appropriate devices to make the simulation correspond to reality.

1.3.3   Compressible Flow

For compressible flow, the integral in the mechanical energy balance in Equation (1.5) must be evaluated, since the density is not constant. There are two limiting cases for frictional flow through a pipe section: isothermal flow and adiabatic flow. For isothermal flow of an ideal gas, the density is expressed as

$$\rho=\frac{PM}{RT} \tag{1.25}$$

where M is the molecular weight, and the integral can be evaluated. For adiabatic, reversible flow of an ideal gas, the temperature in Equation (1.25) is expressed in terms of pressure to evaluate the integral in Equation (1.5) using a relationship obtained from thermodynamics:

$$T=T_1\left(\frac{P}{P_1}\right)^{\frac{\gamma-1}{\gamma}} \tag{1.26}$$

where

$$\gamma=\frac{C_p}{C_v} \tag{1.27}$$

where C_p and C_v are the constant pressure and constant volume heat capacities, respectively. The results are expressed in terms of the superficial mass velocity, G. For isothermal, turbulent flow, the result, presented without derivation, is

$$\frac{M}{2RT}\left(P_2^2 - P_1^2\right) + G^2 \ln\left(\frac{P_1}{P_2}\right) + \frac{2fL_{eq}G^2}{D} = 0 \tag{1.28}$$

which can be solved for an unknown pressure, superficial mass velocity (G), diameter (by expressing superficial mass velocity in terms of diameter), or length. For isothermal, laminar flow, the result is

$$\frac{M}{4RT}\left(P_2^2 - P_1^2\right) + G^2 \ln\left(\frac{P_1}{P_2}\right) + \frac{16\mu GL_{eq}}{D^2} = 0 \tag{1.29}$$

Equation (1.29) is a quadratic in G, or if G is known, any other variable can be found. For adiabatic, turbulent flow, the result is

$$\frac{\gamma}{\gamma+1}\frac{M}{RT_1}P_1^2\left[1 - \left(\frac{P_2}{P_1}\right)^{\frac{\gamma+1}{\gamma}}\right] - \frac{2fL_{eq}G^2}{D} - \frac{1}{\gamma}\ln\left(\frac{P_1}{P_2}\right) = 0 \tag{1.30}$$

For compressible flow in packed beds, the Ergun equation, Equation (1.19), is used for the friction term, and the pressure term in the mechanical energy balance is integrated assuming either isothermal or adiabatic flow. For isothermal flow, the result is

$$\frac{M}{2G^2RT}\left(P_2^2 - P_1^2\right) + \frac{L(1-\varepsilon)}{D_p\varepsilon^3}\left[\frac{150\mu(1-\varepsilon)}{D_p u_s \rho} + 1.75\right] = 0 \tag{1.31}$$

where subscript 1 is upstream and subscript 2 is downstream. Quite often, it is stated that the mechanical energy balance for packed beds, the Ergun equation in Equation (1.19) can be used for gases as long as the pressure drop is less than 10% of the average pressure. However, with the computational tools now available, there is really no need for that approximation.

In Equations (1.28) through (1.33), it is assumed that the flow is in a pipe; therefore, there is no work term. The potential energy term is neglected because it is generally negligible for gases due to their low density.

1.3.4 Choked Flow

In evaluating the flow of compressible fluids, there exists a limit for the maximum velocity of the fluid (gas), that is, the speed of sound in the fluid. As an example, consider a pressurized gas in a supply tank (Tank 1) that is connected to a destination tank (Tank 2) via a pipe. Initially, Tank 1 and Tank 2 are at the same pressure, so no gas flows between them. Gradually, the pressure in Tank 2 is reduced and gas starts to flow from Tank 1 to Tank 2. It seems logical that the lower the pressure in Tank 2, the higher the gas flow rate is and the higher is the velocity of gas entering Tank 2. However, at some critical pressure for Tank 2, P_2^*, the flow of gas into Tank 2 reaches sonic velocity (the speed of sound). Decreasing the tank pressure below this critical pressure has no effect on the exit velocity of the gas entering Tank 2; that is, it remains constant at the speed of sound. This phenomenon of choked flow occurs because the change in downstream pressure must propagate upstream for the change in flow to occur. The speed at which this propagation occurs is the speed of sound. Thus, when the gas velocity is at the speed of sound, any further decrease in downstream pressure cannot be propagated upstream, and the flow cannot increase further. Therefore, there is a critical (maximum) superficial mass velocity of gas, G^*, that can be transferred from Tank 1 to Tank 2 through the pipe. The relationships for critical flow in pipes under turbulent flow conditions are as follows:

Isothermal flow

$$G^* = \frac{P_2^*}{P_1}\sqrt{P_1\rho_1} \tag{1.32}$$

and

$$\frac{4\,fL_{eq}}{D} = \left(\frac{P_1}{P_2^*}\right)^2 - 2\ln\left(\frac{P_1}{P_2^*}\right) - 1 \tag{1.33}$$

adiabatic flow

$$G^* = \sqrt{\gamma P_1\rho_1}\left(\frac{P_2^*}{P_1}\right)^{(\gamma+1)/2\gamma} \tag{1.34}$$

and

$$\frac{4\,fL_{eq}}{D} = \frac{2}{\gamma+1}\left[\left(\frac{P_1}{P_2^*}\right)^{(\gamma+1)/\gamma} - 1\right] - \frac{2}{\gamma}\ln\left(\frac{P_1}{P_2^*}\right) \tag{1.35}$$

When evaluating compressible flows, a check for critical flow conditions in the system should always be done. Usually, critical flow is not an issue when $P_2 > 0.5P_1$, but it is always a good idea to check. The use of Equations (1.32) through (1.35) is illustrated in Example 1.10.

Example 1.10

A fuel gas has an average molecular weight of 18, a viscosity of 10^{-5} kg/m s, and a γ value of 1.4. It is sent to neighboring industrial users through 4-in, schedule-40, commercial-steel pipe. One such pipeline is 100 m long. The pressure at the plant exit is 1 MPa, and the required pressure at the receiver's plant is 500 kPa. It is estimated that the gas maintains a constant temperature of 75°C over the entire length of 100 m. Estimate the volumetric flowrate of the fuel gas, metered at 1 atm and 60°C.

Solution

The conditions for critical flow should be checked first, and this requires the simultaneous solution of Equations (1.32) and (1.33) to find P_2^*. An approximation can be made by assuming that the flow is fully developed turbulent and then checking this assumption. For fully developed turbulent flow, from Equation (1.16),

$$\frac{1}{f^{0.5}} = -4\log_{10}\left[\frac{e}{3.7D}\right] = -4\log_{10}\left[\frac{4.6\times10^{-5}\text{ m}}{3.7(0.0123\text{ m})}\right] \Rightarrow f = 0.00408 \tag{E1.10a}$$

Substituting in Equation (1.33) gives

$$\frac{4\,fL_{eq}}{D} = \frac{4(0.00408)(100)}{(0.10226)} = \left(\frac{P_1}{P_2^*}\right)^2 - 2\ln\left(\frac{P_1}{P_2^*}\right) - 1 \qquad \text{(E1.10b)}$$

Solving gives $P_2^* = 223.9$ kPa < 500 kPa; therefore, the flow is not choked. The actual friction factor is within a few percent of that calculated in Equation (E1.10a), and this difference does not affect the result regarding whether the flow is choked.

Equation (1.28) can now be solved for the superficial mass velocity:

$$G = \left[\frac{\dfrac{M}{2RT}\left(P_1^2 - P_2^2\right)}{\dfrac{2\,fL_{eq}}{D} + \ln\left(\dfrac{P_1}{P_2}\right)}\right]^{0.5} \qquad \text{(E1.10.c)}$$

All terms in Equation (E1.10c) are given other than the friction factor, which must be calculated. So,

$$G = \left[\frac{\dfrac{18\ \text{kg/kmol}}{2(8314\ \text{m}^3\text{Pa/kmol/K})(348\ \text{K})}\left(\left(10^6\ \text{Pa}\right)^2 - \left(5\times10^5\,\text{Pa}\right)^2\right)}{\dfrac{2f(100\ \text{m})}{0.10226\ \text{m}} + \ln\left(\dfrac{1\ \text{MPa}}{0.5\ \text{MPa}}\right)}\right]^{0.5} \qquad \text{(E1.10d)}$$

The friction factor, using the e value for commercial steel at the top of Figure 1.6, is

$$\frac{1}{f^{0.5}} = -4\log_{10}\left[\frac{4.6\times10^{-5}\ \text{m}}{3.7(0.10226\ \text{m})} + \left(\frac{6.81(10^{-5}\ \text{kg/m/s})}{(0.10226\ \text{m})G}\right)^{0.9}\right] \qquad \text{(E1.10e)}$$

where the Reynolds number is expressed as DG/μ. Equations (E1.10d) and (E1.10e) can be solved simultaneously for G. A possible approximation is to assume fully turbulent flow, as was done when checking for choked flow. In that case, the Reynolds number in the Pavlov equation is assumed to be large, so the friction factor asymptotically approaches a value calculated from only the roughness term. In this case, $f = 0.004077$. Then, from Equation (E1.10d), $G = 518.8$ kg/m²s. Simultaneous solution of Equations (E1.10d) and (E1.10e) yields $f = 0.00411$ and $G = 516.7$ kg/m²s, so the fully turbulent approximation is reasonable, even though an exact solution is possible. The Reynolds number is $DG/\mu = 5.28 \times 10^6$, which, from Figure 1.6, is in the fully turbulent, constant-friction-factor region.

Since $G = \dot{m}/A = \rho\dot{v}/A$, using the exact solution, with the density calculated using the ideal gas law $\rho = PM/RT$,

$$\dot{v} = \frac{(516.7\ \text{kg/m}^2/\text{s})(0.0082124\ \text{m}^2)}{\left(\dfrac{101{,}325\ \text{Pa}(18\ \text{kg/kmol})}{8314\ \text{m}^3\text{Pa/kmol/K}(333\ \text{K})}\right)} = 6.44\ \text{m}^3/\text{s} \qquad \text{(E1.10f)}$$

It is observed that the temperature and pressure used to calculate the density in Equation (E1.10d) are not the conditions in the pipeline, because the flowrate required is at 1 atm and 60°C. Since the density of gases is a function of temperature and pressure obtained through an equation of state, a volumetric flowrate must have temperature and pressure specified. In the gas industry, where

American Engineering units are common, the standard conditions, known as *standard cubic feet (SCF)*, are at 1 atm and 60°F.

If the second tank was at a pressure of $P_2^* = 223.9$ kPa or less, then the superficial mass velocity would be at its maximum value, given by Equation (1.32) (where $\rho_1 = 6.5016$ kg/m³):

$$G^* = \frac{P_2^*}{P_1}\sqrt{P_1\rho_1} = \frac{223,900}{10^6}\sqrt{(10^6)(6.5016)} = 570.9 \text{ kg/m}^2/\text{s}$$

1.4   OTHER FLOW SITUATIONS

1.4.1   Flow Past Submerged Objects

Objects moving in fluids and fluids moving past stationary, submerged objects are similar situations that are described by the force balance. When an object is released in a stationary fluid, it will either fall or rise, depending on the relative densities of the object and the fluid. The object will accelerate and reach a terminal velocity. The period of acceleration is found through an unsteady-state force balance, which is

$$m\frac{du}{dt} = -(\rho_s - \rho)gV + F_{drag} \tag{1.36}$$

where ρ_s is the object density, and ρ is the fluid density. For solid objects, the density difference most likely will be positive, so the object moves downward due to gravity and the drag force resists that motion—hence the opposite signs of the two terms on the right-hand side of Equation (1.36). However, for a gas bubble in a liquid, for example, the density difference is negative, so the bubble rises and the drag force resists that motion. Since velocity is generally defined as being positive moving away from gravity, because that is the positive direction of the coordinate system, the signs reconcile.

For a sphere, the mass is

$$m = \rho_s V = \rho_s \frac{\pi D_s^3}{6} \tag{1.37}$$

where D_s is the sphere diameter, and the volume is defined in Equation (1.37). The drag force on an object is defined as

$$F_{drag} = C_D \frac{\rho u^2}{2} A_{proj} = C_D \frac{\rho u^2}{2}\frac{\pi D_s^2}{4} \tag{1.38}$$

where C_D is a drag coefficient that may be thought of as an analog to the friction factor, A_{proj} is the projected area normal to the direction of flow, and u is the velocity of the object relative to the fluid. For a sphere, the projected area is that of a circle, as shown in the second equality of Equation (1.38). For a cylinder with transverse flow, this area is that of a rectangle. Equation (1.37), Equation (1.38), and the volume of a sphere may be substituted into Equation (1.36), and integration between the limits of zero velocity at time zero and velocity u at time t yields the transient velocity. The transient velocity approaches the terminal velocity at $t \rightarrow \infty$, which can also be obtained by solving for velocity in Equation (1.36) when $du/dt = 0$, that is, at steady state, when the sum of the forces on the object equal zero. The terminal velocity is

$$u_t^2 = \frac{4(\rho_s - \rho)gD_s}{3C_D\rho} \tag{1.39}$$

An expression for the drag coefficient is now needed, just as an expression for the friction factor was needed for pipe flow. Similar to pipe flow, there are different flow regimes with different drag coefficients. The Reynolds number for a sphere is defined as $Re = D_s u_t \rho/\mu$, where the density and viscosity are always that of the fluid, and if $Re \ll 1$, which is called *creeping flow*, this is the Stokes flow regime. Stokes law, which is a theoretical result, states that the drag force in Equation (1.36) is defined as

$$F_d = 3\pi\mu D_s u_t \tag{1.40}$$

which yields

$$C_D = \frac{24}{Re} \tag{1.41}$$

Stokes law must be applied only when it is valid, even though its use makes the mathematical results much simpler. In addition to the Reynolds number constraint, the assumptions involved in Stokes law are a rigid sphere and that gravity is the only body force. An example of another body force is electrostatic force; therefore, Stokes law may fail for charged objects. Theoretically, there are two drag force components for flow past an object. This is based on the concept that drag is manifested as a pressure drop. Form drag is caused by flow deviations due to the presence of the object. Since the fluid must change direction to flow around the object, energy is "lost," which is manifested as a pressure drop. Frictional drag is analogous to that in a pipe and is due to the contact between the fluid and the object. In Equation (1.40), two-thirds of the total is due to frictional drag and one-third is due to form drag.

Experimental data are usually used as a means to determine the drag coefficient. There are curve fits for the intermediate region, between creeping flow and the constant value observed for $1000 < Re < 200,000$. Haider and Levenspiel (1989) provide a curve fit to the data for all values of $Re < 200,000$:

$$C_D = \frac{24}{Re} + 3.3643Re^{-0.3471} + \frac{0.4601Re}{Re + 2682.5} \tag{1.42}$$

and these results are plotted in Figure 1.8.

Equation (1.42) is not convenient for solving the terminal velocity of a sphere falling in a fluid because an iterative solution is required (see Example 1.11). However, this equation may be reformulated in terms of two other dimensionless variables, u_t^* and D^*:

$$u_t^* = \left(\frac{4\,Re_t}{3\,C_D}\right)^{1/3} = u_t\left[\frac{\rho_f^2}{\mu(\rho_s - \rho_f)g}\right]^{1/3} \tag{1.43}$$

$$D^* = \left(\frac{3}{4}C_D Re_t^2\right)^{1/3} = D_{sph}\left[\frac{\rho_f(\rho_s - \rho_f)g}{\mu^2}\right]^{1/3} \tag{1.44}$$

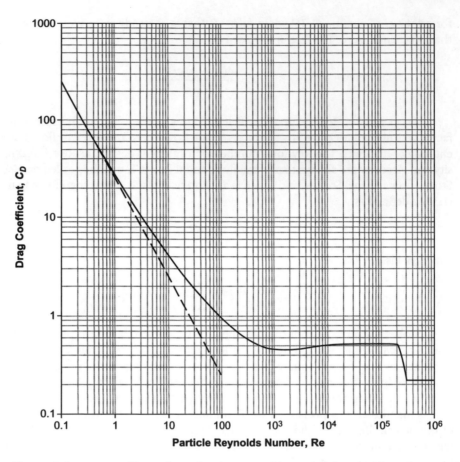

Figure 1.8   Drag coefficient dependence on Reynolds number from Haider and Levenspiel (1989), Equation (1.42). The dotted, straight line is the creeping flow asymptote.

and

$$u_t^* = \left[\frac{18}{(D^*)^2} + \frac{0.591}{(D^*)^{0.5}} \right]^{-1}$$

(1.45)

If the properties of the fluid and particle are known, then D^* can be calculated using Equation (1.44), and then Equation (1.45) can be used to determine u_t^*, and finally u_t can be calculated from Equation (1.43). This is illustrated in Example 1.11.

Example 1.11

In a particular sedimentation vessel, small particles (SG = 1.2) are settling in water. The particles have a diameter of 0.2 mm. What is the terminal velocity of the particles?

Solution

Since the particles are small, creeping flow will be assumed initially. Substituting Equation (1.41) into Equation (1.39) yields

$$u_t = \frac{gD^2(\rho_s - \rho)}{18\mu} = \frac{(9.8 \text{ m/s}^2)(2\times10^{-4} \text{ m})^2(200 \text{ kg/m}^3)}{18(10^{-3} \text{ kg/m/s})} = 0.0044 \text{ m/s} \qquad \text{(E1.11a)}$$

Checking the Reynolds number,

$$\text{Re} = \frac{(2\times10^{-4} \text{ m})(0.0044\text{m/s})(1000 \text{ kg/m}^3)}{10^{-3} \text{ kg/m/s}} = 0.87 \qquad \text{(E1.11b)}$$

which is not in the creeping flow regime. Therefore, simultaneous solution of Equations (1.39) and (1.42) is required, and the result is $u_t = 0.0039$ m/s and Re = 0.78.

Alternatively, using Equations (1.43), (1.44), and (1.45),

$$D^* = D_{sph}\left[\frac{\rho_f(\rho_s - \rho_f)g}{\mu^2}\right]^{1/3}$$

$$D^* = (2\times10^{-4}\text{ m})\left[\frac{(1000)(1200-1000)(9.81)}{(1\times10^{-3})^2}\right]^{1/3} = 2.504 \qquad \text{(E1.11c)}$$

$$u_t^* = \left[\frac{18}{(D^*)^2} + \frac{0.591}{(D^*)^{0.5}}\right]^{-1} = \left[\frac{18}{(2.504)^2} + \frac{0.591}{(2.504)^{0.5}}\right]^{-1} = 0.3082 \qquad \text{(E1.11d)}$$

$$u_t^* = u_t\left[\frac{\rho_f^2}{\mu(\rho_s - \rho_f)g}\right]^{1/3} \Rightarrow u_t = u_t^*\left[\frac{\mu(\rho_s - \rho_f)g}{\rho_f^2}\right]^{1/3}$$

$$u_t = (0.3082)\left[\frac{(1\times10^{-3})(200)(9.81)}{(1000)^2}\right]^{1/3} = 0.00385 \text{ m/s} \qquad \text{(E1.11e)}$$

For Re > 200,000, the phenomenon called *boundary layer separation* occurs. The drag coefficient in this region is $C_D = 0.22$.

With the exception of the boundary layer separation region, Figure 1.8 has about the same shape as Figure 1.6. For low Reynolds numbers, the friction factor and drag coefficient are both inversely proportional to the Reynolds number, though the exact proportionality is different. For large Reynolds numbers, what is generally called *fully turbulent flow*, the friction factor and drag coefficient both approach constant values.

For nonspherical particles, the determination of the drag coefficient and terminal velocity is more complicated. A major challenge is how to account for particle shape. One method is to define the shape in terms of sphericity. Sphericity is defined as

$$\text{Sphericity} = \Psi = \left(\frac{\text{surface area of sphere}}{\text{surface area of particle}}\right)_{\text{same volume}} \qquad (1.46)$$

Then, the diameter of a sphere with the same volume as the particle, d_v, is calculated and used in place of the diameter in Equations (1.37) through (1.42). Care is needed when using sphericity, since particles with quite different shapes but similar sphericities may behave quite differently when falling in a fluid.

Example 1.12

Determine the sphericity and D_v of a cube.

Solution

Call the dimension of the cube x. Therefore, D_v is obtained from

$$\frac{\pi D_v^3}{6} = x^3 \tag{E1.12a}$$

$$D_v = \left(\frac{6x^3}{\pi}\right)^{1/3} = 1.241x \tag{E1.12b}$$

and the sphericity is

$$\Psi_{cube} = \frac{\pi D_v^2}{6x^2} = \frac{\pi(1.241x)^2}{6x^2} = 0.806 \tag{E1.12c}$$

Haider and Levenspiel (1989) have provided a curve fit for previously published experimental data, which were taken for regular geometric shapes. The drag coefficient for different sphericities is illustrated in Figure 1.9, and the curve-fit equation is

$$C_D = \frac{24}{Re}\left[1 + \left(8.1716e^{-4.0655\Psi}\right)Re^{0.0964+0.5565\Psi}\right] + \frac{73.69e^{-5.0748\Psi}Re}{Re + 5.378e^{6.2122\Psi}} \tag{1.47}$$

where $Re = D_v u_t \rho / \mu$.

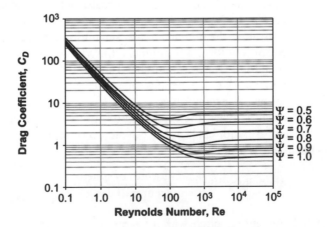

Figure 1.9   Drag coefficient dependence on Reynolds number and sphericity from Haider and Levenspiel (1989), Equation (1.47)

The equivalent expression in terms of D^* and u_t^* is given as

$$u_t^* = \left[\frac{18}{(D^*)^2} + \frac{2.335 - 1.745\Psi}{(D^*)^{0.5}} \right]^{-1} \quad \text{with} \quad D^* = D_v \left[\frac{\rho_f (|\rho_s - \rho_f|)g}{\mu^2} \right]^{1/3} \qquad (1.48)$$

where D_v is the diameter of a sphere with the same volume as the particle.

Equation (1.39) can be solved for one unknown by using either Equation (1.41) or Equation (1.42) for the drag coefficient. For example, the viscosity of a fluid can be determined by measuring the terminal velocity of a falling sphere. Or, the terminal velocity of an object can be determined if all of the fluid and particle physical properties are known. If the Reynolds number is unknown, then the flow regime is unknown. Therefore, depending on the type of problem being solved, judgment may be needed to assume a flow regime, the assumption must be checked, and iterations may be required to get the correct answer.

1.4.2 Fluidized Beds

If fluid flows upward through a packed bed, at a high enough velocity, the particles become buoyant and float in the fluid. For this condition, the upward drag on the particles is equal to the weight of the particles and is called the *minimum fluidization velocity*, and the particles are said to be fluidized. This is one reason why flow through packed beds is usually downward. The benefits of fluidization are that once the particles are fluidized, they can circulate and the bed of solids mixes. If the upward fluid velocity is sufficiently high, then the bed of particles becomes well mixed (like a continuous stirred tank reactor) and approaches isothermal behavior. For highly exothermic reactions, this property is very desirable. Fluidized beds are often used for such reactions and are discussed in Chapter 4, "Reactors." Fluidized beds are also used in drying and coating operations where the movement of solids is desirable to increase heat and/or mass transfer. As the fluid velocity upward through the bed of particles increases, the mixing of particles becomes more vigorous and there is a tendency for particles to be flung upward and elutriate from the bed. Therefore, a cyclone is typically part of a fluidized bed to remove the entrained particles and recirculate them to the fluidized bed. Another desirable feature of fluidized beds is that they can be used with very small catalyst particles without a large pressure drop. For very small catalyst particles in a packed bed, the pressure drop becomes very large. An example of such a catalyst is fluid catalytic cracker used in petroleum refining to make smaller hydrocarbons from large ones.

The general shape of the pressure drop versus superficial fluid velocity in a fluidized bed is shown in Figure 1.10.

The region to the left of u_{mf} is described by the Ergun equation for packed beds because, before fluidization begins, behavior is that of a packed bed. If the particles were restricted, by, say, placing a wire screen on top of the bed, then the bed would continue to behave as a packed bed beyond the u_{mf}. Assuming that the top of the bed is unrestricted, once there is sufficient upward velocity, and hence upward force, the particles begin to lift. This is called *minimum fluidization*. At minimum fluidization, the upward force is equal to the weight of the particles. Hence, the frictional force equals the weight of the bed, and the pressure drop remains constant. Quantitatively,

$$-\Delta P_{ff} A_t = V_{solids} (\rho_s - \rho_f) g = A_t h_{mf} (1 - \varepsilon_{mf})(\rho_s - \rho_f) g \qquad (1.49)$$

where the subscript *mf* signifies minimum fluidization and h_{mf} is the height of the bed at minimum fluidization, which for a packed bed was called the length of the bed, L. At the instant at

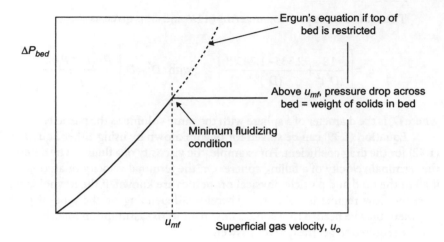

Figure 1.10   Plot illustrating constant value of pressure drop above minimum fluidization velocity

which fluidization begins, the frictional pressure drop is equal to that of a packed bed. Combining Equation (1.19), which is the frictional loss in a packed bed and equals $-\Delta P_{fr}/\rho$, and Equation (1.49) yields

$$h_{mf}(1-\varepsilon_{mf})(\rho_s-\rho_f)g = \frac{\rho_f h_{mf} u_{mf}^2 (1-\varepsilon_{mf})}{D_p \varepsilon_{mf}^3}\left[\frac{150\mu(1-\varepsilon_{mf})}{D_p u_{mf}\rho_f}+1.75\right] \qquad (1.50)$$

Rearranging Equation (1.50) and defining two dimensionless groups that characterize the fluid flow in a fluidized bed,

$$\mathrm{Re}_{mf}=\frac{D_p u_{mf}\rho_f}{\mu} \qquad (1.51)$$

$$\mathrm{Ar}=\frac{D_p^3 \rho(\rho_s-\rho_f)g}{\mu^2} \qquad (1.52)$$

where Equation (1.51) is the particle Reynolds number, which characterizes the flow regime, and Equation (1.52) defines the Archimedes number, which is the ratio of gravitational forces/viscous forces, yields ·

$$\frac{1.75}{\varepsilon_{mf}^3}\mathrm{Re}_{mf}^2+\frac{150(1-\varepsilon_{mf})}{\varepsilon_{mf}^3}\mathrm{Re}_{mf}-\mathrm{Ar}=0 \qquad (1.53)$$

Equation (1.53) is a quadratic in Re_{mf}, so the minimum fluidization velocity can be obtained if the physical properties of the solid and fluid are known. For nonspherical particles, the result is

$$\frac{1.75}{\Psi\varepsilon_{mf}^3}\mathrm{Re}_{mf}^2+\frac{150(1-\varepsilon_{mf})}{\Psi^2\varepsilon_{mf}^3}\mathrm{Re}_{mf}-\mathrm{Ar}=0 \qquad (1.54)$$

If the void fraction at minimum fluidization, which must be measured, and/or the sphericity are not known, Wen and Yu (1966) recommend using

$$\Psi \varepsilon_{mf}^3 = \frac{1}{14} \tag{1.55}$$

$$\frac{(1-\varepsilon_{mf})}{\Psi^2 \varepsilon_{mf}^3} = 11 \tag{1.56}$$

and Equation (1.54) reduces to

$$\mathrm{Re}_{mf} = \left[(33.7)^2 + 0.0408 \mathrm{Ar} \right]^{1/2} - 33.7 \tag{1.57}$$

Since the volume of solid particles remains constant, it is possible to relate the bed height and void fraction at different levels of fluidization.

$$h_{mf}(1-\varepsilon_{mf}) = h_f(1-\varepsilon_f) \tag{1.58}$$

Equation (1.58) is understood by multiplying each side of the equation by A_t, the total bed area, so each side of the equation is the volume of particles because $(1-\varepsilon)$ is the solid fraction, and hA_t is the total bed volume. The operation of fluidized beds above u_{mf} varies considerably on the basis of the size of particles and the superficial velocity of gas. One way to describe the behavior of these beds is through the flow map by Kunii and Levenspiel (1991) in Figure 1.11. In Figure 1.11, u^* and D^* refer to the dimensionless velocity and particle size introduced in Section 1.4.1, except that the superficial velocity of the gas through the bed (not the particle terminal velocity) is used in u^*.

It is clear from this figure that operation of fluidized beds can occur over a wide range of operating velocities from u_{mf} to several times the terminal velocity. For turbulent (lying above

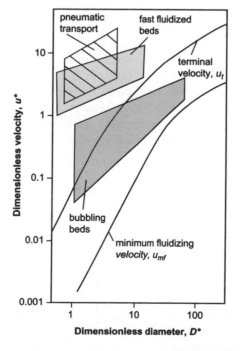

Figure 1.11 Flow regime map for gas-solid fluidization (Modified from Kunii and Levenspiel [1991])

bubbling beds) and fast fluidized beds, internal and external cyclones must be employed, respectively. The gas and solids flow patterns in all these regimes are very complex and can be found only by experimentation or possibly by using complex computational fluid dynamics codes.

1.4.3   Flowrate Measurement

The traditional method for measuring flowrates is to add a restriction in the flow path and measure the pressure drop. The pressure drop can be related to the velocity and flowrate by the mechanical energy balance. More modern instruments include turbine flow meters that measure flowrate directly and vortex shedding devices.

The types of restrictions used are illustrated in Figure 1.12.

The control volume is fluid between an upstream point, labeled 1, and a point in the obstruction, labeled 2. For turbulent flow, the mechanical energy balance written between these two points is

$$\frac{P_2 - P_1}{\rho} + \frac{u_2^2 - u_1^2}{2} + e_f = 0 \qquad (1.59)$$

The friction term is dropped at this point but is incorporated into the problem through a discharge coefficient, C_o. From Equation (1.3), u_1 is expressed in terms of u_2, the cross-sectional areas, and then the diameters; and solving for the velocity in the obstruction yields

$$u_2 = C_o \left[\frac{2(P_1 - P_2)}{\rho(1 - \beta^4)} \right]^{0.5} \qquad (1.60)$$

where

$$\beta = \frac{D_2}{D_1} \qquad (1.61)$$

The flowrate can then be obtained by multiplying the velocity in the restriction by the cross-sectional area of the restriction. The term C_o, a discharge coefficient, is added to account for the frictional loss in the restriction. Figure 1.13 shows C_o as a function of β and the bore (restriction) Reynolds number for an orifice, one of the most common restrictions used. Since C_o is not known, the asymptotic value of 0.61 for high-bore Reynolds number is assumed, and iterations may be required if the bore Reynolds number is not above about 20,000. This calculation method is illustrated in Example 1.13.

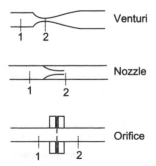

Figure 1.12   Typical devices used to measure flowrate

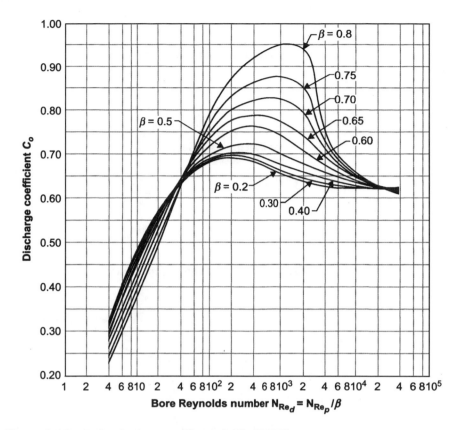

Figure 1.13 Orifice discharge coefficient (Miller [1983])

Other flow measurement devices are used. One such device is the rotameter that has a float that moves within a variable area vertical tube. The level of the float in the device is related to the flowrate, as illustrated in Figure 1.14. As the fluid flow increases, the drag on the float increases and it moves up, but the annular flow area around the float also increases. Consequently, the float comes to a new equilibrium position at which its weight is just balanced by the upward drag force of the fluid. Rotameters are still found in laboratories and provide accurate measurements for both gas and liquid flows. While there is a theoretical description of how a rotameter works, it is typically calibrated by measuring the flowrate versus the height of the float for the given fluid of interest.

Measuring pressure differences is automated in a chemical plant through the use of various devices. However, manometers may still be found in laboratories. Manometers work by having an immiscible fluid of higher density than the flowing fluid in a U-shaped tube, with one end of the tube connected to the pipe at Location 1 and the other end connected as close as possible to Location 2. The height difference between the levels of the immiscible fluid is a measure of the pressure difference between Locations 1 and 2.

Figure 1.15 illustrates a general manometer, where the pipe in which the fluid is flowing may be inclined.

The manometer is an example of fluid statics, so the pressure at any horizontal location must be the same in each manometer leg. For the pressure at height 3 in Figure 1.15,

$$P_1 + \rho_A g(z_1 - z_3) = P_2 + \rho_A g(z_2 - z_4) + \rho_B g(z_4 - z_3) \tag{1.62}$$

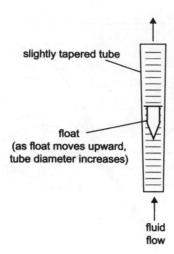

Figure 1.14   Illustration of rotameter

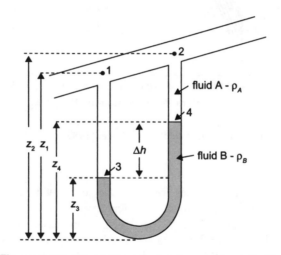

Figure 1.15   Illustration of general manometer situation

Equation (1.62) can be rearranged into the "general" manometer equation

$$P_1 - P_2 + g\Delta h(\rho_A - \rho_B) + \rho_A g(z_1 - z_2) = 0 \tag{1.63}$$

where

$$\Delta h = (z_4 - z_3) \tag{1.64}$$

The third term in Equation (1.63) is zero if the pipe is horizontal. It is important to understand that $z_1 - z_2$ is a difference in vertical distance (height), not a distance along the pipe, and that the coordinate system points upward, so a high height minus a low height is a positive number.

Example 1.13

An orifice having a diameter of 1 in is used to measure the flowrate of an oil (SG = 0.9, μ = 50 cP) in a horizontal, 2-in, schedule-40 pipe at 70°F. The pressure drop across the orifice is measured by a mercury (SG = 13.6) manometer, which reads 2.0 cm. Calculate the volumetric flowrate of the oil.

Solution

Two steps are involved. First, the pressure drop is calculated from the manometer information. Then, the flowrate is calculated.

To calculate the pressure drop, Equation (1.62) is used, but since the pipe is horizontal, the third term on the right-hand side is zero. The result is

$$P_1 - P_2 = g\Delta h(\rho_B - \rho_A) =$$

$$\frac{32.2 \text{ ft/sec}^2}{32.2 \text{ ft lb/lb}_f/\text{sec}^2}\left(2 \text{ cm}\frac{1 \text{ in}}{2.54 \text{ cm}}\right)(13.6-0.9)(62.4 \text{ lb/ft}^3)\left(\frac{\text{ft}}{12 \text{ in}}\right)^3 = 0.361 \text{ psi} \qquad \text{(E1.13a)}$$

Next, the pressure drop is used in Equation (1.60) with the initial assumption that C_o = 0.61. So

$$u_2 = 0.61\left[\frac{2(0.361 \text{ lb}_f/\text{in}^2)(12 \text{ in/ft})^2(32.2 \text{ ft lb/lb}_f/\text{sec}^2)}{0.9(62.4 \text{ lb/ft}^3)\left[1-\left(\frac{1 \text{ in}}{2.067 \text{ in}}\right)^4\right]}\right]^{0.5} = 4.84 \text{ ft/sec} \qquad \text{(E1.13b)}$$

Now, the bore Reynolds number must be checked.

$$\text{Re} = \frac{(1/12 \text{ ft})(4.84 \text{ ft/sec})(62.4 \text{ lb/ft}^3)}{50 \text{ cP }(6.72\times10^{-4} \text{ lb/ft/sec/cP})} = 749 \qquad \text{(E1.13c)}$$

From Figure 1.13, with β = 0.48 and Re = 749, C_o ≈ 0.71. Repeating the calculation in Equation (E1.11b) gives u_2 = 5.63 ft/sec and Re = 872. Within the error of reading Figure 1.12, C_o ≈ 0.71, so the iteration is completed. The volumetric and mass flowrates can now be calculated:

$$\dot{v} = (5.63 \text{ ft/sec})(0.02330 \text{ ft}^2) = 0.131 \text{ ft}^3/\text{sec} \qquad \text{(E1.13d)}$$

$$\dot{m} = (0.131 \text{ ft}^3/\text{sec})(62.4 \text{ lb/ft}^3) = 8.19 \text{ lb/sec} \qquad \text{(E1.13e)}$$

When fluid flows through an orifice, the pressure decreases because the velocity increases through the small cross-sectional area of the orifice. Physically, this is because pressure energy is converted to kinetic energy. This is similar to a nozzle, as illustrated in Example 1.3. Subsequently, when the velocity decreases as the cross-sectional area increases to the total pipe area, the pressure increases again. However, not all of the pressure is "recovered," due to circulating fluid flow at the pipe-orifice diameter. The permanent pressure loss requires incremental pump power, and that is part of the cost of measuring the flowrate using an orifice or nozzle. The amount of recovered pressure has been correlated as a function of β, and for an orifice, and it is illustrated in Figure 1.16.

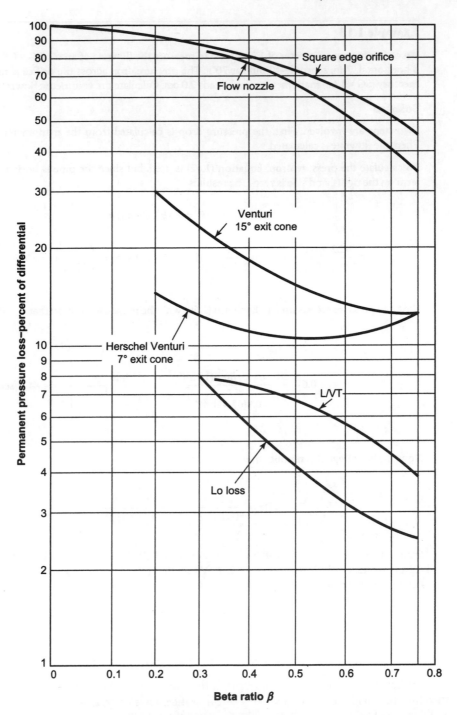

Figure 1.16 Unrecovered frictional loss in different flow measuring devices (Adapted by permission from Cheremisinoff and Cheremisinoff)

Example 1.14

For Example 1.13, how much additional power is needed for the permanent pressure loss through the orifice? The pump is 75% efficient.

Solution

For $\beta \approx 0.5$, from Figure 1.15, the permanent pressure loss is about 73%. From the mechanical energy balance,

$$\frac{0.73(P_1 - P_2)}{\rho} - \frac{\eta_p \dot{W}_s}{\dot{m}} = 0 \tag{E1.14a}$$

$$\frac{(0.73)(0.361\ \text{lb}_f/\text{in}^2)(12\ \text{in}/\text{ft})^2}{0.9(62.4\ \text{lb}/\text{ft}^3)} = \frac{0.75\dot{W}_s}{8.19\ \text{lb}/\text{sec}} \tag{E1.14b}$$

so

$$\dot{W}_s = 7.38\ \text{ft lb}_f/\text{sec} = 0.0134\ \text{hp} \tag{E1.14c}$$

This result shows that, while there is a cost associated with an orifice, it is small.

1.5 PERFORMANCE OF FLUID FLOW EQUIPMENT

In addition to equipment design, the chemical engineer must deal with performance of existing equipment. The differences between the design problem (also called a *rating* problem) (a) and the performance problem (b) are illustrated in Figure 1.17. The use of italics indicates the unknowns in the particular problem. In the design problem, the input and the desired output are specified, and the equipment is designed to satisfy those constraints. In the performance problem, the input and equipment are specified, and the output is determined. The performance problem is what is involved in dealing with day-to-day operations in a chemical plant.

Several different types of problems in frictional fluid flow using the mechanical energy balance were discussed in Section 1.3. Determining the pump power needed for a given situation is a design problem. Similarly, determining the required pipe diameter is a design problem. On the other hand, determining the flowrate when all equipment is specified is a performance problem, as is determining the pressure change for an existing system.

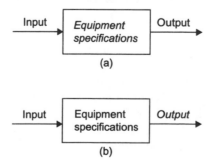

Figure 1.17 Illustrations of (a) design problem, (b) performance problem (unknowns are indicated by italics for each case)

Suppose it is necessary to increase the capacity of a process without adding new equipment. Logically, all flowrates must increase. This is a performance problem, since the input and equipment are specified, and the output must be determined for each unit in the process. Somewhere in the process, the amount of scale-up needed will be limited due to equipment constraints, and this limiting unit is called a *bottleneck*. The process of finding a solution that removes the bottleneck is called *debottlenecking*, which is a performance problem. Similarly, if there is a problem with the output of a process (purity or temperature, for example), the cause of the problem must be determined, which is called *troubleshooting*.

Returning to the situation in which process capacity must be increased, for the fluid flow component, initially, it may appear that problems similar to those in Section 1.3 must be solved from scratch. However, for many situations, not just in fluid flow, very good approximations can be made with a much simpler analysis.

1.5.1 Base-Case Ratios

The ability to predict changes in a process design or in plant operations is improved by anchoring an analysis to a base case. This calculation tool combines use of fundamental relationships with plant operating data to form a basis for predicting changes in system behavior. As will be seen, it is applicable to problems involving all chemical process units when analytical expressions are available.

For design changes, it is desirable to identify a design proven in practice as the base case. For operating plants, actual data are available and are chosen as the base case. It is important to put this base case into perspective. Assuming that there are no instrument malfunctions and these operating data are correct, then these data represent a real operating point at the time the data were taken. As the plant ages, the effectiveness of process units changes and operations are altered to account for these changes. As a consequence, recent data on plant operations should be used in setting up the base case.

The base-case ratio integrates the "best available" information from the operating plant with design relationships to predict the effect of process changes. It is an important and powerful technique with a wide range of applications. The base-case ratio, X, is defined as the ratio of a new-case system characteristic, x_2, to the base-case system characteristic, x_1.

$$X \equiv x_2/x_1 \tag{1.65}$$

Using a base-case ratio often reduces the need for knowing actual values of physical properties (physical properties refer to thermodynamic and transport properties of fluids), equipment, and equipment characteristics. The values identified in the ratios fall into three major groups. They are defined below and applied in Examples 1.15 and 1.16.

1. **Ratios Related to Equipment Sizes** (L_{eq}, equivalent length; diameter, D; surface area, A): Assuming that the equipment is not modified, these values are constant, the ratios are unity, and these terms cancel out.
2. **Ratios Related to Physical Properties** (such as density, ρ; viscosity, μ): These values can be functions of material composition, temperature, and pressure. Only the functional relationships, not absolute values, are needed. For small changes in composition, temperature, or pressure, the properties often are unchanged, and the ratio is unity and cancels out. An exception to this is gas-phase density.
3. **Ratios Related to Stream Properties**: These ratios usually involve velocity, flowrate, concentration, temperature, and pressure.

Using the base-case ratio eliminates the need to know equipment characteristics and reduces the amount of physical property data needed to predict changes in operating systems.

The base-case ratio is a powerful and straightforward tool to analyze and predict process changes. This is illustrated in Example 1.15.

Example 1.15

It is necessary to scale up production in an existing chemical plant by 25%. Your job is to determine whether a particular pump has sufficient capacity to handle the scale-up. The pump's function is to provide enough pressure to overcome frictional losses between the pump and a reactor.

Solution

The relationship for frictional pressure drop is obtained from the mechanical energy balance:

$$\frac{\Delta P}{\rho} = -\frac{2 f L u^2}{D} \tag{E1.15a}$$

This relationship is now written as the ratio of two cases, where subscript 1 indicates the base case, and subscript 2 indicates the new case:

$$\frac{\Delta P_2}{\Delta P_1} = \frac{2 \rho_2 f_2 L_{eq2} u_2^2 D_1}{2 \rho_1 f_1 L_{eq1} u_1^2 D_2} \tag{E1.15b}$$

Because the pipe has not been changed, the ratios of diameters (D_2/D_1) and lengths (L_{eq2}/L_{eq1}) are unity. Because a pump is used only for liquids, and liquids are (practically) incompressible, the ratio of densities is unity. If the flow is assumed to be fully turbulent, which is usually true for process applications, the friction factor is not a function of Reynolds number. This fact should be checked for a particular application. Figure 1.6 illustrates how, for fully turbulent flow in pipes that are not hydraulically smooth, the friction factor approaches a constant value. Since the x-axis is a log scale, changes up to a factor of 2 to 5, which are well beyond the scale-up capability of most equipment, do not represent much of a difference on the graph. Therefore, the friction factor is constant, and the ratio of friction factors is unity. The ratio in Equation (E1.15b) reduces to

$$\frac{\Delta P_2}{\Delta P_1} = \frac{u_2^2}{u_1^2} = \frac{\dot{m}_2^2 / A_2^2 \rho_2^2}{\dot{m}_1^2 / A_1^2 \rho_1^2} = \frac{\dot{m}_2^2}{\dot{m}_1^2} \tag{E1.15c}$$

where the second equality is obtained by substituting for u_i in numerator and denominator using the mass balance $\dot{m}_i = \rho_i A_i u_i$, canceling the ratio of densities for the same reason as above, and canceling the ratio of cross-sectional areas because the pipe has remained unchanged. Therefore, by assigning the base-case mass flow to have a value of 1, for a 25% scale-up, the new case has a mass flow of 1.25, and the ratio of pressure drops becomes

$$\frac{\Delta P_2}{\Delta P_1} = \left(\frac{\dot{m}_2}{\dot{m}_1}\right)^2 = \left(\frac{1.25}{1}\right)^2 = 1.56 \tag{E1.15d}$$

Thus, the pump must be able to deliver enough head to overcome 56% additional frictional pressure drop while pumping 25% more material.

It is important to observe that Example 1.15 was solved without knowing any details of the system. The pipe diameter, length, and number of valves and fittings were not known. The liquid being pumped, its temperature, and its density were not known. Yet the use of base-case ratios along with simple assumptions permitted a solution to be obtained. This illustrates the power and simplicity of base-case ratios.

Example 1.16

It is proposed to improve performance through a section of pipe by adding an identical section in parallel.

- **a.** If the total flowrate remains constant, what parameter changes and by how much, assuming the fluid flow is fully turbulent?
- **b.** If the original pipe is 1.5-in, schedule-40, commercial steel, and the new section is 2-in, schedule-40, commercial steel, answer the same question as in Part (a).

Solution

- **a.** By using the mechanical energy balance and Equation (1.14) for the friction term, with the subscript 1 representing the original case and subscript 2 representing the new case, each being the flow through the original section, the ratio of pressure drops is

$$\frac{\Delta P_2}{\Delta P_1} = \frac{-e_{f2}}{-e_{f1}} = \frac{32\rho_2 f_2 L_{eq2} \dot{v}_2^2}{\pi^2 D_2^5} \frac{\pi^2 D_1^5}{32\rho_1 f_1 L_{eq1} \dot{v}_1^2} \tag{E1.16a}$$

The constants cancel. If the fluid is unchanged, the densities cancel. Since the new and old pipe lengths and diameters are identical, the lengths and diameters cancel. It is assumed that the minor losses due to the elbows and fitting needed to add the parallel pipe are unchanged, so the equivalent lengths cancel. For fully turbulent flow, the friction factor has asymptotically approached a constant value (Figure 1.6), so the friction factors cancel. So, the result is

$$\frac{\Delta P_2}{\Delta P_1} = \frac{\dot{v}_2^2}{\dot{v}_1^2} \tag{E1.16b}$$

Since the two parallel sections are identical, the flowrate splits equally between the two sections, so the flowrate in the original section is half of the original flowrate:

$$\frac{\Delta P_2}{\Delta P_1} = \frac{\dot{v}_2^2}{\dot{v}_1^2} = \frac{(0.5\dot{v}_1)^2}{\dot{v}_1^2} = 0.25 \tag{E1.16c}$$

Therefore, the pressure drop through that section of pipe decreases by 75%.

- **b.** In this case, subscripts 1 and 2 represent the flow though the original and new sections, after the parallel section is installed. The analysis starts identically, but the diameters and friction factors do not cancel. The friction factors do not cancel because the asymptotic value for the friction factor in Figure 1.6 and in the Pavlov equation (Equation [1.16]) depends on the ratio of the roughness factor to the diameter, and that ratio is different for the two sections of pipe. The ratio expression becomes

$$\frac{\Delta P_2}{\Delta P_1} = \frac{f_2 \dot{v}_2^2 D_1^5}{f_1 \dot{v}_1^2 D_2^5} \tag{E1.16d}$$

From the Pavlov equation (Equation [1.16]), using the ratio of the friction factors at an asymptotically large Reynolds number and the schedule pipe diameters, Equation (E1.16d) becomes

$$\frac{\Delta P_2}{\Delta P_1} = \left[\frac{\log_{10}\left(\dfrac{0.0018 \text{ in}}{3.7(1.610 \text{ in})}\right)}{\log_{10}\left(\dfrac{0.0018 \text{ in}}{3.7(2.067 \text{ in})}\right)} \right]^2 \frac{(1.610 \text{ in})^5 \dot{v}_2^2}{(2.067 \text{ in})^5 \dot{v}_1^2} = 0.270 \frac{\dot{v}_2^2}{\dot{v}_1^2} \tag{E1.16e}$$

Since the pressure drops in each parallel section must be equal,

$$\frac{\dot{v}_2}{\dot{v}_1} = \left(\frac{1}{0.270}\right)^{0.5} = 1.92 \qquad (E1.16f)$$

If the flow is laminar, the analysis would be similar, but the results would differ due to the different expression for the friction factor in laminar flow. Examples of this are the subject of problems at the end of the chapter.

1.5.2   Net Positive Suction Head

There is a significant limitation on pump operation called net positive suction head (NPSH). This is the head that is needed on the pump feed (suction) side to ensure that liquid does not vaporize upon entering the pump. Its origin is as follows. Although the effect of a pump is to raise the pressure of a liquid, frictional losses at the entrance to the pump, between the suction pipe and the internal pump mechanism, cause the liquid pressure to drop upon entering the pump. This means that a minimum pressure exists somewhere within the pump. If the feed liquid is saturated or nearly saturated, the liquid can vaporize upon entering due to this internal pressure drop. This causes formation of vapor bubbles. These bubbles rapidly collapse when exposed to the forces created by the pump mechanism, called *cavitation*. This process usually results in noisy pump operation and, if it occurs for a period of time, will damage the pump. As a consequence, regulating valves, which lower fluid pressure, are not normally placed in the suction line to a pump.

Pump manufacturers supply NPSH data with a pump, usually in head units. In this book, both head and pressure units are used. The required NPSH, denoted $NPSH_R$, is a function of the square of velocity because it is a frictional loss and because most applications involve turbulent flow. Figure 1.18(a) shows $NPSH_R$ and $NPSH_A$ curves, which define a region of acceptable pump operation. This is specific to a given liquid. Typical $NPSH_R$ values are in the range of 15 to 30 kPa (2–4 psi) for small pumps and can reach 150 kPa (22 psi) for larger pumps. Figure 1.18 also shows curves for $NPSH_A$, the available NPSH, along with the $NPSH_R$ curve.

The available $NPSH_A$ is defined as

$$NPSH_A = P_{inlet} - P* \qquad (1.66)$$

Equation (1.66) means that the available NPSH ($NPSH_A$) is the difference between the inlet pressure, P_{inlet}, and $P*$, which is the vapor pressure (bubble-point pressure for a mixture). It is required that $NPSH_A \geq NPSH_R$ to avoid cavitation. Cavitation is avoided if operation is to the left of the intersection of the two curves. It is physically possible to operate to the right of the intersection of the two curves, but doing so is not recommended because the pump will be damaged.

All that remains is to calculate or know the pump inlet conditions in order to determine whether sufficient NPSH ($NPSH_A$) is available to equal or exceed the required NPSH ($NPSH_R$). For example, consider the exit stream from a distillation column reboiler, which is saturated liquid. If it is necessary to pump this liquid, cavitation could be a problem. A common solution to this problem is to elevate the column above the pump so that the static pressure increase minus any frictional losses between the column and the pump provides the necessary NPSH to avoid cavitation. This can be done either by elevating the column above ground level using a metal skirt or by placing the pump in a pit below ground level, although pump pits are usually avoided due to safety concerns arising from accumulation of heavier-than-air gases in the pit.

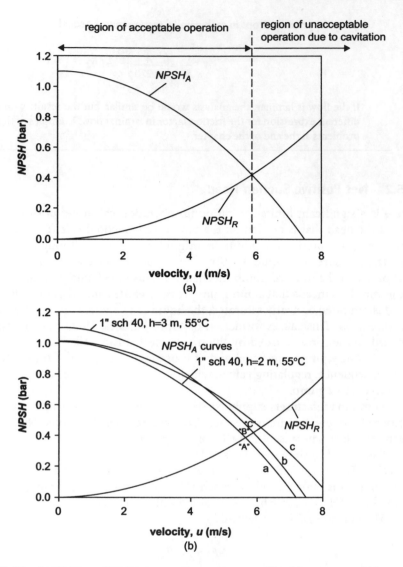

Figure 1.18   (a) $NPSH_A$ and $NPSH_R$ curves showing region of feasible operation, (b) how physical parameters affect shape of $NPSH_A$ curve

In order to quantify NPSH, consider Figure 1.19, in which material in a storage tank is pumped downstream in a chemical process. This scenario is a very common application of the NPSH concept. For NPSH analysis, the only portion of Figure 1.19 under consideration is between the tank and pump inlet.

From the mechanical energy balance, the pressure at the pump inlet can be calculated to be

$$P_{inlet} = P_{tank} + \rho g h - \frac{2\rho f L_{eq} u^2}{D} \tag{1.67}$$

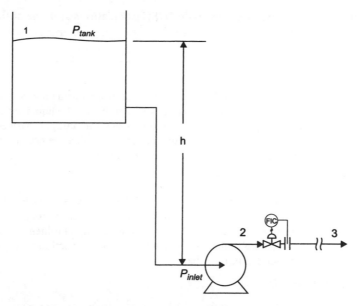

Figure 1.19 Typical situation for application of NPSH principles

which means that the pump inlet pressure is the tank pressure plus the static pressure minus the frictional losses in the suction-side piping. Therefore, by substituting Equation (1.67) into Equation (1.66), the resulting expression for $NPSH_A$ is

$$NPSH_A = P_{tank} + \rho gh - \frac{2\rho f L_{eq} u^2}{D} - P^* \qquad (1.68)$$

This is an equation of a concave downward parabola, of the form $NPSH_A = a - bu^2$, as illustrated in Figure 1.18(b), Curve a. The intercept is $a = P_{tank} + \rho gh - P^*$ and $b = 2\rho f L_{eq}/D$. This analysis does not include the kinetic energy term due to the acceleration of the fluid from the tank into the pipe. Rigorously, this term should also be included in the analysis.

If $NPSH_A$ is insufficient for a particular situation, Equation (1.68) suggests methods to increase the $NPSH_A$:

1. Decrease the temperature of the liquid at the pump inlet. This decreases the value of the vapor pressure, P^*, thereby increasing $NPSH_A$. This increases the intercept of the $NPSH_A$ curve while maintaining constant curvature, as illustrated in Figure 1.18(b), Curve b.

2. Increase the static head. This is accomplished by increasing the value of h in Equation (1.64), thereby increasing $NPSH_A$. As was said earlier, pumps are most often found at lower elevations than the source of the material they are pumping. This increases the intercept of the $NPSH_A$ curve while maintaining constant curvature, as illustrated in Figure 1.18(b), Curve b.

3. Increase the tank pressure. This increases the intercept of the $NPSH_A$ curve while maintaining constant curvature, as illustrated in Figure 1.18(b), Curve b.

4. Increase the diameter of the suction line (feed pipe to pump). This reduces the velocity and the frictional loss term, thereby increasing $NPSH_A$. This decreases the curvature of the $NPSH_A$ curve, as illustrated in Figure 1.18(b), Curve c. It is standard practice to have larger-diameter pipes on the suction side of a pump than on the discharge side.

Example 1.17 illustrates how to do NPSH calculations and one of the preceding methods for increasing $NPSH_A$. The other methods are illustrated in problems at the end of the chapter.

Example 1.17

A pump is used to transport toluene at 10,000 kg/h from a feed tank (V-101) maintained at atmospheric pressure and 55°C. The pump is located 2 m below the liquid level in the tank, and there is 6 m of equivalent pipe length between the tank and the pump. It has been suggested that 1-in, schedule-40, commercial-steel pipe be used for the suction line. Determine whether this is a suitable choice. If not, suggest methods to avoid pump cavitation.

Solution

The following data can be found for toluene: $\ln P^*(\text{bar}) = 10.97 - 4203.06/T(K)$, $\mu = 4.1 \times 10^{-4}$ kg/m s, $\rho = 870$ kg/m³. For 1-in, schedule-40, commercial-steel pipe, the roughness factor is about 0.001 and the inside diameter is 0.02664 m. Therefore, the velocity of toluene in the pipe can be found to be 5.73 m/s. The Reynolds number is about 426,000, and the friction factor is $f = 0.005$. At 55°C, the vapor pressure is found to be 0.172 bar.

From Equation (1.68),

$$NPSH_A = 1.01325 \text{ bar} + 870(9.81)(2)(10^{-5}) \text{ bar}$$
$$-2(870)(0.005)(6)(5.73)^2(10^{-5})/(0.02664) \text{ bar} - 0.172 \text{ bar}$$
$$NPSH_A = 0.37 \text{ bar}$$

This is shown as Point A on Figure 1.18(b). At the calculated velocity, Figure 1.18(b) shows that $NPSH_R$ is 0.40 bar, Point B. Therefore, there is insufficient $NPSH_A$. This means that a 1-in, schedule-40 pipe is unacceptable for this service.

The obvious solution to this problem is to use a larger-diameter pipe for the suction side of the pump. The calculated velocity of 5.73 m/s is far in excess of the typical maximum liquid velocity. The frictional loss in the 6 m of suction piping is approximately 0.64 bar. If, say, a 2-in, schedule-40 pipe was used for the suction line, then the frictional loss would decrease to approximately 0.02 bar and $NPSH_A$ would increase to about 0.99 bar, which is far in excess of $NPSH_R$. Another method for increasing $NPSH_A$ is to increase the height of liquid in the tank. If the height of liquid in the tank is 3 m, with the original 1-in, schedule-40 pipe at the original temperature, $NPSH_A = 0.445$ bar. This is shown as Point C on Figure 1.18(b).

1.5.3 Pump and System Curves

Pumps also have characteristic performance curves, called *pump curves*. Figure 1.20 illustrates a pump curve for a centrifugal pump. Centrifugal pumps are often called *constant head* pumps because, over a wide range of volumetric flowrates, the head produced by the pump is approximately constant. Pump manufacturers provide the characteristic curve, usually in head units. For centrifugal pumps, the shape of the curve indicates that although the head remains constant over quite a wide range of flowrates, eventually, as the flowrate continues to increase, the head produced decreases. Pump curves also include power and efficiency curves, both of which change with flowrate and head; however, these are not shown here.

For a piping system, a system curve can also be defined. Consider the system as illustrated in Figure 1.21. Location 1 is called the source, and Location 2 is the destination. Location 2 may be distant from Location 1, perhaps at the opposite end of a chemical process and at a different elevation from Location 1. Typical processes have only one pump upstream to supply all pressure needed to overcome pressure losses throughout the process. Therefore, the pressure increase

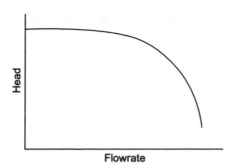

Figure 1.20   Typical shape of pump curve for centrifugal pump

across the pump must be sufficient to overcome all of the losses associated with piping and fittings plus the indicated pressure loss across the control valve. The orifice plate is present to illustrate some type of flowrate measurement, and the flow indicator controller (FIC) illustrates that the measured flowrate is compared to a set point, and deviations from the set point are compensated by adjusting the valve, usually pneumatically. If the flowrate is too large, the valve is partially closed, restricting the flowrate. However, this also increases the frictional pressure loss across the valve, as discussed in Section 1.3.2.

The behavior of the system can be quantified by a *system curve*. The general equation for a chemical process, in terms of pressure, is given by the mechanical energy balance between Points 1 and 2 in Figure 1.21:

$$\Delta P_{pump} = \Delta P_{12} + \rho g \Delta z_{12} + (-\Delta P_f) + (-\Delta P_{cv}) = (P_2 - P_1) + \rho g(z_2 - z_1) + (-\Delta P_f) + (-\Delta P_{cv}) \quad (1.69)$$

where

$$-\Delta P_{ff} = \rho \varepsilon_f = \frac{2\rho f L_{eq} u^2}{D} \quad (1.70)$$

Equation (1.70) is derived from the mechanical energy balance with only the pressure and friction terms. It is important to remember that Δ represents out-in; therefore, the frictional loss term and the pressure loss across the valve are negative numbers before the included negative sign. The system curve is the right-hand side of Equation (1.69) without the term for the control valve:

$$\Delta P_{system} = (P_2 - P_1) + \rho g(z_2 - z_1) + (-\Delta P_{ff}) = (P_2 - P_1) + \rho g(z_2 - z_1) + \frac{32\rho f L_{eq} \dot{v}^2}{\pi^2 D^5} \quad (1.71)$$

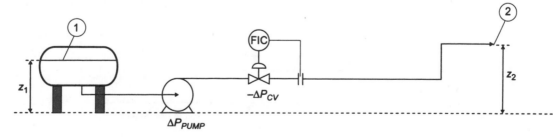

Figure 1.21   Physical situation for system curve

Equation (1.71) is a parabola, concave upward, on a plot of pressure increase versus flow-rate. It is of the form $\Delta P_{system} = a + b\dot{v}^2$, where $a = (P_2 - P_1) + \rho g(z_2 - z_1)$ and $b = 32\rho f L_{eq}/(\pi^2 D^5)$. Since the manufacturer pump curve is usually provided in head units, Equation (1.69) can be rewritten in head units as

$$h_{system} = \frac{\Delta P_{system}}{\rho g} = \frac{(P_2 - P_1)}{\rho g} + (z_2 - z_1) + \frac{32\, f L_{eq} \dot{v}^2}{g\pi^2 D^5} \tag{1.72}$$

Figure 1.22 illustrates the result if the pump curve and the system curve are plotted on the same graph. The indicated pressure changes demonstrate how the head provided by the pump must equal the desired head increase from source to destination, plus the frictional pressure loss, plus the pressure loss across the control valve, as quantified in Equation (1.69). The process of flowrate regulation is also illustrated in Figure 1.22. If the flowrate is to be reduced, the valve is closed, and the operating point moves to the left. At this lower flowrate, the frictional losses are lower, but the pressure loss across the valve is larger. The opposite is true for a higher flowrate. At the intersection of the two curves, the valve is wide open, and the maximum possible flowrate has been reached. This analysis assumes that the pump is operating at constant speed. For a variable speed pump, the pump curve moves up or down as the speed of rotation of the impeller changes. (Note that this simplified explanation omits the very small pressure drop across a wide-open control valve.) Operation to the right of this point is impossible. It is important not to confuse the meanings of the intersection points on the pump-system curve plot and the NPSH plot.

The pump and system curve plot also illustrates the cost of flowrate regulation. The pump must provide sufficient pressure to overcome the losses across the valve over a wide range of flowrates. Additional pump power is required for the possibility of operating at lower flowrates with a very large pressure drop across the valve. In general, this is a small cost for a pump, because the liquid density is high. Variable speed pumps are also available with different pump curves for different speeds. For these, the flowrate is regulated by the rotation speed of the impeller, not by a valve. It is not usually worth their extra cost for small pumps given the low cost of pumping liquids but may be worth considering for larger pumps and flowrates. Pumps with different impeller sizes have different pump curves for each impeller size. However, changing an impeller is not something that can be done while a process is operating.

Pumps (and compressors) are about the only pieces of equipment in a chemical plant with moving parts. Moving parts can fail. Therefore, since pumps are often inexpensive (on the order of

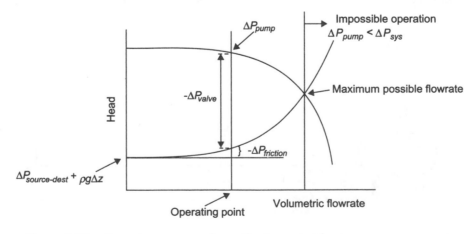

Figure 1.22 Pump (constant speed centrifugal pump) and system curve components

$10,000), a backup pump is typically installed in parallel so the plant can continue operating while the primary pump is maintained. Since shutdown and start-up can take days, it makes sense not to shut down a process that generates profit at a rate of thousands of dollars per minute to avoid purchasing a relatively inexpensive backup pump.

The presence of a backup pump can also be exploited if is necessary to scale-up a process. The piping system can be constructed such that the two pumps can operate simultaneously, either in series or in parallel. If the pumps are in series, the head increase doubles at the same flowrate. If the pumps are in parallel, the flowrate doubles at the same head increase. The pump curves for these situations are illustrated in Figure 1.23. The two system curves illustrate the maximum possible scale-up for two different system curves, indicated by the dots. In one case, the parallel configuration provides more scale-up potential, and in the other case, the series configuration provides more scale-up potential. This demonstrates that it is not possible to make any generalizations about which configuration can produce more scale-up. It all depends on the particular system.

Positive-displacement pumps perform differently from centrifugal pumps. They are usually used to produce higher pressure increases than are obtained with centrifugal pumps. The performance characteristics are represented on Figure 1.24(a), and these are sometimes referred to as *constant-volume pumps*. It can be observed that the flowrate through the pump is almost constant over a wide range of pressure increases, which makes flowrate control using the pressure increase impractical. One method to regulate the flow through a positive-displacement pump is illustrated in Figure 1.24(b). The strategy is to maintain constant flowrate through the pump. By regulating the flow of the recycle stream to maintain constant flowrate through the pump, the downstream flowrate can be regulated independently of the flow through the pump. Therefore, if a higher flow to the process is needed, then the by-pass control valve is closed, and vice versa.

It is observed from Figure 1.21 and Figure 1.24 that, in both cases, flowrate regulation occurs by adjusting a valve. For regulation of temperature, a valve on a cooling or heating fluid is adjusted. For regulation of concentration, valves on mixing streams are adjusted. This emphasizes the concept that about the only way to regulate anything in a chemical process is to adjust a valve position.

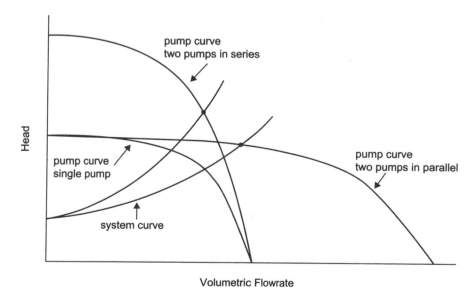

Figure 1.23   Pump and system curves for series and parallel pumps

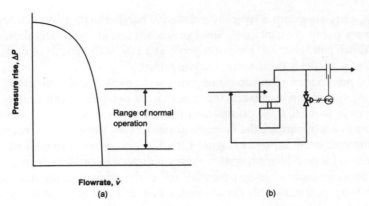

Figure 1.24   (a) Typical pump curve for positive-displacement pump and (b) method for flowrate regulation

> **Process conditions are usually regulated or modified by adjusting valve settings in the plant.**

Example 1.18

Develop the system curve for flow of water at approximately 10 kg/s through 100 m of 2-in, schedule-40, commercial-steel pipe with the source and destination at the same height and both at atmospheric pressure.

Solution

The density of water will be taken as 1000 kg/m^3, and the viscosity of water will be taken as 1 mPa s (0.001 kg/m s). The inside diameter of the pipe is 0.0525 m. The Reynolds number can be determined to be 2.42×10^5. For a roughness factor of 0.001, $f = 0.005$. Equation (1.71) reduces to

$$\Delta P = -19u^2 \tag{E1.18a}$$

since ΔP_{1-3} is zero, with ΔP in kPa and u in m/s. This is the equation of a parabola, and it is plotted in Figure E1.18. Therefore, from either the equation or the graph, the frictional pressure drop is known for any velocity.

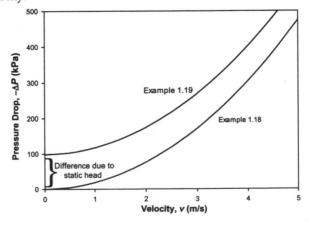

Figure E1.18   System curves for Examples 1.18 and 1.19

Example 1.19

Repeat Example 1.18 for the same length of pipe but with a 10 m vertical elevation change, with the flow from lower to higher elevation, but with the source and destination both still at atmospheric pressure.

Solution

Here, the potential energy term from the mechanical energy balance must be included. The magnitude of this term is 10 m of water, so $\rho g \Delta z$ = 98 kPa. Equation (1.69) reduces to

$$\Delta P = -(98 + 19 u^2)$$ (E1.19)

with ΔP in kPa and u in m/s. This equation is also plotted in Figure E1.18. It is observed that the system curve has the same shape as that in Example 1.18. This means that the frictional component is unchanged. The difference is that the entire curve is shifted up by the constant, static pressure difference.

Example 1.20

The centrifugal pump shown in Figure E1.20 is used to supply water to a storage tank. The pump inlet is at atmospheric pressure, and water is pumped up to the storage tank, which is open to atmosphere, via large-diameter pipes. Because the pipe diameters are large, the frictional losses in the pipes and any change in fluid velocity can be safely ignored.

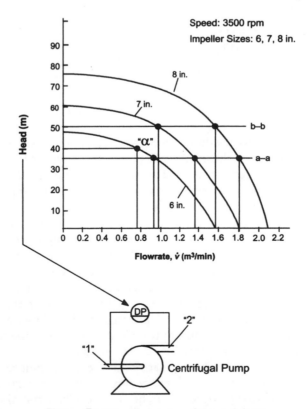

Figure E1.20   Illustration of Example 1.20

a. If the storage tank is located at an elevation of 35 m above the pump, predict the flow using each impeller.

b. If the storage tank is located at an elevation of 50 m above the pump, predict the flow using each impeller.

Solution

a. Figure E1.20 shows the pump curves for three different impeller sizes for the same pump. From Figure E1.20, at Δh_p = 35 m (see line a–a):

6-in Impeller: Flow = 0.93 m³/min
7-in Impeller: Flow = 1.38 m³/min
8-in Impeller: Flow = 1.81 m³/min

Therefore, each impeller can be used, and the larger impeller provides a larger flowrate.

b. From Figure E1.20, at Δh_p = 50 m (see line b–b):

6-in Impeller: Flow = 0 m³/min
7-in Impeller: Flow = 0.99 m³/min
8-in Impeller: Flow = 1.58 m³/min

In this case, the 6-in impeller is not sufficient to provide the desired flowrate, so only the 7-in and 8-in impellers are appropriate choices.

$$-\Delta h_p = 50 \text{ m} = (P_1 - P_2)/\rho g = P_1/\rho g - 1.2 \times 10^5/[750(9.81)] = P_1/\rho g - 16.3 \text{ m}$$
$$P_1 = (50 + 16.3)\rho g = 66.3(750)(9.81) = 4.88 \times 10^5 \text{ Pa} = 4.88 \text{ bar}$$

1.5.4   Compressors

1.5.4.1   Compressor Curves

The performance of centrifugal compressors is somewhat analogous to that of centrifugal pumps. A characteristic performance curve, supplied by the manufacturer, defines how the outlet pressure varies with flowrate. However, compressor behavior is far more complex than that for pumps because the fluid is compressible.

Figure 1.25 shows the performance curves for a centrifugal compressor. It is immediately observed that the *y*-axis is the ratio of the outlet pressure to inlet pressure. This is in contrast to pump curves, which have the difference between these two values on the *y*-axis. Curves for two different rotation speeds are shown. As with pump curves, curves for power and efficiency are often included but are not shown here. Unlike most pumps, the speed is often varied continuously to control the flowrate because the higher power required in a compressor makes it economical to avoid throttling the outlet as in a centrifugal pump.

Centrifugal compressor curves are read just like pump curves. At a given flowrate and revolutions per minute, there is one pressure ratio. The pressure ratio decreases as flowrate increases. A unique feature of compressor behavior occurs at low flowrates. It is observed that the pressure ratio increases with decreasing flowrate, reaches a maximum, and then decreases with decreasing flowrate. The locus of maxima is called the *surge line*. For safety reasons, compressors are operated to the right of the surge line. The surge line is significant for the following reason. Imagine the compressor is operating at a high flowrate and the flowrate is lowered continuously, causing a higher outlet pressure. At some point, the surge line is crossed, lowering the pressure ratio. This means that downstream fluid is at a higher pressure than upstream fluid, causing a backflow. These flow irregularities can severely damage the compressor mechanism, even causing the compressor to vibrate or surge (hence the origin of the term). Severe surging has been known to cause compressors to become detached from the supports keeping them stationary and literally to fly apart, causing great damage. Therefore, the surge line is considered a

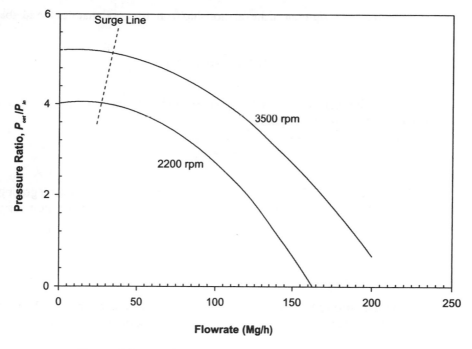

Figure 1.25 Performance curves for a centrifugal compressor

limiting operating condition below which operation is prohibited. Surge control on compressors is usually achieved by opening a bypass valve on a line connecting the outlet to the inlet of the compressor. When the surge point is approached, the bypass valve is opened, and gas flows from the outlet to the inlet, thereby increasing the flow through the compressor and moving it away from the surge condition.

   Positive-displacement compressors also exist and are used to compress low volumes to high pressures. Centrifugal compressors are used to compress higher volumes to moderate pressures and are often staged to obtain higher pressures. Figure 1.5 illustrates the inner workings of a compressor.

1.5.4.2 Compressor Staging

There are two limiting cases for compressor behavior: isothermal and isentropic. An actual compressor is neither isothermal nor isentropic; however, the behavior lies between these two limiting cases. From the general mechanical energy balance, compressor work is

$$\eta_c W_s = \int_{P_1}^{P_2} \frac{dP}{\rho} \tag{1.73}$$

where the subscripts 1 and 2 denote compressor inlet and outlet, respectively. For the isothermal case, assuming ideal gas behavior (which will fail as the pressure increases but is sufficient to illustrate the basic concepts),

$$\eta_c W_{s,isoth} = \int_{P_1}^{P_2} \frac{dP}{\rho} = \frac{RT}{M} \int_{P_1}^{P_2} \frac{dP}{P} = \frac{RT}{M} \ln\left(\frac{P_2}{P_1}\right) \tag{1.74}$$

For isentropic compression, the relationship from thermodynamics for adiabatic, reversible, compression is

$$PV^\gamma = \frac{P}{\rho^\gamma} = \text{constant} \tag{1.75}$$

where $\gamma = C_p/C_v$, the ratio of the constant pressure and constant volume heat capacities. Using the compressor inlet as a reference point,

$$\frac{P_1}{\rho_1^\gamma} = \frac{P}{\rho^\gamma} \tag{1.76}$$

Solving Equation (1.76) for ρ, using that value in Equation (1.73) and integrating, yields a well-known expression from thermodynamics for adiabatic, reversible, compression of an ideal gas:

$$\eta_c W_{s,isen} = \frac{\gamma RT_1}{M(\gamma-1)} \left[\left(\frac{P_2}{P_1} \right)^{\frac{\gamma-1}{\gamma}} - 1 \right] \tag{1.77}$$

Taking the ratio of Equations (1.74) and (1.77), and realizing that $T = T_1$ in Equation (1.74), since the temperature is constant at the inlet value in the isothermal case, yields

$$\frac{W_{s,isoth}}{W_{s,isen}} = \frac{\ln\left(\dfrac{P_2}{P_1} \right)}{\dfrac{\gamma}{\gamma-1} \left[\left(\dfrac{P_2}{P_1} \right)^{\frac{\gamma-1}{\gamma}} - 1 \right]} \tag{1.78}$$

Figure 1.26 is a plot of Equation (1.78), with the dependent variable as the compression ratio, P_2/P_1. Figure 1.26 demonstrates that the reversible, adiabatic work for isothermal compression is always less than that for isentropic compression. As the compression ratio exceeds 3 to 4, the isothermal work is significantly less than the isentropic work, making isothermal compression desirable. Of course, since compressing a gas always increases the gas temperature, isothermal compression cannot be accomplished. However, isothermal compression can be approached by staging compressors with intercooling, as illustrated in Figure 1.27 for a two-stage configuration. Isothermal compression can be reached theoretically with an infinite number of compressors each with an infinitesimal temperature rise, hardly a practical situation. From thermodynamics, it can be shown that the minimum compressor work for staged adiabatic compressors, with interstage cooling to the feed temperature to the first compressor, is accomplished with an equal compression ratio in each compressor stage. This is not necessarily the economic optimum, which would require analysis of the capital cost of the compressor stages and heat exchangers, the operating cost of the compressor, and the utility cost of the cooling medium. However, the preceding analysis explains why compressors are usually staged when the compression ratio exceeds 3 to 4.

1.5.5   Performance of the Feed Section to a Process

A common feature of chemical processes is the mixing of reactant feeds before they enter a reactor. When two streams mix, they are at the same pressure. The consequences of this are illustrated by the following scenario.

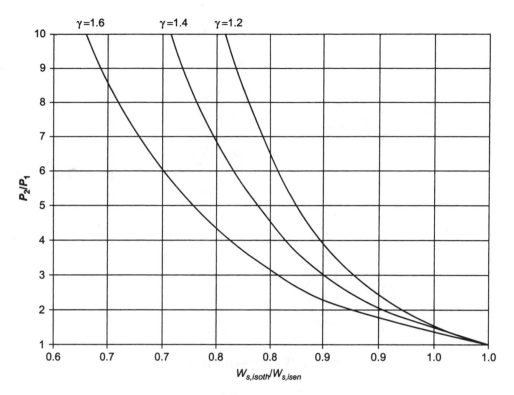

Figure 1.26 Comparison of isothermal and isentropic work for compressors

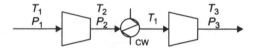

Figure 1.27 Example of two-stage compressor configuration

Phthalic anhydride can be produced by reacting naphthalene and oxygen. The feed section to a phthalic anhydride process is shown in Figure 1.28. The mixed feed enters a fluidized bed reactor operating at five times the minimum fluidization velocity. A stream table is given in Table 1.4. It is assumed that all frictional pressure losses are associated with equipment and that frictional losses in the piping are negligible. It is temporarily necessary to scale down production by 50%. The engineer must determine how to scale down the process and to determine the new flows and pressures.

It is necessary to have pump and compressor curves in order to do the required calculations. In this example, equations for the pump curves are used. These equations can be obtained by fitting a polynomial to the curves provided by pump manufacturers. As discussed in Section 1.5.3, pump curves are usually expressed as pressure head versus volumetric flowrate so that they can be used for a liquid of any density. In this example, pressure head and volumetric flowrate have been converted to absolute pressure and mass flowrate using the density of the fluids involved. Pump P-201 operates at only one speed, and an equation for the pump curve is

$$\Delta P(\text{kPa}) = 500 + 4.663\dot{m} - 1.805\dot{m}^2 \quad \dot{m} \leq 16.00 \text{ Mg/h} \tag{1.79}$$

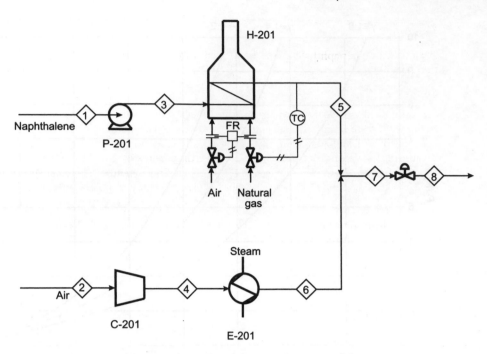

Figure 1.28   Feed section to phthalic anhydride process

Table 1.4   Partial Stream Table for Feed Section in Figure 1.27

	Stream							
	1	2	3	4	5	6	7	8
P (kPa)	80.00	101.33	343.00	268.00	243.00	243.00	243.00	200.00
Phase	L	V	L	V	V	V	V	V
Naphthalene (Mg/h)	12.82	—	12.82	—	12.82	—	12.82	12.82
Air (Mg/h)	—	151.47	—	151.47	—	151.47	151.47	151.47

Compressor C-201 operates at only one speed, and the equation for the compressor curve is

$$\frac{P_{out}}{P_{in}} = 5.201 + 2.662 \times 10^{-3}\,\dot{m} - 1.358 \times 10^{-4}\,\dot{m}^2$$

$$+ 4.506 \times 10^{-8}\,\dot{m}^3 \qquad \dot{m} \le 200\ \text{Mg/h} \tag{1.80}$$

From Figure 1.27, it is seen that there is only one valve in the feed section, after the mixing point. Therefore, the only way to reduce the production of phthalic anhydride is to close the valve to the point at which the naphthalene feed is reduced by 50%. Example 1.21 illustrates the consequences of reducing the naphthalene feed rate by 50%.

Example 1.21

For a reduction in naphthalene feed by 50%, determine the pressures and flows of all streams after the scale-down.

Solution

Because it is known that the flowrate of naphthalene has been reduced by 50%, the new outlet pressure from P-201 can be calculated from Equation (1.79). The feed pressure remains at 80 kPa. At a naphthalene flow of 6.41 Mg/h, Equation (1.79) gives a pressure increase of 455.73 kPa, so P_3 = 535.73 kPa. Because the flowrate has decreased by a factor of 2, the pressure drop in the fired heater decreases by a factor of 4, since $\Delta P \propto L_{eq} \dot{v}^2$. Therefore, P_5 = 510.73 kPa. Consequently, the pressure of Stream 6 must be 510.73 kPa. The flowrate of air can now be calculated from the compressor curve equation.

The compressor curve equation has two unknowns: the compressor outlet pressure and the mass flowrate. Therefore, a second equation is needed. The second equation is obtained from a base-case ratio for the pressure drop across the heat exchanger. The two equations are

$$\frac{P_4}{101.33} = 5.201 + 2.662 \times 10^{-3}\, \dot{m}_{2,new}^2 - 1.358 \times 10^{-4}\, \dot{m}_{2,new}^2 + 4.506 \times 10^{-8}\, \dot{m}_{2,new}^3 \qquad \text{(E1.21a)}$$

$$P_4 - 510.73 = 25\left(\frac{\dot{m}_{2,new}}{151.47}\right)^2 \qquad \text{(E1.21b)}$$

The solution is

$$P_4 = 512.84 \text{ kPa}$$
$$\dot{m}_2 = 43.80 \text{ Mg/h}$$

The stream table for the scaled-down case is given in Table E1.21. Although it is not precisely true, for lack of additional information, it has been assumed that the pressure of Stream 8 remains constant.

It is observed that the flowrate of air is reduced by far more than 50% in the scaled-down case. This is because of the combination of the compressor curve and the new pressure of Streams 5 and 6 after the naphthalene flowrate is scaled down by 50%. The total flowrate of Stream 8 is now 50.21 Mg/h, which is 30.6% of the original flowrate to the reactor. Given that the reactor was operating at five times minimum fluidization, the reactor is now in danger of not being fluidized adequately. Because the phthalic anhydride reaction is very exothermic, a loss of fluidization could result in poor heat transfer, which might result in a runaway reaction. The conclusion is that it is not recommended to operate at these scaled-down conditions.

Table E1.21   Partial Stream Table for Scaled-Down Feed Section in Figure 1.28

	Stream							
	1	2	3	4	5	6	7	8
P (kPa)	80.0	101.33	535.73	512.84	510.73	510.73	510.73	200.00
Phase	L	V	L	V	V	V	V	V
Naphthalene (Mg/h)	6.41	—	6.41	—	6.41	—	6.41	6.41
Air (Mg/h)	—	43.80	—	43.80	—	43.80	43.80	43.80

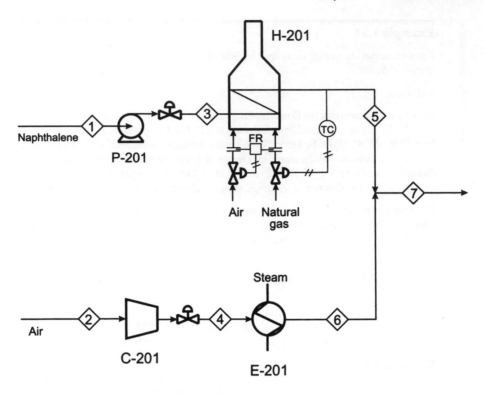

Figure E1.21   Feed section to phthalic anhydride process with better valve placement than shown in Figure 1.28

The question is how the air flowrate can be scaled down by 50% to maintain the same ratio of naphthalene to air as in the original case. The answer is in valve placement. Because of the requirement that the pressures at the mixing point be equal, with only one valve after the mixing point, there is only one possible flowrate of air corresponding to a 50% reduction in naphthalene flowrate. Effectively, there is no control of the air flowrate. A chemical process would not be designed as in Figure 1.28. The most common design is illustrated in Figure E1.21. With valves in both feed streams, the flowrates of each stream can be controlled independently.

WHAT YOU SHOULD HAVE LEARNED:

- How to write the mass balance for pipe flow
- How to apply the mechanical energy balance to pipe flow
- How to apply the force balance to flow around submerged objects
- The types of pipes and pipe sizing
- The types of pumps and compressors and their applicability
- The purpose of including valves in a piping system
- How to design and analyze performance of a system for frictional flow of fluid in pipes
- How to design a system for frictional flow of fluid with submerged objects such as packed and fluidized beds

- Methods for flow measurement
- How to analyze existing fluid flow equipment
- What net positive suction head is and the limitations it places on piping system design
- How to analyze pump and system curves to understand the limitations of pumps
- Why compressors are staged

NOTATION

Symbol	Definition	SI Units
A	cross-sectional area for flow	m^2
A_t	total cross-sectional area of packed bed	m^2
C_D	drag coefficient	
C_p	constant-pressure heat capacity	kJ/kmol/K or kJ/kmol/°C
C_v	constant-volume heat capacity	kJ/kmol/K or kJ/kmol/°C
D	pipe diameter	m
D_p	particle diameter	m
D_s	diameter of sphere	m
D_s^*	dimensionless variable within Haider-Levenspiel equation	m
D_v	diameter of sphere with same surface area as nonsphere	m
e	pipe roughness factor	m
e_f	energy dissipated by friction	J/kg
f	friction factor	
F	force	N
F_d	drag force	N
F_g	gravitational force	N
F_p	pressure force	N
g	acceleration due to gravity	m/s^2
G	superficial mass velocity	$kg/m^2/s$
h	height or head	m
h_{mf}	bed height at minimum fluidization	m
K	loss coefficient for elbows, fittings, etc.	
L	length	m
L_{eq}	equivalent length of pipe	m

Symbol	Definition	SI Units
$\dot{m}$	mass flowrate	kg/s
M	molecular weight	kg/kmol
$NPSH_A$	net positive suction head available	m (or Pa)
$NPSH_R$	net positive suction head required	m (or Pa)
P	pressure	Pa
$P*$	vapor pressure	Pa
R	restoring force to keep elbow stationary	N
Re	Reynolds number	
Re_{mf}	Reynolds number at minimum fluidization	
Re_t	Reynolds number at terminal velocity	
T	absolute temperature	K
u	velocity	m/s
u_t^*	dimensionless variable within Haider-Levenspiel equation	
u_s	superficial volumetric velocity in packed or fluidized bed	m/s
u_t	terminal velocity	m/s
$\dot{v}$	volumetric flowrate	m³/s
V	volume	m³
W_s	shaft work	J/kg
$\dot{W}_s$	shaft power	W
z	coordinate in direction opposite gravity	m

GREEK SYMBOLS

β	orifice diameter/pipe diameter	
γ	Cp/Cv	
ε	void fraction	
η	efficiency	
η_c	compressor efficiency	
η_p	pump efficiency	
η_t	turbine efficiency	
μ	viscosity	kg/m/s
ρ	density (usually of fluid)	kg/m³
ρ_s	solid (particle) density	kg/m³
Ψ	sphericity	

REFERENCES

Bird, R. B., W. E. Stewart, and E. N. Lightfoot. 2006. *Transport Phenomena*, 2nd ed. rev. New York: Wiley, p. 51.

Cheremisinoff, N. P., and P. N. Cheremisinoff. *Instrumentation for Process Flow Engineering*. Lancaster, PA: Technomic.

Couper, J. R., W. R. Penney, J. R. Fair, and S. M. Walas. 2012. *Chemical Process Equipment: Selection and Design*, 3rd ed. New York: Elsevier, Chapter 7.

Darby, R. 2001. *Chemical Engineering Fluid Mechanics*, 2nd ed. New York: Marcel Dekker, p. 209.

Geankoplis, C. 2003. *Transport Processes and Separation Process Principles*, 4th ed. Upper Saddle River, NJ: Prentice Hall, pp. 99–100.

Green, D. W., and R. H. Perry. 2008. *Perry's Chemical Engineers' Handbook*, 8th ed. New York: McGraw-Hill, Section 10.

Haider, A., and O. Levenspiel. 1989. "Drag Coefficient and Terminal Velocity of Spheres and Nonspherical Particles." *Powder Technol* 58: 63–70.

Kittredge, C. P., and D. S. Rowley. 1957. "Resistance Coefficients for Laminar and Turbulent Flow Through One-Half-Inch Valves and Fittings." *Trans ASME* 79, 1759–66.

Kunii, D., and O. Levenspiel. 1991. *Fluidization Engineering*, 2nd ed. Stoneham, MA: Butterworth-Heinemann.

Levenspiel, O. 2014. *Engineering Flow and Heat Exchange*, 3rd ed. New York: Springer, p. 27.

McCabe, W. L., J. C. Smith, and P. Harriott. 1993. *Unit Operations of Chemical Engineering*, 5th ed. New York: McGraw-Hill.

Miller, R.W. 1983. *Flow Measurement Engineering Handbook*. New York: McGraw-Hill.

Pavlov, K. F., P. G. Romankov, and A. A. Noskov. 1981. *Problems and Examples for a Course in Basic Operations and Equipment in Chemical Technology*. Moscow: MIR, translated.

Perry, R. H., and D. Green. 1984. *Perry's Chemical Engineers' Handbook*, 6th ed. New York: McGraw-Hill, Section 5.

Walas, S. 1988. *Chemical Process Equipment: Selection and Design*. Stoneham, MA: Butterworth.

Wen, C. Y., and Y.-H. Yu. 1966. "A Generalized Method for Predicting the Minimum Fluidization Velocity." *AIChE Journal* 12: 610–612.

PROBLEMS

Short Answer Problems

1. Explain the physical meaning of each term in the mechanical energy balance.
2. Fluid flows from a larger-diameter pipe to a smaller-diameter pipe. How does the velocity change?
3. Explain the concept of pressure head.
4. Fluid flows downward in a vertical pipe of uniform diameter. How does the velocity change with position?
5. A liquid flows vertically downward through a pipe of uniform diameter at steady state. Explain how the mass flowrate, volumetric flowrate, and velocity change with vertical position.

6. Explain the meaning of the Reynolds number in terms of forces.

7. Sketch the approximate shape of a graph of frictional losses versus Reynolds number. Discuss two other situations in which the graph has similar shape.

8. There are three key parameters that affect frictional loss in pipe flow. State two of them and explain the effect (i.e., whether the parameter increases or decreases, how the frictional loss is affected).

9. For sections of pipes in series, what is the relationship between the mass flowrate in each section? What is the relationship between the pressure drops in each section?

10. For sections of pipes in parallel, what is the relationship between the mass flowrate in each section? What is the relationship between the pressure drops in each section?

11. How is the mechanical energy balance different for compressible flow compared to incompressible flow?

12. What is the difference between form drag and frictional drag?

13. Define void fraction.

14. Explain the difference between void volume, solid volume, and total volume.

15. Define sphericity.

16. When is mercury a better manometer fluid than water or oil? When is mercury not recommended? Assume the specific gravity of mercury is 13.2, and the specific gravity of oil is 0.8.

17. Explain the physical meaning of the intersection of the $NPSH_R$ and $NPSH_A$ curves.

18. Explain the physical meaning of the intersection of the pump and system curves.

19. Why does a compressor cost more to operate than a pump?

20. Why are compressors often staged with intercooling?

21. For fully developed turbulent flow, assuming all variables not mentioned are held constant:
    a. What is the effect of doubling the flowrate on the pressure drop?
    b. What is the effect of increasing the pipe diameter by 25% on pressure drop?
    c. What is the effect of increasing the pipe diameter on flowrate?
    d. What is the effect of increasing pipe length on pressure drop?
    e. What is the effect of increasing pipe length on flowrate?
    f. What is the effect on pressure drop of replacing one long pipe segment with two equal-sized pipe segments of half the length placed in parallel?

22. Repeat Problem 1.21 for laminar flow.

Problems to Solve

23. Consider the situation depicted in Figure P1.23. The fluid is an oil with a specific gravity of 0.85. Fill in the missing data in Table P1.23.

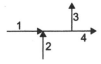

Figure P1.23

Table P1.23

Stream	Pipe	$\dot{m}$ (kg/s)	$\dot{v}$ (m³/s)	u (m/s)
1	2-in, schedule 40	6		
2	3.5-in, schedule 40		0.0106	
3	1.5-in, schedule 40			4.032
4	3-in, schedule 40			

24. For water flowing in the situation shown in Figure P1.24 and the data in Table P1.24, do the following:

   a. Calculate the mass flowrate of Stream 3.

   b. Calculate the velocity of Stream 4.

   c. What schedule-40 pipe size must be used in Stream 3?

Figure P1.24

Table P1.24

Stream	Pipe	v (m/s)
1	1-in, schedule 40	5
2	1.5-in, schedule 40	3
3	?-in, schedule 40	2.18
4	4-in, schedule 40	?

25. Consider the situation depicted in Figure P1.25. The liquid level in the cylindrical tank is increasing at 0.02 ft/sec.

   a. What is the **net** rate of flow into the tank?

   b. What is the velocity in the 3-in pipe?

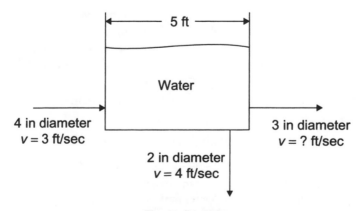

Figure P1.25

26. Water is pumped through a 750-ft length of 6-in, schedule-40 pipe. The discharge at the end of the pipe is 50 ft above the suction end. The pump is 80% efficient and is driven by a 20 hp motor. If the friction loss in the pipe is 50 ft lb$_f$/lb, what is the flowrate through the pipe?

27. A hydroelectric power plant takes 25 m^3/s of water from a large reservoir through its turbine and discharges it to the atmosphere at 1 m/s. The turbine is 50 m below the reservoir surface. The frictional head loss in the system is 10 m. The turbine and electric generator as a whole are 80% efficient. Assuming turbulent flow, calculate the power extracted by the turbine.

28. Water is pumped at a constant rate of 10 m^3/h from an open tank on the floor to an open tank with a level 10 m above the floor. Frictional losses in the 50-mm-diameter pipe between the tanks are 3.5 J/kg. At what height above the floor must the water level be kept if the pump can develop only 0.1 kW? The pump is 75% efficient.

29. Water in a dam on a 75-ft-deep river is passed through a turbine to produce energy. The outlet of the turbine is 15 ft above the river bed. The mass flowrate of water is 65,000 lb/s, and the inside diameter of the discharge pipe is 10 ft. Discharge is to the atmosphere, and frictional losses may be neglected. For a 55% efficient turbine, calculate the power produced.

30. Many potential drugs have low water solubility, hindering their transport in the body's aqueous material distribution medium (blood). One way to improve solubility is to decrease particle size below 1 μm diameter. One method to do this is by high-pressure homogenization, using a homogenizer, which is basically a nozzle. In this process, the drug is dispersed in a solvent and forced through a narrow orifice (nozzle) at high pressure. As the liquid enters the orifice, it experiences a pressure drop so great it partially vaporizes. As it exits the constriction, the vapor bubbles collapse violently and produce local disturbances, breaking up the surrounding solid particles.

    A high-pressure homogenizer is being used to decrease the size of some drug particles. The particles are suspended in water at 25°C and sent through the homogenizer at 250 mL/min. The pressure before the orifice is 34.5 MPa, and the diameter of the pipe is 0.1 m. Determine the diameter orifice (nozzle) that results in an exit pressure at the vapor pressure of water. Neglect friction.

31. A pump operating at 80% efficiency delivers 30 gal/min of water from a reservoir to an open-air storage tank at a chemical plant 1 mi away. A 3-in, schedule-40 pipe is used, and the frictional losses are 200 ft lb$_f$/lb. The elevation of the liquid level in the tank is 873 ft above sea level, and the elevation of the liquid level in the reservoir is 928 ft above sea level.

    a. What is the minimum horsepower required for the pump?

    b. The elevation of the reservoir is fixed. What elevation of the liquid level of the tank would make the pump unnecessary?

32. A pressurized tank situated above ground level contains a liquid with specific gravity of 0.9. The liquid flows down to ground level through 4-in, schedule-40 pipe through a pump (75% efficiency) and into a tank at a level 25 m above the level of the source tank at a pressure of 550 kPa through 2-in, schedule-40 pipe. The pump power is 6.71 kW. A pressure gauge at the pump entrance reads 115.6 kPa, and a pressure gauge at the pump discharge reads 762.6 kPa. The frictional losses in the piping on the suction side of the pump and on the discharge side of the pump are 30 J/kg and 50 J/kg, respectively.

    a. What is the mass flowrate of liquid through the system?

    b. What is the velocity in the 2-in, schedule-40 pipe?

    c. What is the pressure of the liquid in the source tank?

    d. Determine whether the kinetic energy contribution to the mechanical energy balance is small.

33. Water is pumped from one storage tank to a higher tank at a steady rate of 10^{-3} m^3/s. The difference in the elevations of the two water tanks is 50 m. The storage tank, which serves a source, is open to the atmosphere, while the tank receiving the water has a pressure of 170.3 kPa. Pressure

gauges in the pipeline at the inlet and outlet of the pump read 34.5 kPa and 551.6 kPa, respectively. The power supplied by an electric motor to the pump shaft is 1000 W. All piping is 1-in, schedule-40 steel pipe. Find the pump efficiency and friction loss in the pipe per kg of water.

34. Oil (SG = 0.88) flows at 5 ft^3/s from one tank, through a pump, to another tank. The pipe diameter between the source tank and the pump is 12 in, and the pipe diameter between the pump and the destination tank is 6 in. The liquid level in the source tank is 10 ft above the pump, which is at ground level. The liquid level in the destination tank is 12 ft. The source tank is at 25 psia, and the destination tank is open to the atmosphere. A manometer is connected to the upstream and downstream pipes, immediately adjacent to the pump, with a differential height of 36 in of mercury. The pump is 75% efficient. Frictional losses may be neglected.

    a. What is the power rating (fix) of the pump?

    b. What is the maximum possible height of the bottom of the destination tank?

35. A fluid with specific gravity of 0.8 is in a tank, at a pressure of 150 kPa, with a level maintained at 5 m above ground level. The fluid leaves the tank through 4-in, schedule-40 pipe (frictional loss of 30 J/kg) at a mass flowrate of 6.5 kg/s and enters a pump at ground level. The pump power is 1.5 kW and is 70% efficient. The fluid leaving the pump flows through 3.5-in, schedule-40 pipe (frictional loss of 50 J/kg) to a "final" point in the pipe above the original tank level, where the pressure is 200 kPa.

    a. Find the velocity at the final point in the pipe.

    b. Determine the pressure at the pump inlet.

    c. Determine the height above the ground of the final point in the pipe.

36. Consider the problem of how long it takes for a tank to drain. Consider an open-top cylindrical tank with one horizontal exit pipe at the bottom of the tank that discharges to the atmosphere. The tank has a diameter, d, and the height of liquid in the tank at any time is h.

    a. The mass balance is unsteady state. Explain why the mass balance is

    $$\frac{dm}{dt} = -\dot{m}_{out}$$

    where m is the mass of liquid in the tank and $\dot{m}_{out}$ is the mass flowrate out of the tank.

    b. The mass in the tank is the fluid density times the volume of liquid in the tank. The flowrate; $\dot{m}_{out}$, can be related to the velocity and the cross-sectional area of the exit pipe based on what we have already learned. The volume of liquid in the tank can be related to the height of liquid in the tank. Simplify the differential mass balance to obtain an expression for the height of liquid in the tank as a function of the velocity of the liquid through the exit pipe.

    c. Now, write a mechanical energy balance on the fluid in the tank and pipe from the top level in the tank to the pipe outlet, neglecting friction. It is generally assumed the velocity of the tank level (i.e., the fluid level in the tank) is small because of the large diameter. Solve for the velocity, and rearrange the differential equation to look like

    $$\frac{dh}{dt} = af(h)$$

where a is a group of constants and $f(h)$ is a function of the height that you have derived.

d. Solve this differential equation for height as a function of time with the initial condition of a height of h_o at time zero.

e. Rearrange the answer to Part (d) to get an expression for the time for complete drainage.

37. The following equations describe a fluid-flow system. Draw and label the system.

$$\dot{m}_1 + \dot{m}_2 = \dot{m}_3 = \dot{m}_4$$

$$\frac{P_4 - P_3}{\rho} + \frac{v_4^2 - v_3^2}{2} - \eta W_s = 0$$

$$\frac{P_5 - P_4}{\rho} - \frac{v_4^2}{2} + g(z_5 - z_4) + e_f = 0$$

38. The following equations describe a fluid flow system, with friction neglected. Draw and label the system, making sure that your diagram is visually accurate.

$$\dot{m}_3 + \dot{m}_2 = \dot{m}_4 = \dot{m}_5$$

$$\frac{P_2 - P_1}{\rho} + g(z_2 - z_1) + \frac{v_2^2}{2} = 0$$

$$g(z_6 - z_1) - \eta W_s = 0$$

$$\frac{P_5 - P_4}{\rho} + \frac{v_5^2 - v_4^2}{2} - \eta W_s = 0$$

$$\frac{P_6 - P_5}{\rho} + g(z_6 - z_5) - \frac{v_5^2}{2} = 0$$

$$z_6 > z_1$$

39. An aneurysm is a weakening of the walls of an artery causing a ballooning of the arterial wall. The result is a region of larger diameter than a normal artery. If the "balloon" ruptures in a high-blood-flow area, such as the aorta, death is almost instantaneous. Fortunately, there are often symptoms due to slow leakage that can precede rupture. What happens to the blood velocity as it passes through the aneurysm? Justify your answer using equations. In the human body, very small changes in pressure can be significant. Using the mechanical energy balance, neglecting only friction and potential energy effects, analyze the pressure change as blood enters the aneurysm a large distance from the heart, so that the pulse flow is not an issue. The blood is flowing in a region not in the vicinity of the heart. What is the effect of the observed pressure change?

40. A pipeline is replaced by new 2-in, schedule-40, commercial-steel pipe. What power would be required to pump water at a rate of 100 gpm through 6000 ft of this pipe?

41. Hot water at 43°C flows from a constant-level tank through 2-in, schedule-40, commercial-steel pipe, from which it emerges 12.2 m below the level in the tank. The equivalent length of the piping system is 45.1 m. Calculate the rate of flow in m^3/s.

42. Crude oil (μ = 40 cP, SG = 0.87) is to be pumped from a storage tank to a refinery through a series of pump stations via 10-in, schedule-20, commercial-steel pipeline at a flowrate of 2000 gpm. The pipeline is 50 mi long and contains 35 90° elbows and 10 open gate valves. The pipeline exit is 150 ft higher than the entrance, and the exit pressure is 25 psig. What horsepower is required to drive the pumps if they are 70% efficient?

43. A pipeline to carry 1 million bbl/day of crude oil (1 bbl = 42 gal, SG = 0.9, μ = 25 cP) is constructed with 50-in-inside-diameter, commercial-steel pipe and is 700 mi long. The source and destination are at atmospheric pressure and the same elevation. There are 50 wide-open gate valves, 25 half-open globe valves, and 50 45° elbows. There will be 25 identical pumps along this pipeline, each with an efficiency of 70%. What is the power required for each pump?

44. A pump draws a solution of specific gravity 1.2 with the viscosity of water from a ground-level storage tank at 50 psia through 3.5-in, schedule-40, commercial-steel pipe at a rate of 12 lb/s. The pump produces 4.5 hp with an efficiency of 75%. The pump discharges through a 2.5-in, schedule-40 commercial steel pipe to an overhead tank at 100 psia, which is 50 ft above the level of solution in the feed tank. The suction line has an equivalent length of 20 ft, including the tank exit. The discharge line contains a half-open globe valve, two wide-open gate valves, and two 90° elbows. What is the maximum total length of discharge piping allowed for this pump to work?

45. Many chemical plants store fuel oil in a "tank farm" on the outskirts of the plant. To prevent an environmental disaster, there are specific rules regarding the design of such facilities. One such rule is that there be an emergency dump tank with the capacity of the largest storage tank. Should a leak or structural problem occur with a tank, the fuel oil can be pumped into the emergency dump tank.

    Consider the design of the pumping system from a 250 m^3 tank storing #6 fuel oil into a 250 m^3 dump tank. The viscosity of #6 fuel oil is 0.8 kg/m s, and its density is 999.5 kg/m^3. The piping system consists of 43 m of commercial-steel pipe, four 90° flanged regular elbows, a sharp entrance, an exit, and a pump. The oil must be pumped to an elevation 3.35 m above the exit point from the source tank.

    a. If 20-in, schedule-40, commercial-steel pipe is used, and if it is necessary to accomplish the transfer within 45 min, determine the power rating required of the pump. Assume the pump is 80% efficient.

    b. If the pump to be used has 10 kW at 80% efficiency, and the pipe is 20-in, schedule-40, commercial-steel pipe, determine how long the transfer will take.

    c. If the pump to be used has 20 kW at 80% efficiency, and the transfer is to be accomplished in 45 min, determine the required schedule-40 pipe size.

46. Two parallel sections of pipe branch from the same split point. Both branches end at the same pressure and the same elevation. Branch 1 is 3-in, schedule-40, commercial-steel pipe and has an equivalent length of 12 m. Branch 2 is 2-in, schedule-80, commercial-steel pipe and has an equivalent length of 9 m.

    a. Assuming fully turbulent flow, what is the split ratio between the two branches?

    b. Suppose that Branch 2 ends 5 m higher than Branch 1. What is the split ratio between the branches in this case?

47. Consider a two-pipes-in-series system: that is, Pipe 1 is followed by Pipe 2. The liquid is water at room temperature with a mass flowrate of 2 kg/s. The pipes are horizontal. Calculate the pressure drop across these two pipes and the power necessary to overcome the frictional loss. Ignore the minor losses due to the pipe fitting. The pipe data are:

Pipe	L (m)	Pipe Size	Material
1	10	1-in, schedule-40	Commercial steel
2	15	2-in, schedule-40	Cast iron

48. Assume that the same two pipes in Problem 1.47 are now in parallel with the same total pressure drop. Compute the mass flowrate in each section of pipe. Neglect additional frictional losses due to the parallel piping. Explain the reason for the observed split between the parallel pipes.

49. A pipe system to pump #6 fuel oil (μ = 0.8 kg/m s, ρ = 999.5 kg/m^3) consists of 50 m of 8-in, schedule-40, commercial-steel pipe. It has been observed that the pressure drop is 8.79×10^4 Pa.

    a. Determine the volumetric flowrate of the fuel oil.

    b. Extra capacity is needed. Therefore, it has been decided to add a parallel line of the same length (neglect minor losses) using 5-in, schedule-40, commercial-steel pipe. By what factor will the fuel oil volumetric flowrate increase?

50. There are three equal-length sections of identical 3-in, schedule-40, commercial-steel pipe in series. An increased flowrate of 20% is needed. How is the pressure drop affected? It is decided to replace the second section with two equal-length, identical sections of the original pipe in parallel. How is the pressure drop in this system affected relative to the original case? Neglect minor losses due to elbows and fittings and assume fully developed turbulent flow.

51. Consider two parallel arteries of the same length, both fed by a main artery. The flowrate in the main artery is 10^{-6} m^3/s. One branch is stenotic (has plaque build-up due to too many Big Macs, Double Whoppers, etc.). The stenotic artery will be modeled as a rigid pipe with 60% the diameter of the healthy artery (diameter of 0.1 cm). For this problem, blood may be considered to be a Newtonian fluid with the properties of water. What fraction of the blood flows in each arterial branch? Be sure to validate any assumptions made.

52. One of the potential benefits of the production of shale gas is that certain seams of the gas contain significant amounts of ethane, which can be cracked into ethylene, a building block for many other common chemicals (polyethylene, ethylene oxide, which is made into ethylene glycol among many others, and tetrafluoroethylene, the monomer for Teflon). Assume that a cracker plant produces ethylene (C_2H_4) at 5 atm and 70°F. It is to be delivered by pipeline to a neighboring plant, which was built near the ethylene cracker facility, which is 10 miles away. The pressure at the neighboring plant entrance must be 2.5 atm. It has been suggested that 6-in, schedule-40, commercial-steel pipe be used. What delivery mass flowrate is possible with this pipe size? If you need to make an assumption, do so and prove its validity.

53. Natural gas (methane, μ = 10^{-5} kg/m s) must flow in a pipeline between compression stations. The compressor inlet pressure is 250 kPa, and its outlet pressure is 1000 kPa. Assume isothermal flow at 25°C. The pipe is 6-in, schedule-40, commercial steel. The mass flowrate is 2 kg/s. What is the required distance between pumping stations?

54. Your plant produces ethylene at 6 atm and 60°F. It is to be delivered to a neighboring plant 5 miles away via pipeline, and the pressure at the neighboring plant entrance must be 2 atm. The contracted delivery flowrate is 2 lb/sec. It has been suggested that 4-in, schedule-40, commercial-steel pipe be used. Evaluate this suggestion. Be sure to validate any assumptions made.

55. Calculate the terminal velocity of a 2 mm diameter lead sphere (SG = 11.3) dropped in air. The properties of air are ρ = 1.22 kg/m^3 and μ = 1.81×10^{-5} kg/m s.

56. A packed bed is composed of crushed rock with a density of 200 lb/ft^3 with an assumed particle diameter of 0.15 in. The bed is 8 ft deep, has a porosity of 0.3, and is covered by a 3 ft layer of water that drains by gravity through the bed. Calculate the velocity of water through the bed, assuming the water enters and exits at 1 atm pressure.

57. A hollow steel sphere, 5 mm in diameter with a mass of 0.05 g, is released in a column of liquid and attains an upward terminal velocity of 0.005 m/s. The liquid density is 900 kg/m^3,

and the sphere is far enough from the container walls so that their effect may be neglected. Determine the viscosity of the liquid in kg/m s. Hint: Assume Stokes flow and confirm the assumption with your answer.

58. In a particular sedimentation vessel, small particles (SG = 1.1) are settling in water at 25°C. The particles have a diameter of 0.1 mm. What is the terminal velocity of the particles? Validate any assumptions made.

59. At West Virginia University, each Halloween, there is a pumpkin-drop contest. College, high-school, and middle-school students participate. The goal is to drop a pumpkin off the top of the main engineering building (assume about 100 ft) and have it land close to a target without being damaged. Packing and parachutes are commonly used. You have a theory that the terminal velocity at which a pumpkin packed in your newly invented, proprietary bubble wrap can hit the ground and remain intact is 50 m/s. You will use no parachute, and the shape will be approximately spherical. The pumpkin plus wrapping has a diameter of 40 cm. By calculating the actual terminal velocity, determine whether the pumpkin will exceed the desired terminal velocity. Assume that the wrapped pumpkin has the specific gravity of water, and assume the air is at 25°C and 1 atm.

60. Air enters and passes up through a packed bed of solids 1 m in height. Using the data provided, what are the pressure drop and the outlet pressure?

data: $v_s = 1$ m/s $P_{inlet} = 0.2$ MPa

$T = 293$ K $\mu = 1.8 \times 10^{-5}$ kg/m/s

$D_p = 1$ mm $\varepsilon = 0.4$

$\rho_s = 9500$ kg/m^3

61. In the regeneration of a packed bed of ion-exchange resin, hydrochloric acid (SG = 1.2, $\mu = 0.002$ kg/m s) flows upward through a bed of resin particles (particle density of 2500 kg/m^3). The bed is 40 cm in diameter, and the particles are spherical with a diameter of 2 mm and a bed void fraction of 0.4. The bed is 2 m deep, and the bottom of the bed is 2 m off of the ground. The acid is pumped at a rate of 2×10^{-5} m^3/s from an atmospheric pressure, ground-level storage tank through the packed bed and into another atmospheric pressure, ground-level storage tank, in which the filled height is 2 m. The complete piping system consists of 75 equivalent meters of 4-in, schedule-40, commercial-steel pipe.

   a. Determine the required power of a 75% efficient pump for this duty. Remember that a pump must be sized for the maximum duty needed.

   b. What do you learn from the numbers in Part (a) regarding the relative magnitudes of the maximum duty and the steady-state duty?

   c. What is the pressure rise needed for the pump?

62. A gravity filter is made from a bed of granular particles assumed to be spherical. The bed porosity is 0.40. The bed has a diameter of 0.3 m and is 1.75 m deep. The volumetric flowrate of water at 25°C through the bed is 0.006 m^3/s. What particle diameter is required to obtain this flowrate?

63. Calculate the flowrate of air at standard conditions required to fluidize a bed of sand (SG = 2.4) if the air exits the bed at 1 atm and 70°F. The sand grains have an equivalent diameter of 300 μm, and the bed is 3 ft in diameter and 1.5 ft deep, with a porosity of 0.33.

64. Consider a catalyst, specific gravity 1.75, in a bed with air flowing upward through it at 650 K and an average pressure of 1.8 atm ($\mu_{air} = 3 \times 10^{-5}$ kg/m s). The catalyst is spherical with a

diameter of 0.175 mm. The static void fraction is 0.55, and the void fraction at minimum flu-idization is 0.56. The slumped bed height is 3.0 m, and the fluidized bed height is 3.1 m.

   a. Calculate the minimum fluidization velocity.

   b. Calculate the pressure drop at minimum fluidization.

   c. Estimate the pressure drop at one-half of the minimum fluidization velocity assuming incompressible flow.

65. A manometer containing oil with a specific gravity (SG) of 1.28 is connected across an orifice plate in a horizontal pipeline carrying seawater (SG = 1.1). If the manometer reading is 16.8 cm, what is the pressure drop across the orifice? What is it in inches of water?

66. Water is flowing downhill in a pipe that is inclined 35° to the horizontal. A mercury manom-eter is attached to pressure taps 3 in apart. The interface in the downstream manometer leg is 1.25 in higher than the interface in the upstream leg. What is the pressure drop between the two pressure taps?

67. An orifice having a diameter of 1 in is used to measure the flowrate of SAE 10 lube oil (SG = 0.928, μ = 60 cP) in a 2.5-in, schedule-40, commercial-steel pipe at 70°F. The pressure drop across the orifice is measured by a mercury (SG = 13.6) manometer, which reads 3 cm.

   a. Calculate the volumetric flowrate of the oil.

   b. How much power is required to pump the oil through the orifice (not the pipe, just the orifice)?

68. You must install a centrifugal pump to transfer a volatile liquid from a remote tank to a point in the plant 1000 ft from the tank. To minimize the distance that the power line to the pump must be strung, it is desirable to locate the pump as close as possible to the plant. If the liquid has a vapor pressure of 30 psia, the pressure in the tank is 30 psia, the level in the tank is 40 ft above the pump inlet, and the required pump NPSH is 20 ft, what is the closest that the pump can be located to the plant without the possibility of cavitation? The line is 2-in, schedule-40, commercial steel, the flowrate is 75 gpm, and the fluid properties are ρ = 45 lb/ft^3 and μ= 5 cP.

69. Refer to Figure P1.69. Answer the following questions. Explain each answer.

   a. At what flowrate is $NPSH_A$ = 3.2 m? Comment on the feasibility of operating at this flowrate.

   b. At what flowrate does the pump produce 33.5 m of head? What is the system frictional loss at this flowrate? What is the pressure drop across the control valve at this flowrate?

   c. At what flowrate does cavitation become a problem?

   d. What is the maximum possible flowrate?

   e. If the source and destination pressures are identical, what is the elevation difference between source and destination?

   f. At a flowrate of 1 L/s, what is the system frictional head loss?

   g. At a flowrate of 1 L/s, what head is developed by the pump?

   h. At a flowrate of 1 L/s, what is the head loss across the control valve?

70. Benzene at atmospheric pressure and 41°C is in a tank with a fluid level of 15 ft above a pump. The pump provides a pressure increase of 50 psi to a destination 25 ft above the tank fluid level. The suction line to the pump has a length of 20 ft and is 2-in, schedule-40. The discharge line has a length of 40 ft to the destination and is 1.5-in schedule-40. The flowrate of benzene is 9.9 lb/sec.

   a. Derive an expression for the NPSH in head units (ft of liquid) vs. flowrate in ft^3/s.

   b. Derive an expression for the system curve in head units (ft of liquid) vs. flowrate in ft^3/s.

c. Locate the operating point on both plots on Figure P1.70.

d. What is the maximum flowrate before cavitation becomes a problem?

e. What is the pressure drop across the valve at the operating point?

f. What is the maximum flowrate possible with one pump, two pumps in series, and two pumps in parallel?

Data:

$$\log_{10} P^*_{benzene}\,(\mathrm{mm\,Hg}) = 6.90565 - \frac{1211.033}{T(°C)+220.79}$$

$$\rho_{benzene} = 51.9\ \mathrm{lb/ft^3}$$

$$\mu_{benzene} = 0.85\ \mathrm{cP}$$

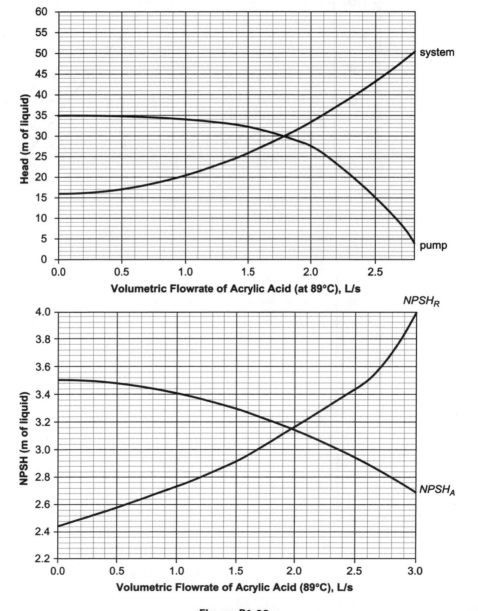

Figure P1.69

71. Acrylic acid at 89°C and 0.16 kPa (ρ = 970 kg/m^3, μ = 0.46 cP) leaves the bottom of a distillation column at a rate of 1.5 L/s. The bottom of a distillation column may be assumed to behave like a tank containing vapor and liquid in equilibrium at the temperature and pressure of the exit stream. The liquid must be pumped to a railroad heading supply tank 4.0 m above the liquid level in the distillation column, where the pressure must be 116 kPa. The liquid level at the bottom of the distillation column is 3.5 m above the pump suction line, and the frictional head loss for the suction line including the tank exit is 0.2 m of acrylic acid. There is a cooler after the pump with a pressure drop of 3.5 m of acrylic acid. The discharge line is 1.5-in,

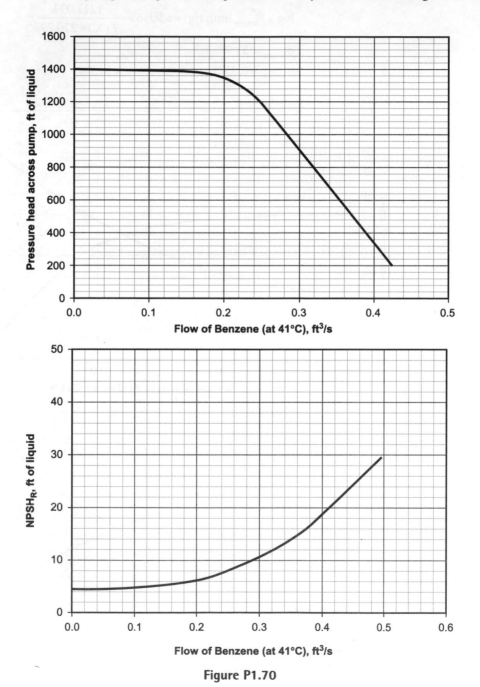

Figure P1.70

schedule-40, commercial-steel pipe, with an equivalent length of 200 m. The entire process may be assumed to be isothermal at 89°C. The problem at hand is whether this system can be scaled up by 20%. The plots required for this analysis are in Figure P1.71.

a. Based on a pump/system curve analysis, can this portion of the process be scaled up by 20%? If not, what is the maximum scale-up percentage?

b. Based on an NPSH analysis, is it good operating policy for this portion of the process to be scaled up by 20%? If not, what is the maximum recommended scale-up percentage?

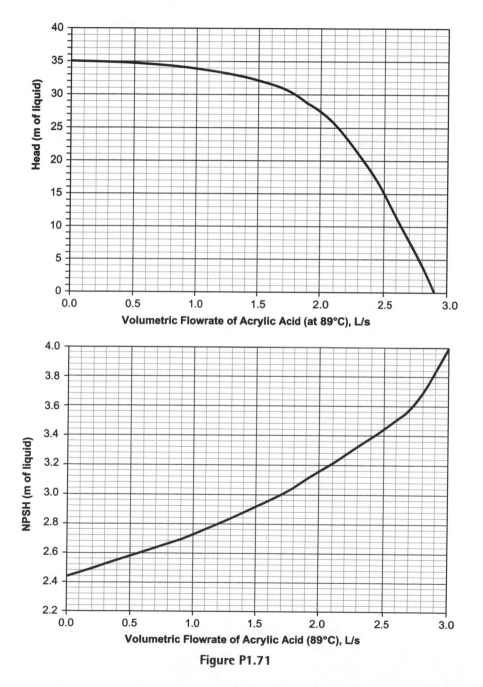

Figure P1.71

72. Consider the pump and system curves indicated by the data in Table P1.72 Answer the following questions.

   a. If the source and destination are at the same height, what is the pressure change from source to destination?

   b. The operating condition is 1.2 L/s. What is $\Delta P_{friction}$ at this point?

   c. At 1.2 L/s, what pressure change does the pump provide?

   d. At 1.2 L/s, what is the pressure drop across the control valve following the pump?

   e. What is the maximum flowrate possible with this (assumed single) pump?

   f. What is the maximum flowrate possible with two identical pumps in series?

   g. What is the maximum flowrate possible with two identical pumps in parallel?

Table P1.72

Pump Curve		System Curve	
Pressure Developed (kPa)	Flowrate (L/s)	Pressure Change (kPa)	Flowrate (L/s)
225.00	0.00	54.00	0.00
225.00	0.40	59.00	0.50
225.00	0.80	80.00	1.00
224.00	1.20	115.00	1.50
220.00	1.50	200.00	2.00
185.00	1.86	559.00	2.50
0.00	2.60		

CHAPTER

2

Process Heat Transfer

WHAT YOU WILL LEARN

- The basics of process heat transfer
- The key relationships for designing and analyzing a heat exchanger
- Analysis of shell-and-tube heat exchangers
- Common correlations for heat transfer coefficients for single-phase and change-of-phase conditions
- The combination of these coefficients with appropriate resistances due to fouling and conduction to determine a single overall heat transfer coefficient
- Equations for extended heat transfer surfaces (fins) for common fin configurations
- Methods to design new heat exchangers
- Predicting the performance of existing exchangers

2.0 INTRODUCTION

The purpose of this chapter is to introduce the concepts needed to design and determine the performance of typical heat-exchange equipment (heat exchangers) used in the process industries. The emphasis is on shell-and-tube (S-T) exchanger design, but other exchanger types, such as air coolers, are mentioned. This chapter does not provide a comprehensive review of the processes of conduction, convection, and radiation, and the design and operation of fuel-fired boilers and fired heaters is not covered. Likewise, a comprehensive review of all heat transfer coefficients is not considered. The emphasis of this chapter is to present a set of useful working equations that can be used to design and evaluate the performance of heat exchangers. Multiple examples are included, and several comprehensive case studies for the design of heat exchangers are given.

2.1 BASIC HEAT-EXCHANGER RELATIONSHIPS

2.1.1 Countercurrent Flow

The flow configuration between process streams and/or utility streams is important in determining the correct driving force for heat transfer. The most common, idealized model is countercurrent

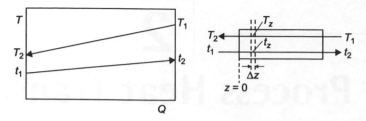

Figure 2.1 Temperature-enthalpy diagram (T-Q diagram) for a countercurrent heat exchanger

flow, which is best illustrated in the sketch and the temperature-enthalpy diagram (T-Q diagram) shown in Figure 2.1.

In Figure 2.1, subscripts 1 and 2 indicate inlet and outlet conditions, respectively. The temperatures of the two streams are identified by uppercase and lowercase letters (T and t). For the simple case shown in Figure 2.1, there are no phase changes in either stream. In addition, the specific heat capacities of both streams are assumed to be constant, which gives rise to the linear temperature profiles shown in Figure 2.1. The energy balances for both streams are

$$Q = \dot{M}C_p(T_1 - T_2) \tag{2.1}$$

$$Q = \dot{m}c_p(t_2 - t_1) \tag{2.2}$$

where capital letters are used for the hot stream and lowercase letters are used for the cold stream. The term Q is the total rate of heat transferred between streams (energy/time), $\dot{M}$ and $\dot{m}$ are the mass flowrates, and C_p and c_p are the heat capacities of the streams. The streams are physically separated, usually by a metal wall, and heat is exchanged through this metal wall. The driving force for the heat exchange is, of course, the temperature difference $(T_z - t_z)$ between the two streams at the given location, z, in the heat exchanger. In general, this driving force changes throughout the exchanger, and some appropriate average should be used. In order to obtain a relationship between Q and the inlet and exit temperatures of each stream a simple differential balance can be written on an element of the heat exchanger, as illustrated in the right-hand sketch of Figure 2.1, to give

$$dQ = U_z(T_z - t_z)dA = U_z\Delta T_z dA \tag{2.3}$$

where $(T_z - t_z) = \Delta T_z$ For many applications, the overall heat transfer coefficient U does not vary substantially with the location in the exchanger and thus $U_z = U$. Since the heat capacities of both streams are assumed to be constant, the temperature difference between the streams $(T_z - t_z)$ or ΔT_z is also linear with respect to the amount of heat transferred, Q, as illustrated in Figure 2.2.

From Figure 2.2, it can be seen that

$$\frac{d\Delta T}{dQ} = \frac{\Delta T_1 - \Delta T_2}{Q} \text{ or } dQ = \frac{Qd\Delta T}{\Delta T_1 - \Delta T_2} \tag{2.4}$$

Substituting Equation (2.4) into Equation (2.3) and dropping the subscript z gives

$$\frac{Qd\Delta T}{\Delta T_1 - \Delta T_2} = U\Delta T dA \tag{2.5}$$

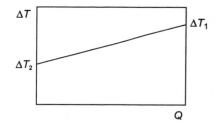

Figure 2.2   Temperature difference between streams, ΔT, as a function of the heat exchanged, Q, in a countercurrent heat exchanger

Integrating Equation (2.5) gives

$$Q\int_{\Delta T_1}^{\Delta T_2}\frac{d\Delta T}{\Delta T}=U(\Delta T_1-\Delta T_2)\int_0^A dA \tag{2.6}$$

$$Q\ln\frac{\Delta T_1}{\Delta T_2}=UA(\Delta T_1-\Delta T_2) \tag{2.7}$$

or

$$Q=UA\Delta T_{lm} \tag{2.8}$$

where ΔT_{lm} is known as the log-mean temperature difference, or *LMTD*, and is given as

$$\Delta T_{lm}=LMTD=\frac{\Delta T_1-\Delta T_2}{\ln\dfrac{\Delta T_1}{\Delta T_2}} \tag{2.9}$$

For heat-exchanger design and performance calculations, the *LMTD* is the correct average temperature to use when the streams flow countercurrently, and Equation (2.8) is the design equation for heat exchangers.

Note: For the special case when the $\dot{M}C_p=\dot{m}c_p$ or when there are constant-pressure phase changes for both streams, then the ΔT between the hot and cold streams is constant throughout the heat exchanger; that is, the *T-Q* lines for both fluids are parallel. For such cases, the expression for *LMTD* gives $0/\ln(1)$, which is indeterminate, but by using L'Hôpital's rule, it can be shown that *LMTD* = ΔT. Since the *LMTD* is simply an average temperature difference between the streams, it should make sense that the average between identical temperature differences is just that temperature difference.

2.1.2   Cocurrent Flow

An alternative flow pattern in heat exchangers is cocurrent flow, which is illustrated in the sketch and *T-Q* diagram shown in Figure 2.3.

A similar analysis to that given for the countercurrent exchanger could be carried out for the cocurrent configuration and would result in the same Equation (2.5). However, now $\Delta T_1 = (T_1 - t_1)$ and $\Delta T_2 = T_2 - t_2$. The cocurrent configuration is less efficient in terms of heat exchange for two reasons. First, the average driving force is smaller than for countercurrent operation; therefore, the heat-exchange area for a given Q will be higher for cocurrent flow than for countercurrent flow. Second, the limiting temperature that either stream can reach is the exit temperature of the other

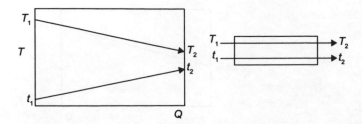

Figure 2.3   T-Q diagram and sketch for a cocurrent heat exchanger

stream, whereas for countercurrent operation, the limiting exit temperature is the inlet temperature of the other stream. Examples 2.1 and 2.2 illustrate these differences.

Example 2.1

A gas is to be cooled from 100°C to 45°C using cooling water that enters the heat exchanger at 30°C and leaves at 40°C. What is the *LMTD* for the heat exchanger if the flow of the streams is

  a. countercurrent?
  b. cocurrent?
  c. How much more heat-exchange area would be required for the cocurrent exchanger compared to the countercurrent exchanger?

Solution

  a. For countercurrent flow, the *T*-Q diagram is shown in Figure E2.1A.

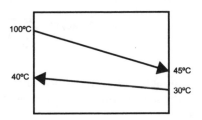

Figure E2.1A   T-Q diagram for countercurrent flow in Example 2.1

$$\Delta T_1 = 60°C, \Delta T_2 = 15°C$$

$$\textbf{LMTD} = (60 - 15)/\ln(60/15) = \textbf{32.46°C}$$

  b. For cocurrent flow, the *T*-Q diagram is shown in Figure E2.1B.

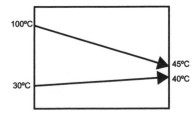

Figure E2.1B   T-Q diagram for cocurrent flow in Example 2.1

$$\Delta T_1 = 70°C, \Delta T_2 = 5°C$$

$$\textbf{LMTD} = (70 - 5)/\ln(70/5) = \textbf{24.63°C}$$

c. Since $A = Q/(U\Delta T_{lm})$

$$\text{then } \frac{A_{cocurrent}}{A_{countercurrent}} = \frac{\Delta T_{lm,countercurrent}}{\Delta T_{lm,cocurrent}} = \frac{32.46}{24.63} = 1.318$$

Thus the cocurrent exchanger area would be 31.8% larger than the countercurrent exchanger area for the same heat duty.

Example 2.2

For the streams and flow configurations used in Example 2.1, what is the lowest temperature to which the gas stream can be cooled assuming the cooling water must leave the exchanger at 40°C?

Solution

The limiting condition for each heat-exchanger configuration is given when the exit gas stream is at the same temperature as the cooling water temperature, this situation is shown for countercurrent and cocurrent flows in Figure E2.2 (a) and (b), respectively.

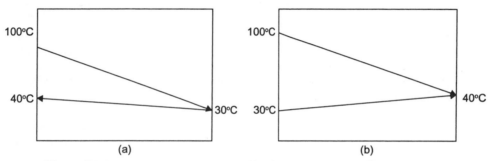

Figure E2.2 Limiting temperature profiles for (a) countercurrent and (b) cocurrent flow

This gives

$$\text{Countercurrent} - \text{limiting temperature} = 30°C$$
$$\text{Cocurrent} - \text{limiting temperature} = 40°C$$

2.1.3 Streams with Phase Changes

Often in chemical processes, heat exchangers are used to boil or condense streams. When there is a change of phase in either or both streams, the temperature-enthalpy diagrams for heat exchangers look different from those shown in Section 2.1.1. Some examples are illustrated in Figure 2.4. In Figure 2.4, the phase change occurs at constant temperature and nearly constant pressure, and there is no subcooling or superheating of streams. Therefore, one fluid is either simply boiling or condensing while the other fluid does not undergo a phase change.

For all these cases, the *LMTD* should be used as the appropriate temperature driving force in heat-exchanger calculations. More complicated examples exist when both temperature and phase changes for one or both fluids occur within the same heat exchanger. This results in sensible heat changes ($\dot{m}c_p$) and phase changes ($\dot{m}\lambda$) occurring within the same heat exchanger. For example, Figure 2.5 illustrates the case when Stream 2 is a subcooled liquid that is heated to saturation, boiled, and then the vapor is superheated using a superheated vapor (Stream 1) that cools, condenses, and subcools.

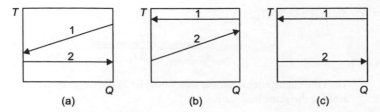

Figure 2.4 *T-Q* diagrams for (a) Stream 1 cooling and Stream 2 boiling, (b) Stream 1 condensing and Stream 2 heating, (c) Stream 1 condensing and Stream 2 boiling

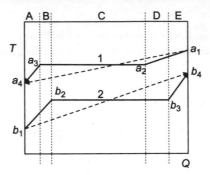

Figure 2.5 *T-Q* diagram in which Stream 1 enters as a superheated vapor and is cooled to saturation (a_1 to a_2), condenses (a_2 to a_3), and then is subcooled (a_3 to a_4) and leaves as a subcooled liquid (a_4) while transferring heat to Stream 2 that enters as a subcooled liquid and is heated to saturation (b_1 to b_2), vaporizes (b_2 to b_3), and is then superheated (b_3 to b_4) and leaves as a superheated vapor (b_4) (the dotted lines between a_1 and a_4 and b_1 and b_4 should not be used to determine the *LMTD*)

From Figure 2.5, it should be clear that using a single *LMTD* (shown by the dotted lines drawn between the end-point temperatures) based on the conditions of the streams entering and leaving the heat exchanger does not provide a realistic or accurate estimate of the correct average driving force within the heat exchanger. In cases such as these, it is necessary to divide the heat exchanger into sections in which the *T* versus *Q* line is a single straight line segment for each fluid in each section. For the case illustrated in Figure 2.5, this would result in five separate sections, known as *zones*, labeled A through E in the figure. The design (and performance) of the heat exchanger would then follow a "zoned analysis" that would consider each section or zone (A–E in Figure 2.5) separately.

2.1.4 Nonlinear *Q* versus *T* Curves

In certain cases, most often when there is a large temperature change for a fluid in a heat exchanger, the specific heat capacity of the fluid may change significantly, and the *T-Q* line for a stream (or both streams) is curved, as shown in Figure 2.6. This situation can also occur when

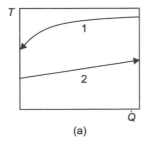

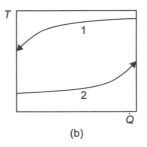

Figure 2.6 *T-Q* diagrams for cases when (a) one fluid or (b) both fluids have nonlinear *T-Q* lines

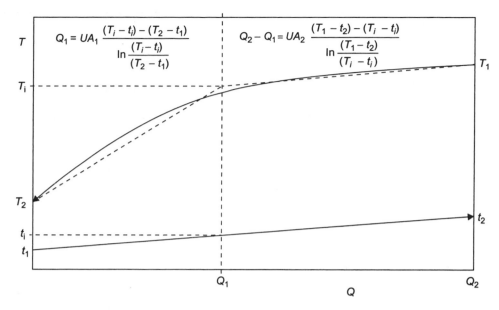

Figure 2.7 *T-Q* diagram for an exchanger with a variable specific heat, where T_i and t_i are intermediate temperatures used in the linear approximation of the temperature profile

partial condensation occurs in a stream containing condensable and noncondensable components; however, this situation is far more complicated and must be considered as a separate case (see Section 2.1.5).

Comparing Figure 2.6 and the analysis carried out in Section 2.1.1, it should be clear that since the *T-Q* lines for at least one of the streams is not straight, the integration of Equation (2.3) will no longer yield the *LMTD* and Equation (2.7). For these cases, the usual practice is to integrate Equation (2.3) numerically to give the appropriate form of Equation (2.8).

An alternative method is to approximate the curved *T-Q* line as straight line segments, and the *LMTD* for each segment can then be found, similar to a zoned analysis. Figure 2.7 illustrates this method.

2.1.5   Overall Heat Transfer Coefficient, *U*, Varies along the Exchanger

In the derivation of *LMTD* given in Section 2.1.1, it was assumed that the overall heat transfer coefficient did not vary within the heat exchanger. Often this assumption is very good. However, for the case when streams undergo large temperature changes, the physical properties used to estimate film heat transfer coefficients, especially liquid viscosity, may change greatly. For this case, the analysis to determine the required heat transfer area becomes complex because the variation in *U* must be included in the integration of Equation (2.3). For the case when the change in *U* is linear or close to linear with the amount of heat transferred, *Q*, a modification of Equations (2.8) and (2.9) can be made:

$$Q = A\frac{U_2\Delta T_2 - U_1\Delta T_1}{\ln\dfrac{U_2\Delta T_2}{U_1\Delta T_1}} = A(\overline{U\Delta T})_{lm} \tag{2.10}$$

where the term $(\overline{U\Delta T})_{lm}$ is the logarithmic mean of the product of *U* and *ΔT*. Example 2.3 illustrates the use of this equation.

Example 2.3

Low-pressure steam at 160°C is used to heat very viscous oil from 30°C to 65°C in a heat exchanger located at the inlet to a pump. The amount of energy needed to heat the oil is 2.65 GJ/h. Over the temperature range of the oil, the physical and transport properties of the oil are all constant except for the viscosity, which decreases significantly as the oil is heated. Using this information, calculations show that the overall heat transfer coefficients at the oil inlet and outlet are 245 and 390 W/m^2/K, respectively. For this exchanger, draw the T-Q diagram and calculate the heat transfer area required in the heat exchanger.

Solution

The information in the problem statement is represented in Figure E2.3.

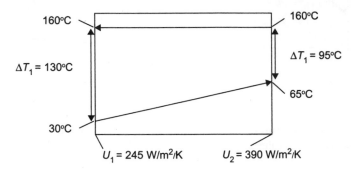

Figure E2.3   Variation of temperature differences and overall heat transfer coefficients over the heat exchanger

Using Equation (2.10),

$$Q = A(\overline{U\Delta T})_{lm} \Rightarrow A = \frac{Q}{(\overline{U\Delta T})_{lm}}$$

$$(\overline{U\Delta T})_{lm} = \frac{U_2\Delta T_2 - U_1\Delta T_1}{\ln\dfrac{U_2\Delta T_2}{U_1\Delta T_1}} = \frac{(95)(390)-(245)(130)}{\ln\dfrac{(95)(390)}{(245)(130)}} = \frac{37,050-31,850}{\ln\dfrac{37,050}{31,850}} = 34,384\,\text{W/m}^2$$

$$A = \frac{2.650\times10^9\,[\text{J/h}]}{(3600)[\text{s/h}](34,384)[\text{W/m}^2]} = 21.4\,\text{m}^2$$

For the case of partial condensation from a gas stream containing noncondensables, the situation is even more complex, because the temperature driving force and U may both vary nonlinearly with Q. Additionally, the mass transfer of the condensable component through the noncondensable component to reach the cold surface must be considered. Such situations require numerical solutions that include the changing physical properties, film heat and mass transfer coefficients, and temperature driving forces throughout the heat-exchange equipment. Many commercial software vendors, including Heat Transfer Research, Inc. (HTRI), Aspen Technology, Inc., Chemstations, and others, provide programs to perform these calculations.

2.2   HEAT-EXCHANGE EQUIPMENT DESIGN AND CHARACTERISTICS

2.2.1   Shell-and-Tube Heat Exchangers

Throughout this chapter, many references are made to S-T heat exchangers. These exchangers are the workhorse of the process industry, and an enormous amount of literature is available about the thermal and mechanical design of this type of equipment. Many of the mechanical considerations are set forth in the *Standards of the Tubular Exchanger Manufacturers Association* (Tubular Exchanger Manufacturers Association [TEMA], 2013). A brief summary of the key design considerations are given here followed by a list of heuristics for choosing which fluid should be placed in the shell or tube side of the heat exchanger. The main components of an S-T exchanger are:

- Shell and shell cover
- Tubes and tubesheet
- Channel and channel cover
- Baffles
- Inlet and outlet nozzles

A good review of each of these components is given by Mukherjee (1998), and a summary of the results of this work are presented in the next section. The TEMA designations for S-T heat exchangers are given in Figure 2.8, the notation for several examples of S-T exchangers are given in Figure 2.9, and some standard designs are given in Figure 2.10.

2.2.1.1   Shell Configurations
There are essentially seven types of shells, designated Types E, F, G, H, J, K, and X in Figure 2.8. The shell type essentially describes the flow pattern in the shell. The simplest and most common configuration is Type E, a one-pass shell, in which the shell-side fluid flows either from left to right (inlet on top left of shell) or from right to left (inlet at bottom right of shell). Two shell passes can be achieved in a single shell by providing a longitudinal baffle down the center of the shell (Type F) that extends to a point short of the tubesheet. If this type of exchanger has only two tube passes, then it is a true counter-current arrangement. Types G and H are characterized by a split flow on the shell side. For Type G, the feed and outlet nozzles are placed at the middle of the shell, and a longitudinal baffle, located at the centerline of the shell, diverts the fluid from the center to both ends and then back to the middle as it exits. The maximum tube length for this type of heat exchanger is ~3 m, since the tube cannot be supported and will lead to unacceptable tube vibration for longer unsupported tube lengths. For Type H, the shell-side fluid is split into two inlet nozzles, longitudinal baffles divert the flow to the left and right, and the flow reverses direction as it passes the baffles and flows back toward the two exit nozzles. This arrangement can be used if tube lengths longer than 3 m are desired. Both Types G and H are used mainly for horizontal thermosiphon reboilers in which the driving force for circulation of the process fluid, which is partially vaporized, is due to the density difference between the liquid entering the reboiler and the two-phase mixture returning to the column. Type J is similar to Type H except there is only a single inlet nozzle and two outlets (1–2 Type J). Alternatively, there may be two inlets and one outlet (2–1 Type J). Type K is for kettle-type reboilers and is characterized by a flared shell that allows significant vapor disengaging space above the boiling liquid and also may contain an overflow weir that maintains a liquid level above the tube bundle. Type X is a cross-flow arrangement, usually containing multiple inlet distribution and collection nozzles inside the shell that approximates a cross-flow movement of the shell-side fluid. The pressure drop for the shell side of these exchangers is extremely low and is preferred for low-pressure and vacuum-vapor coolers and condensers.

2.2.1.2   Tubesheet and Tube Configurations
The tube-side and shell-side fluids are separated by tubesheets, one at either end of the heat exchanger, or, in the case when U-tubes are used, by a single tubesheet at one end of the exchanger. The tubesheet is generally a circular, flat piece of metal with holes drilled in it to accept the tube ends. Examples of tube connections are shown in Figure 2.11. Tubes may be attached to the tubesheet via a number of methods. For example, the tubesheet may be drilled, reamed, and machined with several grooves, with the tubes forced-fit and rolled into the grooves in the tubesheet, Figure 2.11(a). In addition, a strength- or seal-weld may be added, Figure 2.11(b). These welds provide complete separation between the two fluids (shell side and tube side) and ensure no intermixing of fluids occurs.

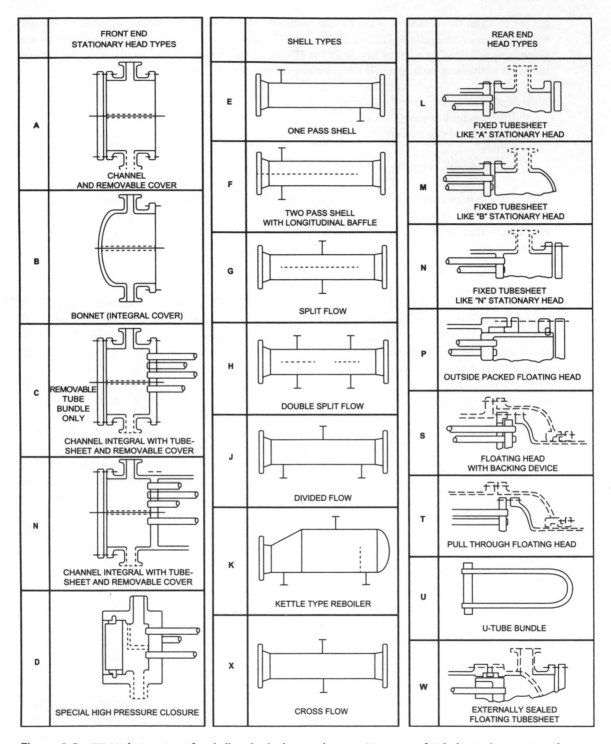

Figure 2.8   TEMA designations for shell-and-tube heat exchangers (Courtesy of Tubular Exchanger Manufacturers Association, Inc. [TEMA, 2013])

1. Stationary Head-Channel
2. Stationary Head-Bonnet
3. Stationary Head Flange-Channel or Bonnet
4. Channel Cover
5. Stationary Head Nozzle
6. Stationary Tubesheet
7. Tubes
8. Shell
9. Shell Cover
10. Shell Flange-Stationary Head End
11. Shell Flange-Rear Head End
12. Shell Nozzle
13. Shell Cover Flange
14. Expansion Joint
15. Floating Tubesheet
16. Floating Head Cover
17. Floating Head Cover Flange
18. Floating Head Backing Device
19. Split Shear Ring
20. Slip-on Backing Flange
21. Floating Head Cover-External
22. Floating Tubesheet Skirt
23. Packing Box
24. Packing
25. Packing Gland
26. Lantern Ring
27. Tierods and Spacers
28. Transverse Baffles or Support Plates
29. Impingement Plate
30. Longitudinal Baffle
31. Pass Partition
32. Vent Connection
33. Drain Connection
34. Instrument Connection
35. Support Saddle
36. Lifting Lug
37. Support Bracket
38. Weir
39. Liquid Level Connection
40. Floating Head Support

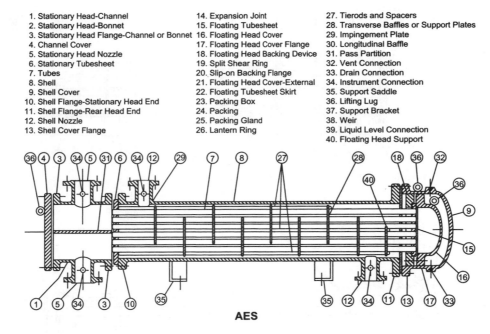

Figure 2.9 Notation for shell-and-tube heat exchangers (Courtesy of Tubular Exchanger Manufacturers Association, Inc. [TEMA, 2013])

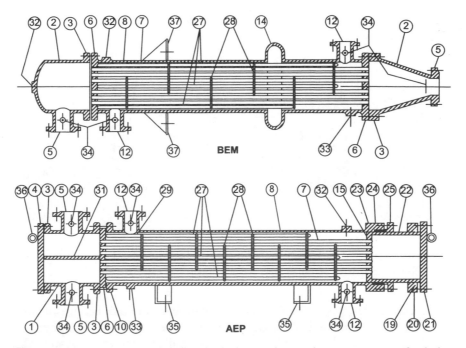

Figure 2.10 Some standard shell-and-tube heat exchanger designs (Courtesy of Tubular Exchanger Manufacturers Association, Inc. [TEMA, 2013]) (Continued)

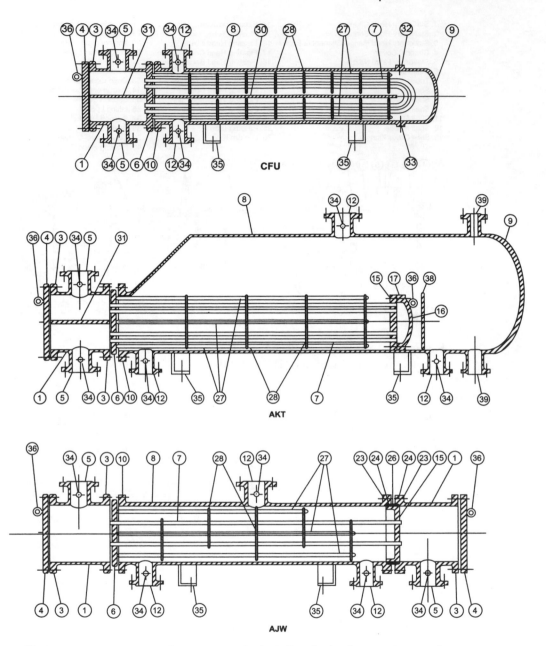

Figure 2.10 (Continued) Some standard shell-and-tube heat exchanger designs (Courtesy of Tubular Exchanger Manufacturers Association, Inc. [TEMA, 2013])

For critical services, when it is vital to avoid intermixing of fluids because of violent cross-reaction and/or degradation of valuable product, a double tubesheet may be employed, Figure 2.11(c). The gap between the tubesheets would typically be vented and monitored for any leaks.

The arrangement or layout of tubes in an S-T exchanger is determined by the rotation or orientation of the tubes relative to the shell-side fluid, the spacing between the tube centers (the tube pitch), and the layout pattern (triangular or square). Some variations of these three factors are illustrated in Figure 2.12. In general, a given shell diameter accommodates more tubes if a triangular arrangement is used (Figure 2.12[c] and 2.12[d]), and this leads to more turbulence on the shell

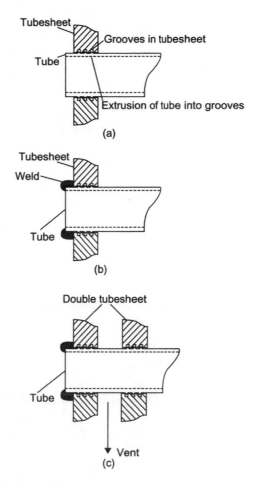

Figure 2.11   Different connections between tubesheets and tubes

side and a correspondingly higher shell-side heat transfer coefficient. On the down side, when the typical tube pitch of 1.25 times the tube outer diameter is used, a triangular arrangement will not permit the mechanical cleaning of the tubes, because there are no access lanes between the tubes to allow for a brush or similar device to clean the accumulated scale from the outside surfaces of the tubes. Mechanical cleaning on the shell side using a triangular arrangement of tubes is possible if a larger tube pitch is used, but this requires a larger shell diameter and greater expense. When mechanical cleaning is desired on the shell side, a square or rotated square pitch is preferred.

2.2.1.3   Fixed Tubesheet and Floating Tubesheet (Head) Designs

For a "standard" 1–2, S-T heat exchanger, the tubesheet at the front end of the exchanger, which contains the feed and discharge nozzles (left column of Figure 2.8), is fixed or anchored to the exchanger. This is referred to as a *stationary head*. For the case when the tubesheet at the other end of the exchanger is also fixed, as illustrated in Types L, M, and N in Figure 2.8, the exchanger is referred to as a *fixed-tubesheet design*. Since the average temperatures on the shell side and tube side will be different, thermal expansion of the tubes and shell will also be different, and this differential expansion will cause mechanical stress at the points where the tubes and shells are connected or anchored. For small differences between the average shell- and tube-side temperatures, this mechanical stress is not excessive, and no damage to the exchanger will result. However, as the average temperature difference between the two fluids in the exchanger becomes larger, the differential expansion becomes excessive and could cause buckling and

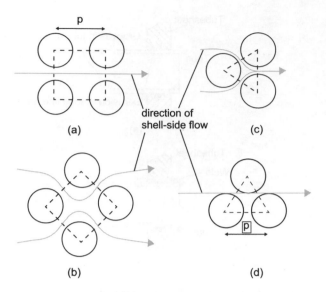

Figure 2.12   Layout patterns for tubes (a) square pitch, (b) rotated square pitch, (c) triangular pitch, and (d) rotated triangular pitch

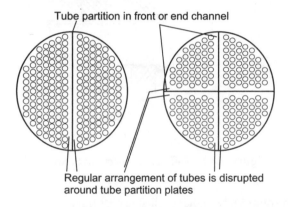

Figure 2.13   Arrangement of tubes for different tube pass configurations

failure for a fixed tubesheet exchanger. To alleviate this problem, a floating head or floating tubesheet is used, as shown in Figure 2.8 (types P, S, T, and W). It should also be noted that the use of U-tubes (Type U) eliminates the use of a second tubesheet and, therefore, eliminates any problems associated with differential thermal expansion. However, it is difficult to clean the "U-bend" in these tubes mechanically, so the use of U-tubes should be avoided when the tube-side fluid fouls.

2.2.1.4   Shell-and-Tube Partitions
The standard 1–2, S-T design can be stacked to form any combination of N-2N shell-tube designs. However, it is often economical to provide multiple tube-pass and shell-pass configurations in a single shell by the use of longitudinal partitions in the shell and by partitions in the tube channel or head. The longitudinal shell partitions are usually welded to the shell to avoid leakage and bypass. Likewise, the partitions in the tube channels at either end of the exchanger have plates welded in place to direct and redirect the tube-side flow through the multiple tube passes. In virtually all cases of S-T exchangers, one or more tube partitions will be present. Therefore, the arrangement of the tubes in the tube bundle along the line of the partition plates will result in areas without tubes, even when a shell-side partition is not used. This situation is shown in Figure 2.13.

2.2.1.5 Baffles

Shell-side baffles are used for two main reasons. First, they support the tubes at regular intervals, and the shorter the unsupported length of a tube, the less damage there is due to vibration. Second, the baffles direct the shell-side fluid back and forth over the tube bundle, and the perpendicular flow increases the shell-side heat transfer coefficient. There are essentially two types of baffles (plate and rod types), and they are available in several different arrangements. Figure 2.14 shows the three common arrangements for plate baffles and the notation used to describe

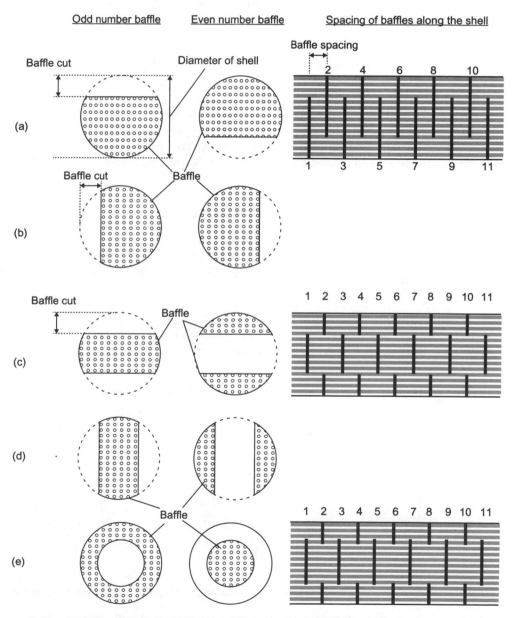

Figure 2.14 Typical arrangements for plate baffles: (a) single segmented baffle, horizontal arrangement; (b) single segmented baffle, vertical arrangement; (c) double segmented baffle, horizontal arrangement; (d) double segmented baffle, vertical arrangement; (e) disk and donut baffles

the baffle arrangements. The baffle-cut is the fraction of the diameter of the baffle plate that is totally open to flow. The baffle cut is usually expressed as a percentage, such as 20%. The maximum range of values for baffle cut is between 15% and 45%, although a narrower range of 20% to 35% is strongly recommended by Mukherjee (1998). Other arrangements of plate baffles are also possible, such as triple segmented baffles and baffles without tubes in the baffle windows. For rod-type baffles, the tubes are held in place by a series of rods placed at right angles to other rods and the tubes. One arrangement is shown in Figure 2.15. The advantages of rod baffles are their low pressure drop and reduction in tube vibrations. For rod-type baffles, the direction of flow is essentially parallel to the tube bundle. The shell-side fluid flows through the gaps between the rods and tubes, which promotes local turbulence and good heat transfer while reducing the shell-side pressure drop.

2.2.1.6 Shell-Side Flow Patterns
From the previous section, it is clear that baffles are usually used to promote the ordered movement of shell-side fluid from the inlet to the outlet. The actual flow pattern that the fluid takes in a design using plate baffles is very complicated, because, in addition to the intended cross flow movement of the shell-side fluid, there are multiple alternative leakage paths that allow fluid to bypass the intended flow direction. An analysis of the flow patterns in plate baffle exchangers was made by Tinker (1958), who identified four leakage or bypass streams in addition to the main cross-flow stream. The main and desired flow direction for the shell-side fluid is shown as B in Figure 2.16.

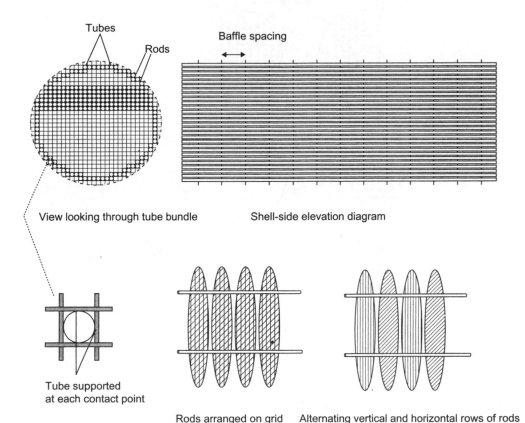

Figure 2.15 Typical arrangement of rod-baffle systems

Stream B moves as a cross-flow stream across the tube bundle and then reverses at the shell wall and flows back across the tube bundle before reversing at the opposite shell wall and so on   movement provides very efficient heat transfer. Accompanying this flow are four alternative streams (A, C, E, and F), which do not follow the desired flow pattern. Flow pattern A is caused by the clearance between the baffle plates and the tubes that allows fluid to pass between the gaps in the baffle plate. Flow pattern C is a bundle bypass stream, which does not fully penetrate the depth of the tube bundle but rather slipstreams around the next baffle. Flow pattern E is caused by fluid flowing parallel to the tube bundle at the walls of the shell. This fluid essentially provides little or no heat transfer between the shell and tube fluids. The final flow pattern, F, is a longitudinal flow created by areas in the tube bundles that have a disrupted tube pattern. Such areas occur when there are shell-side partitions or tube-side partitions. For the case when there is a shell-side partition, a gap exists next to the partition that has fewer tubes in it compared with the main tube bundle. Even for the case when there is no shell-side partition, the arrangement of tubes to accommodate the number of tube passes causes a disruption in the pattern of tubes in the shell giving rise to by-passing lanes (see Figure 2.13). In both these cases, a low resistance path for the shell-side fluid exists, creating a bypass stream, F. Figure 2.17 shows the effect that baffle cut has on side flow patterns. If the baffle cut is too large, then significant shortcutting of fluid (Flow Pattern C in Figure 2.16) occurs that lowers the heat transfer efficiency. As the baffle cut decreases, the shell-side pressure drop increases significantly. As stated previously, the recommended baffle (BC) cut is between 20% and 35%.

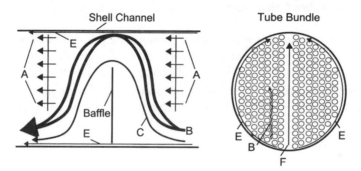

Figure 2.16   Flow patterns on shell-side of an S-T exchanger: B, desired; A, baffle plate leakage; C, flow short circuiting around baffle; E, flow short circuiting around shell wall; F, flow short circuiting through tube bundle (Adapted from Tinker [1958])

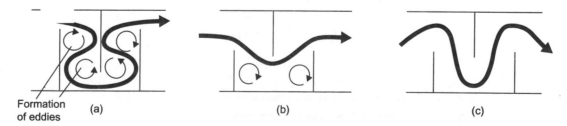

Figure 2.17   The effect of baffle-cut on the flow pattern of shell-side fluid: (a) baffle-cut too small causing high pressure drop, (b) baffle-cut too large giving significant flow bypassing and poor penetration of tube bundle, (c) baffle-cut just right showing good penetration of shell-side flow through tube bundle (Adapted from Mukherjee [1998])

2.2.1.7   Heuristics for Shell-and-Tube Exchanger Designs

The following is a list of heuristics for the design of S-T heat exchangers that is modified from Bell and Mueller (2014). Like any set of heuristics, these should be used as guidelines and not as immutable laws. Some heuristics may seem to be contradictory, and the final choice is often a compromise between conflicting phenomena, with the deciding factor being which design is the least expensive.

Choice of Tube-Side Fluid

- If one stream has a much lower film coefficient that the other, the lower coefficient stream should be placed in the shell so that extended surfaces (fins) can be added to the outside of the tubes to increase the surface area (see Section 2.6). Therefore, if a gas and a liquid are to exchange heat, the gas would normally go on the shell side. For high- and low-viscosity liquids, the low-viscosity fluid is normally placed in the tubes.
- If one fluid is highly corrosive (compared to the other), this fluid should be placed in the tubes. This heuristic is due to the corrosive fluid needing to be in contact with a corrosion-resistant alloy, which is usually expensive. It makes the best economic sense to have the tubes (and connecting piping) made from this alloy. On the other hand, the shell (and baffles and tie rods, etc.) can be made out of a less expensive material (like carbon steel) that is only in contact with the shell-side (less corrosive) fluid.
- If the temperature and pressure conditions for one of the fluids require the use of specialty alloys, that fluid is placed in the tubes for the same reasons given for highly corrosive fluids.
- When one fluid is at much higher pressure than the other fluid, that fluid is placed in the tubes. Since higher pressure requires thicker walls, it makes sense to place the high-pressure fluid inside the tubes, while the shell can be designed for the lower-pressure fluid.
- If one fluid causes severe fouling or scaling, that fluid is placed in the tubes, but a U-tube (Type U) design is not used. Since straight tubes can be mechanically cleaned relatively easily, it makes sense to put the fouling fluid in the tubes.

Choice of Shell Type

- Fixed tubesheet arrangements are not designed to be able to relieve thermally induced stresses. One heuristic is to use fixed tubesheet exchangers only when the inlet temperatures of the two streams differ by less than 55°C (100°F).
- Fixed tubesheet exchangers have been used for temperature differences up to 110°C (200°F) with rolled expansion joints for moderate pressures in the shell (<10 bar).
- U-tube bundles provide a comprehensive solution to the thermal stress problem on the tube side, because they contract and expand independently of the shell, but tubesheet stresses still exist.
- If one fluid is very viscous, this fluid should be placed on the shell side, provided that by doing so the fluid is in the turbulent regime ($Re_s > 100$). If turbulent flow conditions cannot be achieved in the shell, the viscous fluid should be placed in the tubes.
- If one fluid requires a low-pressure drop through the exchanger, this fluid is placed on the shell side of the exchanger. In general, the shell-side pressure drop is lower than the tube-side pressure drop. In addition, as discussed in Section 2.2.1.1, many different shell configurations are available that lower the shell-side pressure drop.

2.3 *LMTD* CORRECTION FACTOR FOR MULTIPLE SHELL AND TUBE PASSES

2.3.1 Background

In the previous sections, the ideal case of pure countercurrent flow and the configuration of typical S-T heat exchangers were investigated. For nonreacting systems, pure countercurrent flow is the best flow configuration (the configuration that requires the smallest heat-exchange area) for a given system. In practice, exchangers with pure countercurrent flow are not the usual choice. For reasons of compactness and cost, heat exchangers are configured with flow patterns that usually fall somewhere between the extremes of pure countercurrent and pure cocurrent flow.

To account for flow patterns that are not purely countercurrent, Equation (2.8) can be rewritten to include an *LMTD* correction factor, F_y, where the subscript refers to a type of flow such as cocurrent or cross-current or to the type of heat exchanger being used. Thus, the generalized working design equation for exchangers becomes:

$$Q = UA\Delta T_{lm} F_y \tag{2.11}$$

By convention, the *LMTD* for pure countercurrent flow is taken as the reference configuration, and the correction factor, F_y, is always the correction relative to pure countercurrent flow. In this way, F_y is always less than or equal to 1.0. For pure cocurrent flow, the *LMTD* correction factor, $F_{cocurrent}$, is given by

$$F_{cocurrent} = \frac{(R+1)\ln\left[\dfrac{1-P}{1-RP}\right]}{(R-1)\ln\left[\dfrac{1}{1-RP-P}\right]} \tag{2.12}$$

where P is the ratio of the temperature change of the tube-side stream to the maximum temperature difference in the exchanger and R is the ratio of heat capacities of the tube-side to shell-side streams,

$$P = \frac{(t_2 - t_1)}{(T_1 - t_1)} \text{ and } R = \frac{(T_1 - T_2)}{(t_2 - t_1)} = \frac{\dot{m}c_p}{\dot{M}C_P}$$

Example 2.4

Revisit Example 2.1 and show that the same answer to Part (c) can be found using Equation (2.12).

Solution

$$t_1 = 30°C, t_2 = 40°C, T_1 = 100°C, T_2 = 45°C$$

For cocurrent flow, $P = (40 - 30)/(100 - 30) = 10/70 = 0.1429$ and $R = (100 - 45)/(40 - 30) = 5.5$. Substituting into Equation (2.12),

$$F_{cocurrent} = \frac{(R+1)\ln\left[\dfrac{1-P}{1-RP}\right]}{(R-1)\ln\left[\dfrac{1}{1-RP-P}\right]} = \frac{(5.5+1)\ln\left[\dfrac{1-(0.1429)}{1-(5.5)(0.1429)}\right]}{(5.5-1)\ln\left[\dfrac{1}{1-(5.5)(0.1429)-(0.1429)}\right]} = 0.7588$$

Applying Equation (2.11) to the cocurrent and countercurrent exchangers and taking ratios give

$$\frac{A_{cocurrent}}{A_{countercurrent}} = \frac{\Delta T_{lm}}{F_{cocurrent}\Delta T_{lm}} = \frac{1}{F_{cocurrent}} = \frac{1}{0.7588} = 1.318$$

This is the same result as given in Part (c) of Example 2.1. It should be noted that the choice of which stream to assign T and t is not important, and this point will be illustrated Example 2.5.

In the process industries, the workhorse of industrial heat exchangers is the S-T design. Many of the important design issues for these heat exchangers were covered in Section 2.2. However, in the next section, the effect of different, nonideal flow patterns on the *LMTD* correction factor (F_y) is covered.

2.3.2   Basic Configuration of a Single–Shell–Pass, Double–Tube–Pass (1–2) Exchanger

As pointed out in Section 2.2, the most common form of heat exchanger in the process industries is the S-T heat exchanger. The most common form of S-T exchanger is the single-shell-pass, two-tube-pass or 1–2 design, which is illustrated in Figure 2.18.

Figure 2.18 is a schematic diagram illustrating the flow pattern for a 1–2, S-T heat exchanger. Many of the technical details of S-T heat exchangers are covered in Section 2.2, but a further, brief description of the fluid flow paths is given here. Referring to Figure 2.18, the tube-side fluid enters from the top of the diagram and passes into a head-channel. The fluid then flows horizontally (from left to right) through a series of parallel tubes into a second head-channel on the right of the diagram. The tube-side stream then reverses direction and passes from right to left through the bottom set of parallel tubes into the left-hand head-channel where it finally leaves the heat exchanger. The shell-side fluid enters the shell, which is separated from the two head-channels by a tubesheet at either end of

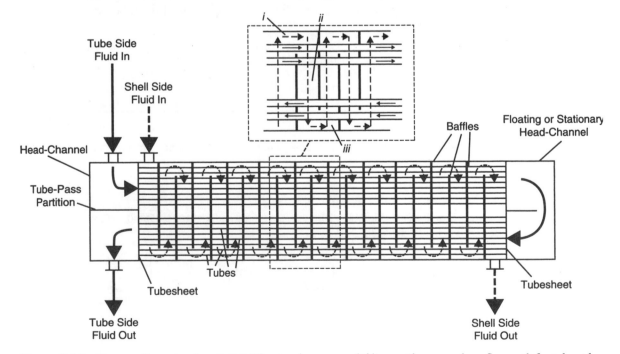

Figure 2.18   Flow configuration for a 1–2, S-T heat exchanger—solid lines with arrows show flow path for tube-side fluid, and dotted lines with arrow show flow path for shell-side fluid

the exchanger. The shell-side fluid passes from left to right in the shell and is guided by a set of baffles that direct the flow. This fluid finally leaves the shell at the bottom right in the diagram. An insert is shown at the top of Figure 2.18 that shows in more detail the flow patterns of the different streams. The insert identifies three regions (labeled *i*, *ii*, and *iii*). In Region *i*, the shell-side and tube-side fluids flow cocurrently with each other. In Region *ii*, the shell-side fluid flows perpendicular or crosscurrently with the fluid in the tubes. Finally, in Region *iii*, the shell- and tube-side fluids flow countercurrent to one another. From this description, it should be clear that the flow pattern in this type of heat exchanger does not fit either the countercurrent or cocurrent models described previously.

Despite the apparently complicated flow pattern shown in Figure 2.18, it has been shown by Bowman, Mueller, and Nagle (1940) that the flow pattern is closely approximated by half the exchanger with cocurrent flow (top half of tubes in Figure 2.18) and the other half with countercurrent flow. With this assumption, an analytical expression for F_{1-2} can be found, and is given as Equation (2.13) and Figure 2.19.

$$F_{1-2} = \frac{\sqrt{(R^2+1)}\ln\left[\dfrac{1-P}{1-PR}\right]}{(R-1)\ln\left[\dfrac{2-P\left(R+1-\sqrt{(R^2+1)}\right)}{2-P\left(R+1+\sqrt{(R^2+1)}\right)}\right]} \quad \text{for } R \neq 1$$

$$F_{1-2} = \frac{P\sqrt{2}}{(1-P)\ln\dfrac{2-2P+P\sqrt{2}}{2-2P-P\sqrt{2}}} \quad \text{for } R = 1$$

(2.13)

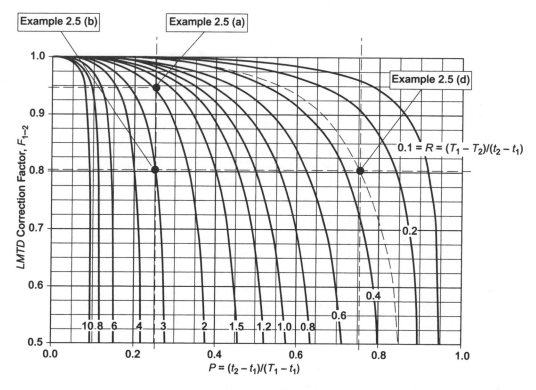

Figure 2.19 *LMTD* correction factor, F_{1-2}, for a 1–2, S-T heat exchanger (this figure is also good for one shell pass and any even number of tube passes, i.e., F_{1-4}, F_{1-6}, F_{1-8}, etc.).

Figure 2.19 illustrates the correction factor, F_{1-2}, as a function of P and R for a 1–2, S-T exchanger. It should be noted that the correction factor for a 1–2, S-T exchanger is essentially the same (within 1% to 2%) for a single shell pass with any number of even tube passes, and the results from Equation (2.13) and Figure 2.19 may be used. Therefore, $F_{1-2} = F_{1-4} = F_{1-6} = F_{1-8}$, and so on.

From Figure 2.19, it should be clear that for any given value of R, there exists a maximum value of P, P_{max}. The value of P_{max} is given by Equation (2.14):

$$P_{max} = \frac{2}{R+1+\sqrt{(R^2+1)}} \tag{2.14}$$

For values of $P > P_{max}$ no solution to Equation (2.13) exists, or to put it another way, a 1–2, S-T exchanger cannot always be used to make the temperature of the streams change as desired. The uses of the *LMTD* correction factor, Figure 2.19, and the limitations of 1–2, S-T exchangers are illustrated in Example 2.5.

Example 2.5

a. A 1–2, S-T heat exchanger is used to cool a process stream from 70°C to 50°C using cooling water at 30°C heated to 40°C. For this heat exchanger, determine the value of F_{1-2}.

b. Repeat this calculation, using the same cooling water temperatures for the case when the exit process temperature is 40°C.

c. Repeat this calculation, using the same cooling water temperatures for the case when the exit process temperature is 35°C.

d. Repeat Part (b) but switch the tube and shell side fluids.

Solution

a. $P = (t_2 - t_1)/(T_1 - t_1) = (40 - 30)/(70 - 30) = 0.25$
   $R = (T_1 - T_2)/(t_2 - t_1) = (70 - 50)/(40 - 30) = 2.0$
   From Figure 2.19, values for P and R are shown as dotted lines, and $F_{1-2} = 0.942$.
   Alternatively, using Equation (2.13),

$$F_{1-2} = \frac{\sqrt{(R^2+1)}\ln\left[\dfrac{1-P}{1-PR}\right]}{(R-1)\ln\left[\dfrac{2-P(R+1-\sqrt{(R^2+1)})}{2-P(R+1+\sqrt{(R^2+1)})}\right]} = \frac{\sqrt{(2^2+1)}\ln\left[\dfrac{1-0.25}{1-(0.25)(2)}\right]}{(2-1)\ln\left[\dfrac{2-(0.25)(2+1-\sqrt{(2^2+1)})}{2-(0.25)(2+1+\sqrt{(2^2+1)})}\right]} = 0.942$$

b. $P = (t_2 - t_1)/(T_1 - t_1) = (40 - 30)/(70 - 30) = 0.25$
   $R = (T_1 - T_2)/(t_2 - t_1) = (70 - 40)/(40 - 30) = 3.0$
   From Figure 2.19, $F_{1-2} = 0.81$ and using Equation (2.13),

$$F_{1-2} = = \frac{\sqrt{(3^2+1)}\ln\left[\dfrac{1-0.25}{1-(0.25)(3)}\right]}{(3-1)\ln\left[\dfrac{2-(0.25)(3+1-\sqrt{(3^2+1)})}{2-(0.25)(3+1+\sqrt{(3^2+1)})}\right]} = 0.809$$

c. $P = (t_2 - t_1)/(T_1 - t_1) = (40 - 30)/(70 - 30) = 0.25$
   $R = (T_1 - T_2)/(t_2 - t_1) = (70 - 35)/(40 - 30) = 3.5$

From Figure 2.19 or using Equation (2.11), there is no solution; that is, it is not possible to design a 1–2, S-T exchanger for this situation.

The situation can be verified by applying Equation (2.14):

$$P_{max} = \frac{2}{R+1+\sqrt{(R^2+1)}} = \frac{2}{3.5+1+\sqrt{(3.5^2+1)}} = 0.2457 < P = 0.25$$

This confirms that only values of $P < P_{max}$ result in feasible solutions.

d. $P = (t_2 - t_1)/(T_1 - t_1) = (40 - 70)/(30 - 70) = 0.75$
   $R = (T_1 - T_2)/(t_2 - t_1) = (30 - 40)/(40 - 70) = 0.333$
   From Figure 2.19, $F_{1-2} = 0.81$ and using Equation (2.13),

$$F_{1-2} = \frac{\sqrt{(0.333^2+1)}\ln\left[\dfrac{1-0.75}{1-(0.75)(0.333)}\right]}{(0.333-1)\ln\left[\dfrac{2-(0.75)(0.333+1-\sqrt{(0.333^2+1)}}{2-(0.75)(0.333+1+\sqrt{(0.333^2+1)}}\right]} = 0.809$$

The result is the same as in Part (b).

Two important conclusions can be drawn from Example 2.5. First, by comparing the results for Parts (b) and (d), it can be seen that it does not matter what stream is chosen for the shell-side or tube-side fluid. Equation (2.13) and Figure 2.19 are essentially symmetrical in R and P, so that the choice of which fluid to put in the shell and which to put in the tubes does not affect the result and value of F_{1-2}. Other factors do affect which fluid goes where, but these were covered in Section 2.2.1.5. Secondly, it was seen that, as the exit temperature of the process fluid decreased, F_{1-2} started to decrease. Specifically, when the outlet temperature of the hot stream drops below the outlet temperature of the cold stream ($T_2 < t_2$), the efficiency drops very rapidly, and for Part (c) no solution existed. These results can be explained by the fact that as the temperature between the hot exit and cold exit streams approaches zero, the efficiency of the 1–2, S-T design is reduced significantly. This effect is illustrated in Figure 2.20. For pure cocurrent flow, the highest temperature to which the cold stream can be heated is equal to the hot stream exit temperature. However, in a 1–2, S-T exchanger only half of the flow is cocurrent, so not surprisingly, the efficiency (F_{1-2}) of the heat exchanger starts to decrease rapidly when this condition is reached. Heat exchangers with multiple shell passes more closely mimic pure countercurrent flow, and these are covered next.

2.3.3   Multiple Shell-and-Tube-Pass Exchangers

For a 1–2, S-T exchanger with the condition $T_2 = t_2$, using the relationships between P and R as follows,

$$T_2 = T_1 - R(t_2 - t_1) \text{ and } t_2 = t_1 + P(T_1 - t_1)$$
$$\therefore T_1 - R(t_2 - t_1) = t_1 + P(T_1 - t_1)$$
$$T_1 - t_1 = R(t_2 - t_1) + P(T_1 - t_1)$$
$$\text{divide by } (T_1 - t_1)$$
$$1 = R\frac{(t_2-t_1)}{(T_1-t_1)} + P\frac{(T_1-t_1)}{(T_1-t_1)} = RP + P$$

$$\therefore P + PR - 1 = 0 \text{ or } P = \frac{1}{1+R} \tag{2.15}$$

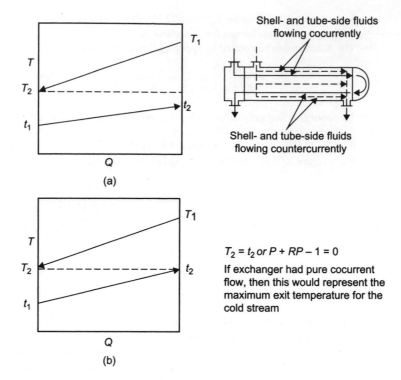

Figure 2.20   T-Q profiles in 1–2, S-T exchanger: (a) satisfactory temperature approach, (b) exit temperatures of both fluids are equal ($F_{1-2} \cong 0.8$)

The line representing Equation (2.15) can be approximated by the condition $F_{1-2} = 0.8$ on Figure 2.19. The usual design criterion for exchanger design is to limit the *LMTD* correction factor to a value ≥ 0.8. If the condition of $F = 0.8$ is taken as the reasonable practical limit of operation of a 1–2, S-T exchanger, the number of shells needed for any design can be calculated graphically using a McCabe-Thiele–type construction, illustrated in Figure 2.21. The analytical expression for the number of shells using this criterion if the specific heats of hot and cold streams are constant (lines on the T-Q diagram are straight) is given by

$$N_{shells} = \frac{\ln\left[\dfrac{1-PR}{1-P}\right]}{\ln\left[\dfrac{1}{R}\right]} \quad \text{for } R \neq 1$$

$$N_{shells} = \left[\frac{P}{1-P}\right] \quad \text{for } R = 1$$

(2.16)

Analytical expressions for the *LMTD* correction factor, *F*, for higher shell and tube passes exist, and many of these are given by Bowman et al. (1940). For a 2–4, S-T exchanger, Equation (2.17) applies (this expression can also be used for a 2–4N, S-T exchanger, where $N = 1, 2, 3,...$):

$$F_{2-4N} = \frac{\sqrt{R^2+1}\left[\ln\dfrac{1-P}{1-PR}\right]}{2(R-1)\ln\left[\dfrac{2-P-PR+2\sqrt{(1-P)(1-PR)}+P\sqrt{R^2+1}}{2-P-PR+2\sqrt{(1-P)(1-PR)}-P\sqrt{R^2+1}}\right]}$$

(2.17)

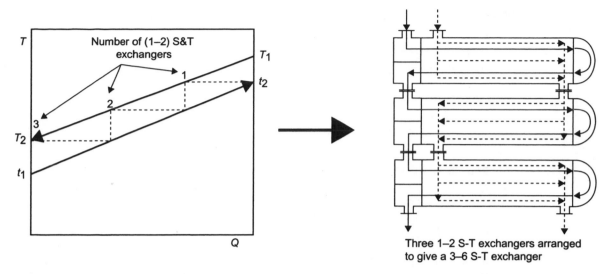

Figure 2.21   McCabe-Thiele–type construction for estimating the number of shell passes for an S-T heat exchanger and the arrangement of three 1–2, S-T exchangers to give a 3–6 S-T configuration

For the case when $R = 1$, Equation (2.17) reduces to Equation (2.18):

$$F_{2-4N} = \frac{P\sqrt{2}}{2(1-P)\ln\left[\dfrac{4(1-P)+P\sqrt{2}}{4(1-P)-P\sqrt{2}}\right]} \qquad (2.18)$$

For higher shell passes, the analytical expressions become more complicated. However, Bowman (1936) has shown that for any given values of F and R, the value of P for an exchanger with N shell-side and $2N$ tube-side passes can be related to P for a 1–2, S-T exchanger through Equation (2.19):

$$P_{N-2N} = \frac{1-\left[\dfrac{1-P_{1-2}R}{1-P_{1-2}}\right]^N}{R-\left[\dfrac{1-P_{1-2}R}{1-P_{1-2}}\right]^N} \qquad (2.19)$$

Equation (2.19) was used to generate the *LMTD* correction factors for 3–6, 4–8, 5–10, and 6–12 S-T exchangers, and the results are presented in Appendix 2.A. *LMTD* correction factors for many types of heat exchanger can also be found at http://checalc.com/solved/LMTD_Chart.html. It is observed that as the number of shell passes increases, the curve for the same value of R moves up, resulting in a higher F value. Example 2.6 illustrates the use of multiple shell heat exchangers.

Example 2.6

   a. How many shell passes would be needed to cool a heavy oil stream from 230°C to 150°C using another process stream that is heated from 130°C to 190°C?

   b. What is the *LMTD* correction factor for this arrangement?

Solution

a. $P = (t_2 - t_1)/(T_1 - t_1) = (190 - 130)/(230 - 130) = 60/100 = 0.60$

$R = (T_1 - T_2)/(t_2 - t_1) = (230 - 150)/(190 - 130) = 80/60 = 1.333$

Applying Equation (2.16),

$$N_{shells} = \frac{\ln\left[\dfrac{1-PR}{1-P}\right]}{\ln\left[\dfrac{1}{R}\right]} = \frac{\ln\left[\dfrac{1-(0.6)(1.333)}{1-0.6}\right]}{\ln\left[\dfrac{1}{1.3333}\right]} = 2.41$$

Rounding up, to $N_{shells} = 3$ means that a 3–6, S-T exchanger is required.

b. From Figure A.3, $F_{3-6} = 0.88$ this value is >0.8, which satisfies the criterion for acceptable efficiency. From Figure A.2, $F_{2-4} = 0.67$, which is below the acceptable limit of 0.8. It should be noted that 4–8, 5–10, and higher numbers of S-T passes would all be acceptable designs, but the complexity and cost of the exchanger increases with the addition of shell passes; therefore, the design with the fewest number of shell passes should be chosen. The effect of the number of shell and tube passes for this example is further illustrated in Figure E2.6 where the curves for different S-T configurations for $R = 1.33$ are plotted.

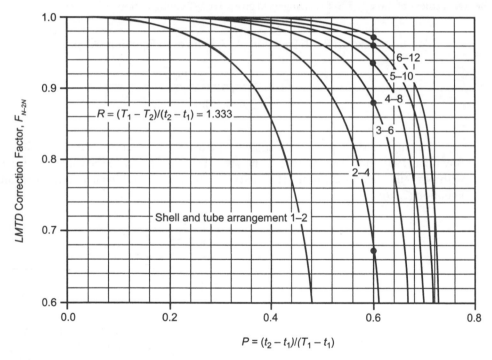

Figure E2.6   The effect of the number of passes for an S-T exchanger for $R = 1.33$; a 3–6, S-T exchanger has the least number of shell passes with an $F \geq 0.8$.

2.3.4   Cross–Flow Exchangers

For some heat exchangers, the two fluids exchanging energy flow are at right angles to each other. This flow pattern is termed *cross-flow*, and there are several important examples of this type of flow pattern in commercial heat exchangers. One example is the plate-and-frame heat exchanger,

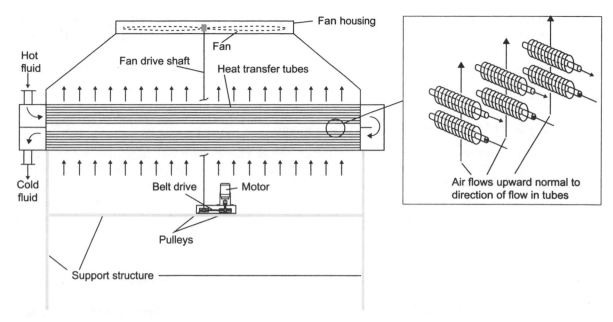

Figure 2.22   Schematic of an air cooler: Air is pulled up and across heat transfer tubes with external fins by an induction fan and hot fluid is transported through the tubes. The insert shows detail of the cross-flow pattern for the air and process fluid.

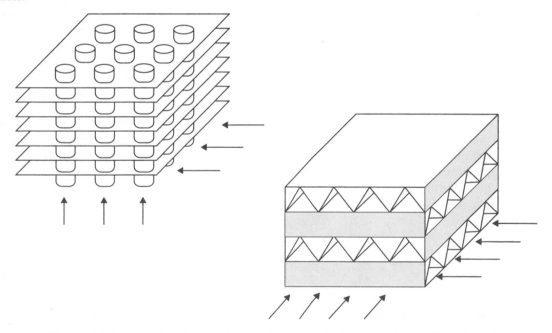

Figure 2.23   Examples of gas-gas heat-exchanger configurations with cross-flow channels

another example is the air-cooled heat exchanger, where air flows perpendicular to a set of finned tubes in which a liquid or condensing vapor flow. The purpose of finned tubes is to provide extended heat transfer surface to fluids with a low heat transfer coefficient; this topic is covered in detail in Section 2.6. A typical arrangement of an air cooler (sometimes termed a *fin-fan*) is illustrated in Figure 2.22. Other examples of cross-flow occur in gas-gas exchangers for which the film heat transfer coefficients for both gases are low. In such situations, each gas may pass through a channel that contains regular geometric fins, as shown in Figure 2.23. Values for the *LMTD* correction factor for one case of cross-flow is given in Appendix 2.A, Figure A.7. Other practical examples of cross-flow exchangers are car radiators and heat sinks on circuit boards and electronic power handling devices.

2.3.5 *LMTD* Correction and Phase Change

For cases when one or both fluids experience a phase change, with *T-Q* diagrams similar to those shown in Figure 2.4, the *LMTD* correction factor can be taken as unity; that is, no correction to the *LMTD* is needed. This should be apparent to the reader, since the fluid that is not changing phase is contacted everywhere with the phase change material that is at constant temperature (assuming constant pressure operation). Therefore, the flow pattern has no effect on the driving forces within the exchanger.

2.4 OVERALL HEAT TRANSFER COEFFICIENTS—RESISTANCES IN SERIES

Up to this point, there has been no discussion on the value or determination of the overall heat transfer coefficient, U. U is the combination of several heat transfer resistances that act in series, and these resistances are illustrated in Figure 2.24 for both planar and circular geometries. In Figure 2.24(a), a wall is shown separating two fluids at different temperatures. For generality, it is assumed that the solid wall is composed of multiple layers. These layers might comprise a composite wall or a single wall with scale deposits on both sides. Either way, each layer acts as a separate resistance to heat transfer. In the case of a deposit of scale, the term *fouling resistance* is used. The heat transfer resistance between each fluid and the surface is given in terms of a film heat transfer coefficient (h_i, h_o), while the resistances of the solid layers are given in terms of their thermal conductivity (k) and thickness (Δx). In Figure 2.24(b), a similar situation is shown for cylindrical or tubular geometry.

For steady-state operations, the amount of heat transferred through each layer or resistance must the same. By writing the basic heat flux equation across each resistance and eliminating the intermediate temperatures, an overall expression of the heat transferred in terms of the bulk temperature driving force (which is used in the exchanger design equation) is obtained. This process is given for the planar geometry as follows:

$$\frac{Q}{A} = h_{hf}(T_{hot} - T_1) = \frac{k_h}{\Delta x_h}(T_1 - T_2) = \frac{k_w}{\Delta x_w}(T_2 - T_3) = \frac{k_c}{\Delta x_c}(T_3 - T_4) = h_{cf}(T_4 - T_{cold})$$

$$\therefore T_{hot} - T_1 = \frac{Q}{Ah_{hf}}, T_1 - T_2 = \frac{Q\Delta x_h}{Ak_h}, T_2 - T_3 = \frac{Q\Delta x_w}{Ak_w}, T_3 - T_4 = \frac{Q\Delta x_c}{Ak_c}, T_4 - T_{cold} = \frac{Q}{Ah_{cf}}$$

Adding all the left-hand sides together gives

$$T_{hot} - T_{cold} = \frac{Q}{Ah_{hf}} + \frac{Q\Delta x_h}{Ak_h} + \frac{Q\Delta x_w}{Ak_w} + \frac{Q\Delta x_c}{Ak_c} + \frac{Q}{Ah_{cf}} = \frac{Q}{A}\left[\frac{1}{h_{hf}} + \frac{\Delta x_h}{k_h} + \frac{\Delta x_w}{k_w} + \frac{\Delta x_c}{k_c} + \frac{1}{h_{cf}}\right]$$

$$\therefore Q = \left[\frac{1}{h_{hf}} + \frac{\Delta x_h}{k_h} + \frac{\Delta x_w}{k_w} + \frac{\Delta x_c}{k_c} + \frac{1}{h_{cf}}\right]^{-1} A(T_{hot} - T_{cold}) = UA(T_{hot} - T_{cold}) = UA\Delta T_{Bulk}$$

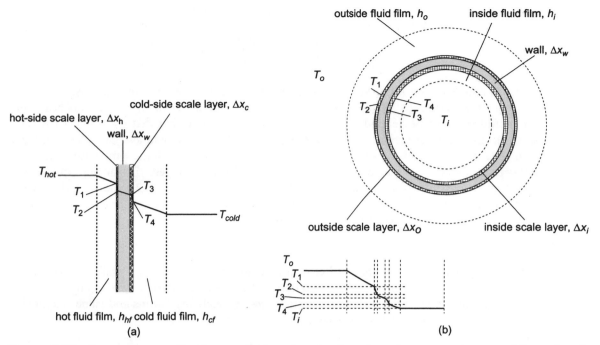

Figure 2.24 Temperature profiles for two fluids exchanging heat through a composite or multilayered wall: (a) planar geometry, (b) radial or tubular geometry.

where

$$\frac{1}{U}=\frac{1}{h_{hf}}+\frac{\Delta x_h}{k_h}+\frac{\Delta x_w}{k_w}+\frac{\Delta x_c}{k_c}+\frac{1}{h_{cf}} \quad \text{or} \quad U=\left[\frac{1}{h_{hf}}+\frac{\Delta x_h}{k_h}+\frac{\Delta x_w}{k_w}+\frac{\Delta x_c}{k_c}+\frac{1}{h_{cf}}\right]^{-1} \tag{2.20}$$

For circular (tubular) geometry with a single wall and scale formation on the inner and outer surfaces, a similar analysis gives

$$Q = A_o h_o (T_o - T_1) = \frac{A_o}{R_{fo}}(T_1 - T_2) = \frac{k_w}{\Delta x_w}(T_2 - T_3) = \frac{A_i}{R_{fi}}(T_3 - T_4) = A_i h_i (T_4 - T_i) \tag{2.21}$$

where the subscripts o and i refer to the outside and inside of the tube, respectively, and the resistances due to fouling are given in terms of an effective fouling heat transfer resistance, $R_{fo} = 1/h_{of}$ and $R_{fi} = 1/h_{if}$, respectively. The form of Equation (2.21) differs from that derived for planar geometry, because for circular geometry, the cross-sectional area through which the heat flows changes from the outside to the inside of the tube. By eliminating the intermediate temperatures and rearranging, the following working equations for circular geometry can be derived:

$$Q = U_o A_o \Delta T_{Bulk} = U_i A_i \Delta T_{Bulk} \tag{2.22}$$

where

$$U_o = \left[\frac{1}{h_o}+R_{fo}+\frac{D_o \ln(D_o/D_i)}{2k_w}+\frac{D_o}{D_i}R_{fi}+\frac{D_o}{D_i}\frac{1}{h_i}\right]^{-1} \tag{2.23}$$

and

$$U_i = \left[\frac{D_i}{D_o} \frac{1}{h_o} + \frac{D_i}{D_o} R_{fo} + \frac{D_i \ln(D_o / D_i)}{2k_w} + R_{fi} + \frac{1}{h_i} \right]^{-1} \qquad (2.24)$$

where D_i and D_o are the inside and outside diameters of the tube (or pipe), respectively.

From Equations (2.23) and (2.24), it is clear that for circular geometry, the overall heat transfer coefficient must be defined with respect to a specific surface but that the product UA is the same no matter what surface is chosen; that is, $U_i A_i = U_o A_o$. It should also be noted that Equation (2.22) was derived for a specific location where ΔT_{Bulk} is known. If the temperature difference between fluids varies with location, as it would in most heat exchangers, then an appropriate average temperature difference should be used. The analysis given in Section 2.1 and all the limitations discussed in Section 2.1 still apply. Thus the general working equation for heat transfer between fluids exchanging heat through tube walls becomes

$$Q = U_o A_o \Delta T_{lm} F_y = U_i A_i \Delta T_{lm} F_y \qquad (2.25)$$

In summary, for heat transfer between fluid streams separated by a wall, the overall heat transfer resistance is a function of the two film heat transfer coefficients, two fouling resistances, and a resistance due to the wall. In the next section, typical correlations and values required to determine all these resistances are presented.

2.5   ESTIMATION OF INDIVIDUAL HEAT TRANSFER COEFFICIENTS AND FOULING RESISTANCES

In Section 2.4, the terms needed to evaluate the overall heat transfer coefficient (U_o or U_i) were identified. In this section, correlations and other data for calculating each of these heat transfer resistances are presented.

2.5.1   Heat Transfer Resistances Due to Fouling

In designing heat exchangers, it is important to make allowance for the deposition of scale, that is, the process of fouling, on heat-exchanger surfaces and the subsequent reduction in overall heat transfer due to the fouling process. The a priori determination of fouling through some mechanistic model is not possible because of the highly complex nature of the fouling process. Moreover, fluctuations in process temperatures and other conditions (pH, total dissolved solids, etc.) due to abnormal plant operations may significantly increase the fouling process and are impossible to predict with any accuracy. With this in mind, the designer must resort to the use of typical fouling resistances for similar services (fluids) obtained through experience.

Fouling resistances for water and chemical process streams as suggested by TEMA (2013) are shown in Tables 2.1 and 2.2.

2.5.2   Thermal Conductivities of Common Metals and Tube Properties

Thermal conductivities for metals usually used as heat-exchanger tubes or in extended surface (compact) heat exchangers are given in Table 2.3. For low-temperature service such as cryogenic air separation, aluminum and stainless steel are the preferred materials of construction. It should be noted that the thermal conductivities of most metals do not vary widely over quite large temperature changes, and usually the resistance to conduction through the wall is not the limiting resistance, so minor changes in thermal conductivity rarely have a significant effect on heat-exchanger design or performance. From Table 2.3, it is clear that copper is the material with the

Table 2.1   Typical Fouling Factors for Water Streams in Heat Exchangers

Water Service	Water Temperature <52°C				Water Temperature >52°C			
Water Velocity	<1 m/s		>1 m/s		<1 m/s		>1 m/s	
	R_f (m^2K/W)	h_f (W/m^2/K)	R_f (m^2K/W)	h_f (W/m^2/K)	R_f (m^2K/W)	h_f (W/m^2/K)	R_f (m^2K/W)	h_f (W/m^2/K)
Sea Water	88.1×10^{-6}	11,350	88.1×10^{-6}	11,350	1.76×10^{-4}	5675	1.76×10^{-4}	5675
Boiler Feed Water	1.76×10^{-4}	5675	88.1×10^{-6}	11,350	1.76×10^{-4}	5675	1.76×10^{-4}	5675
Cooling Water	1.76×10^{-4}	5675	1.76×10^{-4}	5675	3.52×10^{-4}	2840	3.52×10^{-4}	2840
Hard Water	5.28×10^{-4}	1895	5.28×10^{-4}	1895	8.81×10^{-4}	1135	8.81×10^{-4}	1135

Source: Reproduced by permission from Tubular Exchanger Manufacturers Association, Inc. (TEMA, 2013).

Table 2.2   Typical Fouling Factors for Chemical Process Streams in Heat Exchangers

Chemical/Process Stream	R_f (m^2K/W)	h_f (W/m^2/K)
Inorganic		
Gases (oil-bearing or dirty)	3.52×10^{-4}	2840
Liquids (heating or vaporization)	3.52×10^{-4}	2840
Refrigerant brines	1.76×10^{-4}	5675
Molten heat transfer salts	88.1×10^{-6}	11,350
Organic		
Process gas	1.76×10^{-4}	5675
Utility gas (oil bearing, refrigerant, etc.)	3.52×10^{-4}	2840
Condensing vapors	1.76×10^{-4}	5675
Process Liquid	1.76×10^{-4}	5675
Vaporizing liquid	3.52×10^{-4}	2840
Refrigerant liquid	1.76×10^{-4}	5675
Heat transfer media	3.52×10^{-4}	2840
Polymer forming liquid	8.81×10^{-4}	1135
Oils (vegetable and heavy gas oil)	5.28×10^{-4}	1895
Asphalt and residuum	8.81×10^{-4}	1135

Source: Reproduced by permission from Tubular Exchanger Manufacturers Association, Inc. (TEMA, 2013).

Table 2.3   Thermal Conductivity of Metals Typically Used in Heat–Exchanger Construction

	Thermal conductivity, k, W/m/K or W/m/°C				
	$T = -150°C$	$T = 0°C$	$T = 100°C$	$T = 200°C$	$T = 400°C$
Metal					
Admiralty (71% Cu, 28% Zn, and 1% Sn)[1]		111			
Alloys					
Carbon-moly (0.5% Mo)[1]		43			
Chrome-moly steel (1% Cr and 0.5% Mo)[1]		42			
Chrome-moly steel (2¼% Cr and 0.5% Mo)[1]		38			
Chrome-moly steel (5% Cr and 0.5% Mo)[1]		35			
Chrome-moly steel (12% Cr and1% Mo)[1]		28			
Aluminum[5]	100	202	206	215	249
Brass (70% Cu and 30% Zn)[1]		99			
Carbon Steel[4]		45	45.8	43.5	39.0
Copper[1, 6]		386	377		
Inconel® 625 (21.5% Cr, 9% Mo, 5% Fe, and Ni balance)[2]	7.3	9.9	10.7	12.6	15.6 (20.0@700°C)
Hasteloy® N (71% Ni, 16% Mo, 5% Fe, and 1% Si)[3]			11.5	13.1	16.5 (23.6@700°C)
Nickel[1]		62			
Stainless Steel (304)[4]		16	16.4	17.7	20.3
Stainless Steel (316)[1]		16			
Titanium[1]		19			

[1]Engineering Toolbox, Thermal Conductivities of Heat Exchanger Materials, http://www.engineeringtoolbox.com/heat-exchanger-material-thermal-conductivities-d_1488.html.

[2]Azo Materials, http://www.azom.com/article.aspx?ArticleID=4461.

[3]Haynes International Data Sheet, http://haynesintl.com/docs/default-source/pdfs/new-alloy-brochures/corrosion-resistant-alloys/n-brochure.pdf?sfvrsn=6.

[4]"Thermal Conductivity of Some Irons and Steels over the Temperature Range 100 to 500°C," S. M. Shelton, U.S. Department of Commerce Bureau of Standards, Research Paper RP669, part of *Bureau of Standards Journal of Research*, vol. 12, April 1934.

[5]*Perry's Chemical Engineers Handbook*, 8th ed., D. W. Green and R. H. Perry, New York, McGraw-Hill, 2008.

[6]*Transport Processes and Separation Process Principles*, 4th ed, C. Geankoplis, Upper Saddle River, NJ, Prentice Hall, 2003.

highest thermal conductivity (lowest heat transfer resistance) and is favored as long as it is compatible with the fluids it contacts. Similarly, aluminum is also a good conductor but may suffer from low strength, although it is often used in cryogenic operations along with stainless steel.

For S-T heat exchanges, the sizes of standard tubes are given in Table 2.4. The third column heading is BWG, which stands for Birmingham Wire Gauge, which is a standard method, albeit a

Table 2.4   Heat-Exchanger Tube Data

Outside Diameter		Wall Thickness			Inside Diameter	
in	mm	BWG	in	mm	in	mm
⅝ (0.625)	15.875	12	0.109	2.769	0.407	10.338
	15.875	14	0.083	2.108	0.459	11.659
	15.875	16	0.065	1.651	0.495	12.573
	15.875	18	0.049	1.245	0.527	13.386
¾ (0.750)	19.050	12	0.109	2.769	0.532	13.513
	19.050	14	0.083	2.108	0.584	14.834
	19.050	16	0.065	1.651	0.620	15.748
	19.050	18	0.049	1.245	0.652	16.561
⅞ (0.875)	22.225	12	0.109	2.769	0.657	16.688
	22.225	14	0.083	2.108	0.709	18.009
	22.225	16	0.065	1.651	0.745	18.923
	22.225	18	0.049	1.245	0.777	19.736
1	25.400	10	0.134	3.404	0.732	18.593
	25.400	12	0.109	2.769	0.782	19.863
	25.400	14	0.083	2.108	0.834	21.184
	25.400	16	0.065	1.651	0.870	22.098
1¼ (1.25)	31.750	10	0.134	3.404	0.982	24.943
	31.750	12	0.109	2.769	1.032	26.213
	31.750	14	0.083	2.108	1.084	27.534
	31.750	16	0.065	1.651	1.120	28.448
1½ (1.50)	38.100	10	0.134	3.404	1.232	31.293
	38.100	12	0.109	2.769	1.282	32.563
	38.100	14	0.083	2.108	1.334	33.884
2	50.800	10	0.134	3.404	1.732	43.993
	50.800	12	0.109	2.769	1.782	45.263

Source: Adapted from Geankoplis (2003).

very old standard, of specifying tube wall thickness. Unlike schedule pipe sizes discussed in Chapter 1, where the nominal pipe size does not correspond to any actual dimension of the pipe, the stated diameter of BWG tubing is the outside diameter. Standard tube sizes are virtually always used in heat-exchanger design, because the cost of customized tubes and the associated fittings and tooling would be prohibitively expensive. Standard lengths for tubes used in heat-exchanger design are from 8 ft to 20 ft (2.438–6.096 m) in 2 ft increments. The number of tubes that can fit in standard shell diameters is given in the tube-count tables shown later in Section 2.7.

2.5.3   Correlations for Film Heat Transfer Coefficients

The following sections cover the estimation of the convective film heat transfer coefficients. The cases for convective flow without phase change for inside and outside tubes (internal and external flows) are covered first. The coefficients for a change of phase are then introduced.

2.5.3.1   Flow Inside Tubes

In heat-exchanger design, the most common geometry for the equipment uses circular tubes, and the bulk of correlations and experimental work has been done for this geometry. The type of flow, turbulent, laminar, or transition, has a strong influence on the form of the correlation used to determine the film heat transfer coefficient. Each flow regime is considered separately in the following sections.

Turbulent Flow
For a fluid flowing inside a tube or pipe, the heat transfer coefficient is a function of fluid properties, fluid velocity, and the diameter of the tube. Many correlations exist for estimating the heat transfer coefficients for flow in tubes, and one of the most common is the Seider-Tate (1936) equation:

$$\text{Nu} = \frac{h_i D_i}{k} = 0.023 \left[1 + \left(\frac{D_i}{L}\right)^{0.7} \right] \left(\frac{D_i u \rho}{\mu}\right)^{0.8} \left(\frac{c_p \mu}{k}\right)^{1/3} \left(\frac{\mu}{\mu_w}\right)^{0.14} \quad \text{Re} \geq 6000$$

$$\text{Nu} = \frac{h_i D_i}{k} = 0.023 \left[1 + \left(\frac{D_i}{L}\right)^{0.7} \right] \text{Re}^{0.8} \text{Pr}^{1/3} \left(\frac{\mu}{\mu_w}\right)^{0.14} \quad \text{Re} \geq 6000 \tag{2.26}$$

where Nu is the Nusselt number, Re is the Reynolds number, and Pr is the Prandtl number. The Prandtl number is the ratio of the rate of viscous diffusion to the rate of thermal diffusion. The Seider-Tate correlation, Equation (2.26), is valid for all fluids (except molten metals) for Re ≥ 6000. All the fluid properties (ρ, μ, c_p, k) should be evaluated at the average bulk temperature with the exception of μ_w, which is evaluated at the average wall temperature. The first term in parentheses on the right-hand side takes account of the entrance effects for short tubes. For the case when $L/D > 72$, the error in neglecting this term is <5%. The last term in parentheses on the right-hand side takes account of the viscosity change near the tube wall that affects the thickness of the thermal boundary layer. Usually this term is close to unity, partly due to the small exponent. However, the term may become important for highly viscous fluids, where large changes in viscosity may take place because of the temperature variation between the bulk fluid and the wall. The heat transfer coefficient determined from Equation (2.26) changes along the length of the heat exchanger as the viscosity at the wall changes with temperature and position; therefore, use of the Seider-Tate equation with the viscosity correction requires an iterative procedure.

Another common correlation is the Dittus-Boelter (1930) equation that is given by

$$\text{Nu} = \frac{h_i D_i}{k} = 0.023 \text{Re}^{0.8} \text{Pr}^n \tag{2.27}$$

where the value of the exponent n is 0.3 when the fluid is being cooled and 0.4 when the fluid is being heated. This equation does not require an iterative solution but is generally less accurate than Equation (2.26), especially for fluids undergoing large temperature changes.

For an annulus, the following equation can be used:

$$\text{Nu} = \frac{h_{an} D_H}{k} = 0.023 \left(\frac{D_H v \rho}{\mu} \right)^{0.8} \left(\frac{\mu c_p}{k} \right)^{0.4} \left(\frac{D_{i-o}}{D_{o-i}} \right)^{0.45} \tag{2.28}$$

where D_{i-o} is the inside diameter of the outer pipe and D_{o-i} is the outside diameter of the inside pipe, and D_H is hydraulic diameter of the annular opening = $(D_{i-o} - D_{o-i})$, where D_H is defined in Equation (2.29).

For noncircular ducts (square, rectangle, triangular, and ellipsoidal), Equations (2.26) and (2.27) can still be used except that the hydraulic diameter, D_H, should be substituted for D_i in the equation (and in the definition of Re), where

$$D_H = 4 \frac{\text{flow area}}{\text{wetted perimeter}} \tag{2.29}$$

The use of Equations (2.26) and (2.27) is illustrated in Example 2.7.

Example 2.7

Use Equation (2.26) to determine the inside heat transfer coefficient for a fluid flowing inside a 12 ft (3.6576 m) long, ¾-in tube (16 BWG) at a velocity of 2 ft/s (0.6096 m/s) that is being cooled and has an average temperature of 212°F (100°C) and an average wall temperature of 104°F (40°C). Consider the following three fluids:

    a. Liquid water

    b. n-cetane

    c. Air at 2 atm pressure—use a velocity of 20 ft/s (6.609 m/s)

    d. Compare the results using Equation (2.27).

Solution

From Table 2.4, $D_i = 0.620$ in (15.748 mm)

    a. For water at 100°C,

$\rho = 961$ kg/m^3, $\mu = 0.000282$ kg/m/s, $c_p = 4216$ J/kg/K, $k = 0.6804$ W/m/K
At 40°C,
$\mu = 0.000654$ kg/m/s
Applying Equation (2.26),

$$\text{Re} = \frac{D_i u \rho}{\mu} = \frac{(15.748 \times 10^{-3})(0.6096)(961)}{(2.82 \times 10^{-4})} = 32{,}715 \text{ and } \text{Pr} = \frac{c_p \mu}{k} = \frac{(4216)(2.82 \times 10^{-4})}{(0.6804)} = 1.7474$$

$$\text{Nu} = \frac{h_i D_i}{k} = 0.023 \left[1 + \left(\frac{D_i}{L} \right)^{0.7} \right] \left(\frac{D_i \mu \rho}{\mu} \right)^{0.8} \left(\frac{c_p \mu}{k} \right)^{1/3} \left(\frac{\mu}{\mu_w} \right)^{0.14}$$

$$\text{Nu} = \frac{h_i D_i}{k} = 0.023 \left[1 + \left(\frac{15.748 \times 10^{-3}}{3.6576} \right)^{0.7} \right] (32{,}715)^{0.8} (1.7474)^{1/3} \left[\frac{2.82 \times 10^{-4}}{6.54 \times 10^{-4}} \right]^{0.14}$$

$$= 0.023(1.0221)(4090.7)(1.2045)(0.8889) = 102.95$$

$$h_i = \text{Nu} \left[\frac{k}{D_i} \right] = (102.87) \left[\frac{0.6804}{15.748 \times 10^{-3}} \right] = 4448 \, \text{W/m}^2/\text{K}$$

b. For n-cetane at 100°C,

$\rho = 717.6 \, \text{kg/m}^3, \mu = 0.0009207 \, \text{kg/m/s}, c_p = 2392 \, \text{J/kg/K}, k = 0.1236 \, \text{W/m/K}$
At 40°C,
$\mu = 0.002266 \, \text{kg/m/s}$
Applying Equation (2.26),
$h_i = 534 \, \text{W/m}^2/\text{K}$ (Re = 7482, Pr = 17.81)

c. For air at 2 atm pressure at 100°C,

$\rho = 1.8907 \, \text{kg/m}^3, \mu = 2.183 \times 10^{-5} \, \text{kg/m/s}, c_p = 1008 \, \text{J/kg/K}, k = 0.0313 \, \text{W/m/K}$,
At 40°C,
$\mu = 1.916 \times 10^{-5} \, \text{kg/m/s}$
Applying Equation (2.26),
$h_i = 57.9 \, \text{W/m}^2/\text{K}$ (Re = 8315, Pr = 0.703)

d. Using the Dittus-Boelter equation, Equation (2.27) gives

Water	Re = 32,715, Pr = 1.7474	$\text{Nu} = 0.023 \text{Re}^{0.8} \text{Pr}^{0.3} = 111.2 \rightarrow h_i = 4804 \, \text{W/m}^2/\text{K}$
n-cetane	Re = 7842, Pr = 17.81	$\text{Nu} = 71.20 \rightarrow h_i = 559 \, \text{W/m}^2/\text{K}$
Air	Re = 8315, Pr = 0.703	$\text{Nu} = 28.30 \rightarrow h_i = 56.2$

Summarizing the results:

Fluid	h_i (W/m²/K) Eq. (2.26)	h_i (W/m²/K) Eq. (2.27)	% difference
Water	4448	4804	7.4
n-cetane	534	559	4.7
Air	57.9	56.2	−2.9

From the results of Example 2.7, three important points emerge. First, heat transfer coefficients for liquid water are generally significantly higher than for other liquids. Second, heat transfer coefficients for gases are normally much lower (~5–10 times) than for liquids. Third, while the Seider-Tate and Dittus-Boelter equations give different heat transfer coefficients, the differences are not that large and may be within typical design tolerances.

Transition Flow

For the transition between laminar and fully developed turbulent flow, the equation of Hausen (1943) satisfies both the upper and lower limits of Reynolds numbers and is recommended.

$$\text{Nu} = 0.116 \, [\text{Re}^{2/3} - 125] \text{Pr}^{1/3} \left[1 + \left(\frac{D_i}{L} \right)^{2/3} \right] \left(\frac{\mu}{\mu_w} \right)^{0.14} \quad 2100 < \text{Re} < 6400 \qquad (2.30)$$

Laminar Flow

When the flow of fluid inside the tube is laminar (Re < 2100), then theoretical analyses for certain cases are possible and may be used to determine the heat transfer coefficient. Two limiting cases, constant heat flux at the wall and constant wall temperature, are normally considered. Neither case applies directly to normal operating conditions in process heat exchangers. However, the case for constant wall temperature is often used as a starting point for the development of suitable correlations. The appropriate equation to use is determined by the magnitude of the Graetz number, Gz, where $Gz = RePr\dfrac{D_i}{L}$. The correlations attributed to Hausen (1943),

$$Nu = \left[3.66 + \frac{0.085Gz}{1+0.047Gz^{2/3}}\right]\left(\frac{\mu}{\mu_w}\right)^{0.14} \quad Gz < 100 \tag{2.31}$$

and Seider and Tate (1936),

$$Nu = 1.86Gz^{1/3}\left(\frac{\mu}{\mu_w}\right)^{0.14} \quad Gz > 100 \tag{2.32}$$

are recommended.

It should be noted that Gz is a function of $1/L$, so the heat transfer coefficient is strongly affected by the tube length. As the tube length becomes very long and Gz tends to zero, the Nusselt number tends to a limiting value of Nu_∞. When viscosity corrections are negligible, it can be seen that $Nu_\infty = 3.66$ assuming constant wall temperature.

Example 2.8 demonstrates the use of the equations for laminar flow.

Example 2.8

Consider the same conditions and fluids in Example 2.7, except use a liquid velocity of 0.1 ft/s (0.03048 m/s) and a gas velocity of 2 ft/s (0.6096 m/s).

Solution

a. For water,

$$Re = \frac{D_i u \rho}{\mu} = \frac{(15.748\times10^{-3})(961)(0.03048)}{(2.82\times10^{-4})} = 1636, \quad Pr = 1.7474,$$

$$\frac{D_i}{L} = \frac{(15.748\times10^{-3})}{(3.6576)} = 4.306\times10^{-3} \text{ and } Gr = RePr\frac{D_i}{L} = (1636)(1.7474)(4.306\times10^{-3}) = 12.31$$

Applying Equation (2.31),

$$Nu = \left[3.66 + \frac{0.085Gz}{1+0.047Gz^{2/3}}\right]\left(\frac{\mu}{\mu_w}\right)^{0.14} = \left[3.66 + \frac{0.085(12.31)}{1+0.047(12.31)^{2/3}}\right]\left[\frac{2.82\times10^{-4}}{6.54\times10^{-4}}\right]^{0.14} = 3.997$$

$$h_i = Nu\left[\frac{k}{D_i}\right] = (3.997)\left[\frac{0.6804}{15.748\times10^{-3}}\right] = 172.7 \text{ W/m}^2\text{/K}$$

b. For n-cetane,
Re = 374.1, Pr = 17.8181, D_i/L = 4.306 × 10⁻³, Gz = (374.1)(17.8181)(4.306 × 10⁻³) = 28.70, Nu = 4.719, h_i = 37.04 W/m²/K

c. For air,

Re = 415.7, Pr = 0.7030, D_i/L = 4.306 × 10^{-3}, Gz = (415.7)(0.7030)(4.306 × 10^{-3}) = 1.258, Nu = 3.831, h_i = 7.61 W/m^2/K

Summarizing the results:

Fluid	h_i (W/m^2/K)
Water	172.7
n-cetane	37.04
Air	7.61

Clearly, the heat transfer coefficients for laminar flow are much lower than for turbulent flow. In general, laminar flow should be avoided except for highly viscous liquids where pumping costs needed to give turbulent flow conditions may become prohibitively large.

For laminar flow in annuli, the heat transfer correlation of Chen, Hawkins, and Solberg (1946) is recommended:

$$\text{Nu}_{an} = \frac{h_{an}D_H}{k} = 1.02\,\text{Re}^{0.45}\,\text{Pr}^{0.5}\,\text{Gz}^{0.05}\left(\frac{D_H}{L}\right)^{0.4}\left(\frac{D_{i-o}}{D_{o-i}}\right)^{0.8}\left(\frac{\mu}{\mu_w}\right)^{0.14} \qquad 200 < \text{Re} < 2000 \quad (2.33)$$

where Re and Gz are based on the hydraulic diameter, D_H, μ_w is the viscosity at the inner wall of the annulus (at D_{o-i}), and D_{i-o} and D_{o-i} are the inside diameter of the outside tube and the outside diameter of the inside tube, respectively.

For flow in rectangular channels, the limiting value of Nu for a constant wall temperature and very long channels (Nu$_\infty$) is a function of the ratio of the shorter to longer dimensions of the channel ($0 < a/b < 1$). The equivalent limiting value for tubes is 3.66, which is the first term on the right-hand side of Equation (2.31). The relationship between Nu$_\infty$ and (a/b) for rectangular channels attributed to Clark and Kays (1953) is shown in Figure 2.25 and Equation (2.34).

$$\text{Nu}_\infty = 7.57 - 17.71\left(\frac{a}{b}\right) + 28.80\left(\frac{a}{b}\right)^2 - 24.04\left(\frac{a}{b}\right)^3 + 8.26\left(\frac{a}{b}\right)^4 \qquad (2.34)$$

It is suggested that Equation (2.31) be modified to use with rectangular channels as

$$\text{Nu} = \left[\text{Nu}_\infty + \frac{0.085\text{Gz}}{1 + 0.047\text{Gz}^{2/3}}\right]\left(\frac{\mu}{\mu_w}\right)^{0.14} \qquad (2.35)$$

where Nu$_\infty$ is taken from Equation (2.34) and the hydraulic diameter is used in determining Re in Gz.

2.5.3.2 Flow Outside of Tubes (Shell-Side Flow)

The estimation of the heat transfer coefficient for the flow of a fluid over a bundle of tubes is very complicated. Referring back to Figure 2.18, in an S-T exchanger, the flow on the shell side is parallel with the tube axis for some portion of the flow path and perpendicular to the tube axis for the remainder of the time. Moreover, the flow path of the shell-side fluid is quite tortuous, since it

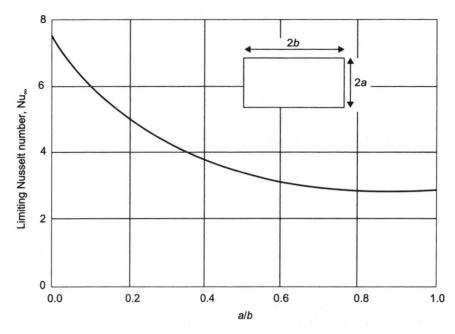

Figure 2.25 Relationship between Nu_∞ and the ratio of the channel side lengths for rectangular channels (Clark and Kays [1953])

flows around and between the tubes, and the arrangement of the tubes (square or triangular pitch) significantly affects the mixing and turbulence of the fluid. Needless to say, analytical expressions for heat transfer coefficients do not exist, and the number of correlations for overall shell-side heat transfer coefficient are numerous. In this section, some of the different phenomena occurring on the shell side of the heat exchanger are discussed, and a simplified method for calculating shell-side coefficients is presented.

Flow Normal to the Outside of a Single Cylinder
The flow of fluid normal to a single tube or cylinder is a complex but well-studied process. A review of data over a wide range of fluid properties (Pr) and flow conditions (Re) has been performed by Žukauskas (1972) and Churchill and Bernstein (1977). The equation recommended by Žukauskas is

$$\mathrm{Nu}_f = \frac{hD_o}{k_f} = C\,\mathrm{Re}_f^m\,\mathrm{Pr}_f^{0.365}\left[\frac{\mathrm{Pr}_{bulk}}{\mathrm{Pr}_w}\right]^{1/4} \tag{2.36}$$

where subscripts *w*, *f*, and *bulk* refer to the wall, film, and bulk fluid conditions, respectively.

$\mathrm{Re} = \dfrac{D_o u_{bulk}\rho_f}{\mu_f}$ is evaluated using the fluid approach velocity upstream of the tube, and the film

properties are evaluated at the average film temperature, $T_f = (T_{bulk} + T_w)/2$. Finally, the different Prandtl numbers in Equation (2.36) are evaluated at the average film temperature, wall temperature, or bulk fluid temperature depending on the subscript.

Table 2.5   Values for Parameters in Equation (2.36)

Range of Re	C	m
<2000	0.677	0.453
$2000 - 1 \times 10^6$	0.260	0.600

Equation (2.36) is valid for values of Pr in the range 0.7 to 500. The values of C and m depend on the Reynolds number and are given in Table 2.5.

Flow Normal to Banks of Tubes
The heat transfer coefficient for a fluid flowing normal to a bank of tubes is very complicated, and the average coefficient for a given tube will vary depending on the location of the tube in the bank. A good review of data on tube banks is again given by Žukauskas (1972). In this chapter, a simpler method for estimating an average shell-side heat transfer coefficient attributed to Kern (1950) is presented.

Kern's Method for Shell-Side Heat Transfer
A popular method for making a preliminary estimate of the shell-side heat transfer coefficient and shell-side pressure drop is that attributed to Kern (1950). The correlations used in Kern's method depend on an equivalent hydraulic diameter for the shell side, $D_{H,s}$. Figure 2.26 shows the basic tube arrangements, and equations for the hydraulic diameter are
Square pitch

$$D_{H,s} = \frac{4 \, (\text{flow area})}{\text{wetted perimeter}} = \frac{4(p^2 - \frac{\pi}{4}D_o^2)}{\pi D_o} = \frac{1.273 p^2 - D_o^2}{D_o} \tag{2.37}$$

Triangular pitch

$$D_{H,s} = \frac{4 \, (\text{flow area})}{\text{wetted perimeter}} = \frac{4(0.866 p \frac{p}{2} - \frac{1}{2}\frac{\pi}{4}D_o^2)}{\frac{\pi D_o}{2}} = \frac{1.103 p^2 - D_o^2}{D_o} \tag{2.38}$$

The term $(p - D_o)$ is sometimes referred to as the *clearance*, C. The shell-side fluid changes velocity as it passes through the baffle window and travels across the bank of tubes. This situation is illustrated in Figure 2.27, where the inside diameter of the shell is D_s, and the baffle spacing is L_b.
Assuming that the mass flowrate of fluid on the shell side is $\dot{m}_s$, then using the notation in Figures 2.26 and 2.27, the following parameters are defined:
Shell-side superficial mass velocity,

$$G_s = \frac{\dot{m}_s}{A_s} \tag{2.39}$$

Shell-side velocity,

$$u_s = \frac{\dot{m}_s}{\rho A_s} = \frac{G_s}{\rho} \tag{2.40}$$

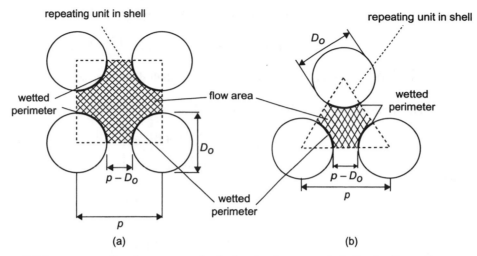

Figure 2.26 Notation for determining the hydraulic diameter of shell-side flow: (a) square pitch, (b) triangular pitch

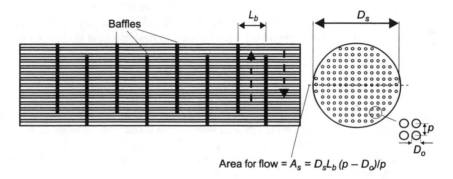

Area for flow = $A_s = D_s L_b (p - D_o)/p$

Figure 2.27 Sketch illustrating the flow area, A_s, used in Kern's method

where the A_s is the maximum flow area on the shell side given as

$$A_s = D_s L_b \frac{(p - D_o)}{p}$$ (2.41)

Shell-side Reynolds number,

$$\text{Re}_s = \frac{G_s D_{H,s}}{\mu} = \frac{u_s \rho D_{H,s}}{\mu}$$ (2.42)

The average heat transfer coefficient for the shell side of the exchanger is given by

$$\text{Nu} = \frac{h_s D_{H,s}}{k_f} = j_h \text{Re} \text{Pr}^{1/3} \left(\frac{\mu}{\mu_w} \right)^{0.14}$$ (2.43)

where

$$j_h = 1.680 (BC)^{-0.263} \text{Re}_s^{-0.581} \quad \text{for Re}_s < 100$$

and

$$j_h = 1.249(BC)^{-0.329} \mathrm{Re}_s^{-0.470} \quad \text{for } 100 < \mathrm{Re}_s < 1 \times 10^6$$

where BC is the baffle cut as a percentage of the shell diameter, with typical values from 20% to 35%.

The frictional pressure drop through the shell side of the exchanger is then given by

$$-\Delta P_f = \frac{4 j_f G_s^2 D_s (N_B + 1)}{2 \rho D_{H,s}} \left[\frac{\mu}{\mu_w} \right]^{-0.14} \tag{2.44}$$

where

$$j_f = 456.6 (BC)^{-0.739} \mathrm{Re}_s^{-0.928} \quad \text{for } \mathrm{Re}_s < 300$$

and

$$j_f = 5.460 (BC)^{-0.674} \mathrm{Re}_s^{-0.189} \quad \text{for } \mathrm{Re}_s \geq 300$$

It should be noted that the original method attributed to Kern (1950) uses a series of charts to determine the values of j_h and j_f as functions of Re and BC. However, the equations given for j_h and j_f were found by regressing data from these charts and are accurate enough for preliminary designs.

Example 2.9 illustrates the use of Kern's method for estimating the heat transfer coefficient for a bank of tubes. This method will also be used to design a heat exchanger in Section 2.7.

Example 2.9

Water flows across the outside of a bank of tubes. It enters at 30°C and leaves at 40°C. The water enters at a flowrate of 34.27 kg/s, the shell diameter is 20 in, the baffle spacing is half (½) the shell diameter, the baffle cut (BC) is 15%, and ¾-in outside diameter (OD) tubes on a 1-in pitch are used. The fluid inside the tubes may be assumed to be at a constant temperature of 90°C (condensing organic), the inside coefficient is expected to be much higher than the shell-side coefficient, and thus the wall temperature may be taken as 90°C.

Use Kern's method to determine the average heat transfer coefficient for the shell side for the following arrangements:

a. Square pitch
b. Equilateral triangular pitch

Solution

a. Tubes on a square pitch

Table E2.9   Properties of Water at Different Temperatures

Temperature	T_{ave} = 35°C	T_f = 62.5°C	T_w = 90°C
C_p (J/kg/K)	4178	4185	4206
ρ (kg/m³)	996	984	968
μ (kg/m/s)	742×10^{-6}	457×10^{-6}	319×10^{-6}
k (W/m/K)	0.6205	0.6535	0.6746
$Pr = c_p\, \mu/k$	5.00	2.92	1.99

From Equations (2.37), (2.41), (2.42), and (2.43),

$$D_{H,s} = \frac{1.273\,p^2 - D_o^2}{D_o} = \frac{1.273(1)^2 - (0.75)^2}{(0.75)} = 0.9473 \text{ in} = 0.02406 \text{ m}$$

$$D_s = (20)(0.0254) = 0.508 \text{ m}$$

$$L_b = \tfrac{1}{2} D_s = 0.254 \text{ m}$$

$$A_s = D_s L_b \frac{(p - D_o)}{p} = (0.508)(0.254)\frac{(1 - 0.75)}{(1)} = 0.03226 \text{ m}^2$$

$$G_s = \frac{\dot{m}_s}{A_s} = \frac{(34.27)}{(0.03226)} = 1062.4 \text{ kg/m}^2/\text{s}$$

$$\text{Re}_s = \frac{G_s D_{H,s}}{\mu} = \frac{(1062.4)(0.02406)}{(742 \times 10^{-6})} = 34{,}448$$

$$j_h = 1.249(BC)^{-0.329}\,\text{Re}^{-0.470} \quad \text{for } 100 < \text{Re} < 1 \times 10^6$$

$$BC = 15\%$$

$$j_h = 1.249(15)^{-0.329}(34{,}448)^{-0.470} = 0.003777$$

$$\text{Nu} = \frac{h_s D_{H,s}}{k_f} = j_h \text{Re}\,\text{Pr}^{1/3}\left(\frac{\mu}{\mu_w}\right)^{0.14} = (0.003777)(34{,}448)(5.00)^{1/3}\left(\frac{742 \times 10^{-6}}{319 \times 10^{-6}}\right)^{0.14} = 250.4$$

$$h_s = 250.4\frac{(0.6535)}{(0.02406)} = 6801 \text{ W/m}^2/\text{K}$$

b. Tubes on a triangular pitch
From Equations (2.38), (2.41), (2.42), and (2.43),

$$D_{H,s} = \frac{1.103\,p^2 - D_o^2}{D_o} = \frac{1.103(1)^2 - (0.75)^2}{(0.75)} = 0.7207 \text{ inch} = 0.01830 \text{ m}$$

$$\text{Re}_s = \frac{G_s D_{H,s}}{\mu} = \frac{(1062.4)(0.0183)}{(742 \times 10^{-6})} = 26{,}209$$

$$j_h = 1.249(BC)^{-0.329}\,\text{Re}^{-0.470} \quad \text{for } 100 < \text{Re} < 1 \times 10^6$$

Assume a 15% baffle cut:

$$j_h = 1.249(15)^{-0.329}(26{,}209)^{-0.470} = 0.004295$$

$$\text{Nu} = \frac{h_s D_{H,s}}{k_f} = j_h \text{Re}\,\text{Pr}^{1/3}\left(\frac{\mu}{\mu_w}\right)^{0.14} = (0.004295)(26{,}208)(5.00)^{1/3}\left(\frac{742 \times 10^{-6}}{319 \times 10^{-6}}\right)^{0.14} = 216.6$$

$$h_s = 216.6\frac{(0.6535)}{(0.0183)} = 7736 \text{ W/m}^2/\text{K}$$

It can be seen that a higher heat transfer coefficient is obtained using triangular pitch. However, this will usually come at the expense of a higher shell-side pressure drop.

2.5.3.3   Boiling Heat Transfer

When one or both fluids undergo a change in phase, the heat transfer process is much more complex and, in general, will be less influenced by the rate of flow of material past the heat transfer surface than by the temperature driving force between the surface and the bulk fluid temperature. There is a large body of work in the area of boiling heat transfer that is far too extensive to review here. The approach adopted in this text is to provide a brief overview of boiling phenomena and to present some useful working equations that are applicable to the design of process heat exchangers.

Typical Pool Boiling Curve

A typical boiling curve for water at 1 atm pressure is shown in Figure 2.28 and represents the situation when a heated surface is placed in a pool of water. In the figure, the x-axis represents the temperature difference between the heating surface (T_s) and the saturation temperature (T_{sat}), and the y-axis is the heat flux (Q/A). The behavior of water at other pressures and other liquids below the critical pressure mimics the general trends shown in Figure 2.28.

When the temperature difference between the surface and saturation temperatures of the water $(T_{sat} = 100°C$ for water at 1 atm) is less than 5°C, all vapor is produced by evaporation at the liquid surface, and the process is termed *free convection boiling*. In this region, the heat transfer coefficient is proportional to $(\Delta T_s)^{1/4}$. As ΔT_s increases beyond about 5°C, bubbles of vapor appear

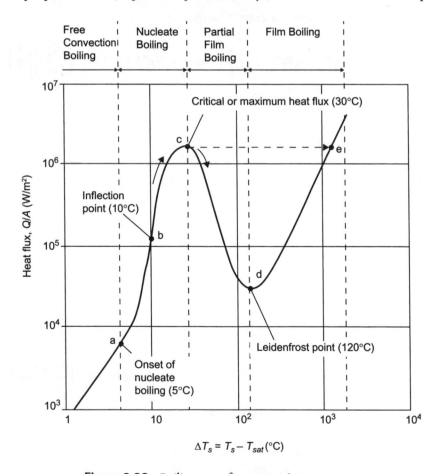

Figure 2.28   Boiling curve for water at 1 atm pressure

on the heating surface, detach, and move to the surface of the liquid. At first, bubbles appear at certain preferred nucleation sites on the surface, but as the ΔT_s increases, the number of nucleation sites increases, and more bubbles are produced. The bubble motion near the surface tends to increase turbulence and promotes higher heat transfer. This process is known as *nucleate boiling*. In this boiling regime (region a–c), the heat transfer coefficient is a strong function of ΔT_s and $h \propto (\Delta T_s)^n$ where n is between 2 and 3. As ΔT_s increases further, more and more bubbles are formed at the surface, and they start to interfere with the movement of liquid toward the surface. This phenomenon tends to reduce the heat transfer coefficient because the gas has a much lower heat capacity and thermal conductivity than the liquid. As a result, an inflection point (Point b) is seen in the boiling curve (at 10°C). With further increase in ΔT_s, the increase in the rate of bubble formation eventually causes the heat flux to reach a maximum (at about 30°C). At Point c on Figure 2.28, the path by which the process proceeds to the right depends on how the experiment is conducted.

First, consider the case when the temperature of the heat transfer surface can be increased independently of the heat flux. Increasing ΔT_s beyond Point c is accompanied by a reduction in heat transfer coefficient (and heat flux) due to the intermittent formation of a vapor film at the heat transfer surface, which occurs because the rate of bubble formation is faster than the rate of bubble detachment from the surface. This region is termed the *partial film boiling* regime (Region c–d), and the heat transfer surface may at any time be completely covered by either gas or a liquid-bubble mixture. In this regime, the surface oscillates between a nucleate boiling and a film boiling condition. This behavior leads to a reduction in heat flux and a corresponding reduction in heat transfer coefficient. This behavior persists until ΔT_s is large enough to maintain a stable gas film at the surface, and then the film boiling regime is reached (Region d–e). The transition to the film boiling regime occurs at the point of minimum heat flux, Point d, which is referred to as the *Leidenfrost point*. As ΔT_s increases beyond the Leidenfrost point, the heat transfer coefficient and heat flux increase because of a combined effect of increasing gas thermal conductivity with increasing temperature and radiation heat transfer that is present at the high temperatures required for film boiling.

Consider now the case when the experiment is conducted with the heat flux as the controlled or independent variable and ΔT_s is the dependent variable, which would be the case if, for example, an electric heater or a direct flame was used to heat the surface in contact with the liquid. Using this experimental procedure, the boiling curve would be identical to that described previously from Points a to c, but at Point c, the only way that the heat flux can be increased is for ΔT_s to jump to Point e. This would be accompanied by a very large increase in T_s, which may lead to permanent damage to the heated surface or in extreme cases could melt the surface.

It is important to recognize that this temperature jump at the critical heat flux may occur in actual processes in which the energy for boiling is supplied by an electric heater or by a burning fuel, as would be the case in a gas or oil fired heater or boiler. This phenomenon generally cannot happen if heat is supplied by a hot process stream, because the temperature of the process stream is bounded by process conditions and the heat flux is, therefore, also bounded. As a result, process heat exchangers used for raising steam such as waste heat boilers or used for reboiling a distillation column do not exhibit this unstable behavior. Nevertheless, operation to the right of the critical heat flux is generally avoided in process heat exchangers to avoid the effect that as ΔT increases the heat flux decreases. Such inverse and counterintuitive behavior may cause problems with control and diagnosis of operations. For process heat exchangers, operation in the nucleate boiling regime is recommended.

Determining the Critical or Maximum Heat Flux in Pool Boiling
Many equations exist to predict the maximum heat flux, denoted by Point c in Figure 2.28. A recommended correlation that gives good predictions is that attributed to Zuber (1958):

$$\left[\frac{Q}{A}\right]_{max} = 0.131\rho_v\lambda\left[\frac{(\rho_l-\rho_v)\sigma g}{\rho_v^2}\right]^{1/4}\left(1+\frac{\rho_v}{\rho_l}\right)^{1/2} \qquad (2.45)$$

where ρ_l and ρ_v are the densities of saturated liquid and vapor, respectively, λ is the latent heat of vaporization, and σ is the surface tension of the boiling liquid.

An alternative relationship that requires only the critical pressure is given by the Cichelli-Bonilla (1945) correlation:

$$\left[\frac{Q}{A}\right]_{max} = 0.3673P_c\left(\frac{P}{P_c}\right)^{0.35}\left(1-\frac{P}{P_c}\right)^{0.9} \qquad (2.46)$$

where $[Q/A]_{max}$ is given in W/m^2 and pressure is in Pa.

Example 2.10

Estimate the critical heat flux for water at 1 atm pressure using Equations (2.45) and (2.46) and compare with the value given in Figure 2.28.

Properties of water at 1 atm pressure ($P = 1.013 \times 10^5$ Pa) and $T = 100°C$
$\rho_l = 958$ kg/m^3, $\rho_v = 0.598$ kg/m^3, $\lambda = 2.257 \times 10^6$ J/kg, $\sigma = 0.060$ N/m, $P_c = 22.06 \times 10^6$ Pa

Solution

From Equation (2.45),

$$\left[\frac{Q}{A}\right]_{max} = 0.131\rho_v\lambda\left[\frac{(\rho_l-\rho_v)\sigma g}{\rho_v^2}\right]^{1/4}\left(1+\frac{\rho_v}{\rho_l}\right)^{1/2}$$

$$\left[\frac{Q}{A}\right]_{max} = (0.131)(0.598)(2.257\times10^6)\left[\frac{(958-0.598)(0.060)(9.81)}{(0.598)^2}\right]^{1/4}\left(1+\frac{0.598}{958}\right)^{1/2} = 1.11\,\text{MW/m}^2$$

From Equation (2.46),

$$\left[\frac{Q}{A}\right]_{max} = 0.3673P_c\left(\frac{P}{P_c}\right)^{0.35}\left(1-\frac{P}{P_c}\right)^{0.9}$$

$$\left[\frac{Q}{A}\right]_{max} = (0.3673)(22.06\times10^6)\left(\frac{1.013\times10^5}{22.06\times10^6}\right)^{0.35}\left(1-\frac{1.013\times10^5}{22.06\times10^6}\right)^{0.9} = 1.23\,\text{MW/m}^2$$

From Figure 2.28, $\left[\dfrac{Q}{A}\right]_{max} \cong 1.2\,\text{MW/m}^2$

Thus both predictions are within ±10%.

Heat Transfer Coefficient for Nucleate (Pool) Boiling
Perhaps the most widely used correlation for the nucleate pool boiling regime is that attributed to Rohsenow (1964)

$$\frac{Q}{A} = \mu_l \lambda \left(\frac{g(\rho_l - \rho_v)}{\sigma} \right)^{1/2} \left(\frac{c_{pl} \Delta T_s}{C_f \lambda \Pr_l^s} \right)^3 \tag{2.47}$$

where l and v refer to saturated liquid and vapor conditions, $s = 1.0$ for water and 1.7 for other materials, and C_f is a constant that depends on the material of the heated surface and varies between 0.006 and 0.013.

This gives the equivalent heat transfer coefficient as

$$h = \mu_l \lambda \left(\frac{g(\rho_l - \rho_v)}{\sigma} \right)^{1/2} \left(\frac{c_{pl}}{C_f \lambda \Pr_l^s} \right)^3 \Delta T_s^2 \tag{2.48}$$

The application of Equations (2.47) and (2.48) is illustrated in Example 2.11.

Example 2.11

Use Equations (2.47) and (2.48) to estimate the heat flux and heat transfer coefficient for water at 1 atm pressure for a temperature driving force of 10°C.

Solution

Properties of water at 1 atm pressure ($P = 1.013 \times 10^5$ Pa) and $T = 100$°C
$\rho_l = 958$ kg/m^3, $\rho_v = 0.598$ kg/m^3, $\lambda = 2.257 \times 10^6$ J/kg, $\sigma = 0.060$ N/m, $c_{pl} = 4216$ J/kg/K, $\mu_l = 2.82 \times 10^{-4}$

kg/m/s, $k_l = 0.6804$ W/m/K, $\Pr_l = \dfrac{c_{p,l}\mu_l}{k_l} = \dfrac{(4216)(2.82 \times 10^{-4})}{(0.6804)} = 1.747$

Applying Equation (2.47) with $s = 1.0$ and $C_f = 0.013$ gives

$$\frac{Q}{A} = \mu_l \lambda \left(\frac{g(\rho_l - \rho_v)}{\sigma} \right)^{1/2} \left(\frac{c_{pl} \Delta T_s}{C_f \lambda \Pr_l^s} \right)^3$$

$$\frac{Q}{A} = (2.82 \times 10^{-4})(2.257 \times 10^6) \left(\frac{(9.81)(958 - 0.598)}{(0.060)} \right)^{1/2} \left(\frac{(4216)(10)}{(0.013)(2.257 \times 10^6)(1.747)^{1.0}} \right)^3$$

$$\frac{Q}{A} = 1.40 \times 10^5 \text{ W/m}^2$$

The equivalent heat transfer coefficient, from Equation (2.48), is $h = 14{,}000$ W/m^2/K.

From Figure 2.28, the value of the heat flux for $\Delta T = 10$°C (Point b) is approximately 1.5×10^5 W/m, or about 10% higher than the predicted value. It should be noted that the factor C_f was chosen as 0.013, which is a conservative estimate. The value of C_f has a very strong influence on the flux, and generally C_f may not be known for practical situations. The resulting heat transfer coefficient is very high, even using this value of C_f. In practice, the heat transfer coefficient for boiling is hardly ever the limiting heat transfer coefficient, and hence using a conservative value of C_f will not lead to significant errors in estimating the overall heat transfer coefficient U.

Effects of Forced Convection on Boiling
The conditions for pool boiling described in the previous sections are generally not those present in process heat exchangers, with the exception of a kettle-type reboiler, in which a vapor is

generated in the shell side of the heat exchanger. Often, boiling is accompanied by a significant bulk flow of fluid. For example, if liquid is pumped through vertical (or horizontal) tubes that are surrounded by fluid (on the shell side) that is hot enough to boil the liquid in the tubes, then the heat transfer process becomes much more complicated. The processes occurring for this situation are illustrated in Figure 2.29.

For the flow in a vertical tube, from Figure 2.29(a), it can be seen that boiling starts at the bottom of the tube at some point above the inlet and results in a bubbly flow. As bubbles start to coalesce and form bigger spherical-cap-shaped bubbles, the flow pattern becomes more chaotic, and large slugs start to form and move rapidly upward. As more of the liquid becomes vaporized, there is a transition region where the continuous phase switches from liquid to vapor. The liquid tends to move up the tube at the walls, and the vapor forms a core at the center of the tube, in which nonvaporized liquid drops travel. Above the core-annular region, all the liquid at the walls has been vaporized, and the entrained drops of liquid in the vapor rapidly evaporate until only vapor exists. If the tube extends further vertically, then superheating of the vapor will occur. The situation for forced boiling in horizontal tubes is illustrated in Figure 2.29(b). The flow pattern in the horizontal flow mimics the vertical flow pattern in many ways. The big difference between the two orientations is that for horizontal flow, gravity tends to force the liquid to the bottom of the tube and the flow pattern is not radially sym-

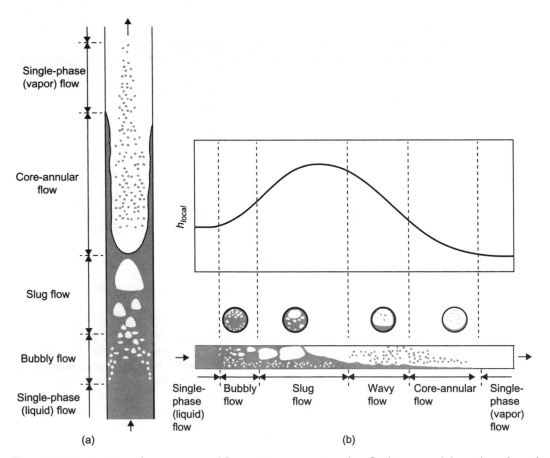

Figure 2.29   Heat transfer processes and flow regimes occurring when fluid is pumped through a tube and vaporizes: (a) vertical tube, (b) horizontal tube

metrical, as shown by the inserts above the figure. It should also be apparent that the local heat transfer coefficient varies widely over the range of conditions in the tube, which is also shown in Figure 2.29(b).

From the description of the flows given for these two orientations, it is clear that the heat transfer coefficient probably changes significantly over the length of the tube. Moreover, the flow patterns are very complex and depend on many parameters. Indeed, the accurate prediction of pressure drop for these two-phase flows is difficult. There have been many attempts to correlate heat transfer coefficients with the flow regimes and also in combination with parameters that describe the two-phase pressure drop. The approach used here is to give the results from a study by Gungor and Winterton (1986), in which correlations were developed for a large set of data taken from a wide variety of boiling liquids in horizontal and vertical tubes. The authors claim that the correlation is able to predict the data within ±25%. The general form of their correlation is

$$h_{cb} = f h_l + s h_{pb} \tag{2.49}$$

where h_{cb} is the overall convective boiling coefficient, h_l is the convective coefficient for liquid flow in the tube, and h_{pb} is the pool boiling coefficient. The factor f, which is greater than 1, accounts for the much higher velocities for two-phase flow compared with single-phase flow. This factor was correlated against the two-phase Martinelli parameter, X_{tp}, and is given by

$$f = 1 + 24{,}000 B_o^{1.16} + \frac{1.37}{X_{tp}^{0.86}} \tag{2.50}$$

where X_{tp} is the Martinelli two-phase flow parameter given by

$$X_{tp} = \left(\frac{1-x}{x}\right)^{0.9} \left(\frac{\rho_v}{\rho_l}\right)^{0.5} \left(\frac{\mu_l}{\mu_v}\right)^{0.1} \tag{2.51}$$

where x is the stream quality (mass fraction of vapor in the stream), and density and viscosity are for the saturated liquid and vapor conditions. B_o is a boiling number and is given by

$$B_o = \frac{Q/A}{\lambda G} \tag{2.52}$$

where Q/A is the heat flux, λ is the latent heat of vaporizations, and G is the superficial mass velocity flowing (axially) past the heating surface. In using Equations (2.51) and (2.52), it should be noted that as $x \to 0$ or 1, the equations are no longer valid. These limits, however, are represented by the pure convective heat transfer coefficients for all liquid flow and all vapor flow, respectively.

The parameter s in Equation (2.49) is a "suppression" factor that is less than 1 and accounts for the lower superheat that is available in convective boiling compared to pool boiling alone. This parameter is given by

$$s = \frac{1}{1 + 1.15 \times 10^{-6} f^2 \text{Re}_l^{1.17}} \tag{2.53}$$

where Re_l is the Reynolds number for the flow inside the tube assuming that the flow is all liquid.

The convective coefficient in Equation (2.49), h_l, is taken to be from the Seider-Tate or Dittus-Boelter correlations (Equation [2.26] or [2.27]) for the liquid. The recommended expression for pool boiling is taken from the correlation attributed to Cooper (1984) and is given by

$$h_{pb} = 55 P_r^{0.12}(-\log_{10} P_r)^{-0.55}(M)^{-0.5}(Q/A)^{0.67} \qquad (2.54)$$

or

$$h_{pb}^{0.33} = 55 P_r^{0.12}(-\log_{10} P_r)^{-0.55}(M)^{-0.5}(T_w - T_{sat})^{0.67} \qquad (2.55)$$

where P_r is the reduced pressure, M is the molecular weight (g/mol) and (Q/A) is the heat flux (W/m²), and the units of h_{pb} are W/m²/K. The pool boiling coefficient from Equation (2.48) could alternatively be used instead of Equation (2.54).

The sequence of Equations (2.49) to (2.54) must be solved to determine the convective boiling heat transfer coefficient and in general will require an iterative method because the heat flux (Q/A) is imbedded in the boiling number term B_o (and Equation [2.52]), which will not be known a priori. Moreover, the value of h_{cb} changes along the length of the tube as the vapor quality changes. An illustration of the use of equations for forced convection boiling is given in Example 2.12.

Example 2.12

An organic liquid (1-propanol at 1 bar) is to be vaporized inside a set of vertical 1-in BWG 16 tubes using condensing steam on the outside of the tubes to provide the energy for vaporization. The major resistance to heat transfer is expected to be on the inside of the tubes, and the wall temperature, as a first approximation, may be assumed to be at the temperature of the condensing steam, which for this case is 110°C. It may be assumed that the value of the vapor quality, x, varies from 0.01 to 0.99 in the tube. The flow and other physical parameters for the organic liquid are

ρ_v = 2.003 kg/m³, ρ_l = 732.5 kg/m³, μ_v = 9.617 × 10⁻⁶ kg/m/s, μ_l = 455.8 × 10⁻⁶ kg/m/s, T_{sat} = 96.9°C, P_c = 52 bar, M = 60 g/mol, k_l = 0.1406 W/m/K, λ = 693.5 kJ/kg, $c_{p,l}$ = 3215.1 J/kg/K, $\dot{m}$ = 0.04 kg/s/tube, D_i = 0.745 in = 18.923 × 10⁻³ m

$$\mathrm{Re}_l = \frac{\rho_l v D_i}{\mu_l} = \frac{4\dot{m}}{\pi \mu_l D_i} = \frac{(4)(0.04)}{\pi(455.8\times10^{-6})(0.018923)} = 5905$$

$$\mathrm{Pr} = c_{pl}\mu_l/k_l = (3215.1)(455.8\times10^{-6})/(0.1406) = 10.42$$

Determine the average heat transfer coefficient and the length of the tube needed to vaporize the liquid in this vertical reboiler.

Solution

From Equation (2.55), the pool boiling coefficient is given by

$$h_{pb}^{0.33} = (55)(0.01923)^{0.12}(-\log_{10}(0.01923))^{-0.55}(60)^{-0.5}(110-96.9)^{0.67} = 18.4056$$

$$h_{pb} = (18.4056)^{1/0.33} = 6810 \ \mathrm{W/m^2/K}$$

The convective heat transfer coefficient is given by Equation (2.26), assuming that $L \gg D$ and $\mu_w \cong \mu_l$, then

$$\mathrm{Nu} = \frac{h_l D_l}{k} = 0.023\left[1+\left(\frac{D_l}{L}\right)^{0.7}\right]\mathrm{Re}^{0.8}\mathrm{Pr}^{1/3} = (0.023)(5905)^{0.8}(10.42)^{1/3} = 52.2$$

$$h_l = (52.2)(0.1406)/(0.018923) = 388 \ \mathrm{W/m^2/K}$$

From Equation (2.51),

$$X_{tp}=\left(\frac{1-x}{x}\right)^{0.9}\left(\frac{\rho_v}{\rho_l}\right)^{0.5}\left(\frac{\mu_l}{\mu_v}\right)^{0.1}=\left(\frac{1-x}{x}\right)^{0.9}\left(\frac{2.003}{732.5}\right)^{0.5}\left(\frac{455.8\times10^{-6}}{9.617\times10^{-6}}\right)^{0.1}=0.07692\left(\frac{1-x}{x}\right)^{0.9}$$

The value of x in the preceding equation varies from 0.01 to 0.99, and hence X_{tp} will also vary over a wide range. To account for the change in x, the problem will be discretized with respect to x and solved for each increment of $\Delta x = 0.1$. The calculations for $x = 0.01$ to 0.1 will be covered, and then the results for the other increments will be summarized. To calculate f in Equation (2.50), the value of B_o must be found from Equation (2.52); however, the heat flux (Q/A) is unknown. Using the discretization scheme shown in Figure E2.12A, the value of B_o may be found in terms of the tube length Δz_i.

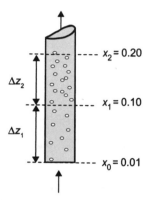

Figure E2.12A   Discretization scheme used in Example 2.12

Therefore, for the first increment is

$$B_o=\frac{Q/A}{\lambda G}=\frac{\dot{m}(x_1-x_0)\lambda}{\pi D_i\Delta z_1}\frac{1}{\lambda}\frac{\pi D_i^2}{4\dot{m}}=\frac{(x_1-x_0)D_i}{4\Delta z_1}=\frac{(0.1-0.01)(0.018923)}{4\Delta z_1}=\frac{4.2577\times10^{-4}}{\Delta z_1}\quad\text{and}$$

$$\overline{x_1}=\frac{(0.1+0.01)}{2}=0.055,\text{ and }X_{tp}=0.07692\left(\frac{1-0.055}{0.055}\right)^{0.9}=0.9945$$

In the equation for B_o, the amount of heat required to vaporize the stream from a vapor fraction of x_0 to x_1 is simply $\dot{m}(x_1-x_0)\lambda$, and the surface area of the tube required for this to happen is $\pi D_i\Delta z_1$. By guessing a value for Δz_1, the values of f and s from Equations (2.50) and (2.53) can be calculated and used in Equation (2.49) to calculate h_{cb}. Now an energy balance on the first increment of tube gives

$$h_{cb}\pi D_i\Delta z_1(T_w-T_{sat})=\dot{m}\lambda(x_1-x_0)\quad h_{cb}\pi D_i\Delta z_1(T_w-T_{sat})=\dot{m}\lambda(x_1-x_0)$$

and rearranging

$$\Delta z_1=\frac{\dot{m}\lambda(x_1-x_0)}{h_{cb}\pi D_i(T_w-T_{sat})}\qquad\qquad\text{(E2.12)}$$

The solution for the first increment is found by iterating between Equations (2.49) and (E2.12) until a constant value of Δz_1 is obtained.

For the first iteration, choose a value of $\Delta z_1 = 0.5$ m:

$$B_o=\frac{4.2577\times10^{-4}}{(0.5)}=8.5154\times10^{-4}$$

$$f = 1 + 24{,}000(B_0)^{1.16} + \frac{1.37}{(X_{tp})^{0.86}} = 1 + 24{,}000(8.5154 \times 10^{-4})^{1.16} + \frac{1.37}{(0.9954)^{0.86}} = 8.97$$

and

$$s = \frac{1}{1 + 1.15 \times 10^{-6} f^2 \, \mathrm{Re}_l^{1.17}} = \frac{1}{1 + 1.15 \times 10^{-6}(8.97)^2(5905)^{1.17}} = 0.2949$$

$$h_{cb} = f h_l + s h_{pb} = (8.97)(388) + (0.2949)(6810) = 5489 \ \mathrm{W/m^2/K}$$

Substitute into Equation (E.2.12) to get

$$\Delta z_1 = \frac{\dot{m}\lambda(x_1 - x_0)}{h_{cb}\pi D_i(T_w - T_{sat})} = \frac{(0.04)(693.5 \times 10^3)(0.1 - 0.01)}{(5489)\pi(0.018923)(110 - 96.9)} = 0.584 \ \mathrm{m}$$

Now recalculate B_0 and iterate until to Δz_1 does not change; finally this gives $\Delta z_1 = 0.635$ m. The results for the whole range of x from 0.01 to 0.99 are given in Table E2.12.

Therefore, a tube length of 3.488 m (11.4 ft) is required. The average heat transfer coefficient for this service is found from

$$\overline{h_{cb}} = \frac{\dot{m}\lambda}{\pi D_i z_{total}(T_w - T_{sat})} = \frac{(0.04)(693.5 \times 10^3)}{\pi(0.018923)(3.488)(110 - 96.9)} = 10{,}210 \ \mathrm{W/m^2/K}$$

From Equation (2.46), the maximum heat flux for pool boiling is given by

$$\left[\frac{Q}{A}\right]_{max} = 0.3673 P_c\left(\frac{P}{P_c}\right)^{0.35}\left(1 - \frac{P}{P_c}\right)^{0.9} = 0.3673(52 \times 10^5)\left(\frac{1}{52}\right)^{0.35}\left(1 - \frac{1}{52}\right)^{0.9} = 471 \ \mathrm{kW/m^2}$$

Table E2.12   Results for Convective Boiling Heat Transfer as a Function of Distance from the Entrance

$x_i - x_{i-1}$	$B_{0,i}$	$\overline{x_i}$	$X_{tp,i}$	f_i	s_i	$h_{cb,i}$ (W/m²/K)	Q/A (kW/m²)	Δz_i (m)
0.01–0.1	0.00067	0.055	0.9945	7.374	0.3823	5465	66.12	0.635
0.1–0.2	0.000764	0.15	0.3665	10.06	0.2495	5603	75.33	0.619
0.2–0.3	0.000833	0.25	0.2068	12.75	0.1716	6114	82.20	0.568
0.3–0.4	0.000969	0.35	0.1343	16.37	0.1116	7110	95.59	0.488
0.4–0.5	0.001197	0.45	0.0921	21.44	0.0682	8783	118.1	0.395
0.5–0.6	0.001563	0.55	0.0642	28.86	0.0388	11,464	154.1	0.303
0.6–0.7	0.002161	0.65	0.0441	40.51	0.0201	15,855	213.2	0.219
0.7–0.8	0.00325	0.75	0.0286	61.30	0.0089	23,843	320.6	0.146
0.8–0.9	0.00581	0.85	0.0161	109.8	0.0028	42,622	573.0	0.081
0.9–0.99	0.01244	0.945	0.0059	261.4	0.0005	101,412	1227.1	0.034
						Total		3.488

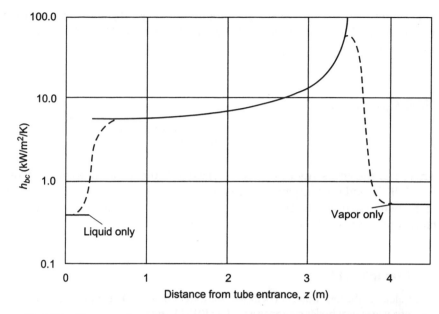

Figure E2.12B   Heat transfer coefficient as a function of distance from the entrance to the tube

The critical heat flux of 471 kW/m^2 is exceeded at the end of the tube ($x > 0.9$), but this critical value is for quiescent pool boiling, and the effect of convection means that flux values above the critical value are possible. Alternatively, the flux could be limited to the value of 471 kW/m^2 but the effect on the required tube length would be very small. Therefore, the estimate of a 3.49 long tube is reasonable.

One last point of interest is the relationship between h_{cb} and distance from the entrance given in Table E2.12, which is shown in Figure E2.12B.

The basic shape of the curve appears to be somewhat different from that given in Figure 2.29. However, as noted previously, for $x = 0$ the liquid convective coefficient should be used, and for $x = 1$ the gas convective coefficient should be used, and these limits are shown on the figure. The transition from the convective boiling correlation of Gungor and Winterton (1986) to these single-phase limits is not predicted, but the general trend in the results from this example do follow the shape of Figure 2.29, and a dashed curve showing possible transitions from the single- to two-phase regimes is shown in Figure E2.12B.

Film Boiling
When the temperature driving force between the tube wall and the boiling liquid exceeds a critical value (Point d, the Leidenfrost point on Figure 2.28), then the heat transfer enters the film boiling regime. This regime is characterized by a coherent continuous film of vapor covering the heat transfer surface. The recommended correlation for the film boiling regime on the outside of a horizontal tube is attributed to Bromley (1950):

$$\overline{Nu}_{fb} = \frac{\overline{h}_{fb} D}{k_v} = 0.62 \left[\frac{g(\rho_l - \rho_v)\lambda D^3}{\mu_v k_v (T_w - T_{sat})} \right]^{1/4} \quad \text{or} \quad \overline{h}_{fb} = 0.62 \left[\frac{g(\rho_l - \rho_v)\lambda k_v^3}{\mu_v (T_w - T_{sat})D} \right]^{1/4} \tag{2.56}$$

where the bar over the symbol represents an average heat transfer coefficient over the length of the tube. Often, the convective heat transfer coefficient in Equation (2.56) is augmented by a radiation component that becomes increasingly important as the absolute temperature difference between the surface and the boiling point of the liquid increases. When this is the case, the overall heat transfer coefficient, $\bar{h}_{fb, total}$, should be calculated from

$$(\bar{h}_{fb, total})^{4/3} = (\bar{h}_{fb})^{4/3} + \bar{h}_{rad}(h_{fb, total})^{1/3} \tag{2.57}$$

where $\bar{h}_{rad}$ is calculated from

$$\bar{h}_{rad} = \frac{\varepsilon\sigma(T_w^4 - T_{sat}^4)}{(T_w - T_{sat})} \tag{2.58}$$

where ε is the emissivity of the tube surface, σ is the Stefan-Boltzmann constant (5.6704×10^{-8} $W/m^{-2}/K^{-4}$), and the temperatures used in Equation (2.58) are absolute (Kelvin). It should be clear that when radiation is important, the calculation of the boiling film heat transfer coefficient using Equations (2.56), (2.57), and (2.58) is an iterative process.

Example 2.13

Calculate the heat transfer coefficient for water boiling at 42 bar pressure (253.3°C) in the shell side of a waste heat boiler equipped with 1.25 BWG 12 tubes.
Assume that the tube wall temperature is constant at 400°C, and include radiation from the wall that has an emissivity of 0.40.

Solution

$\rho_v = 21.14 \, kg/m^3$, $\rho_l = 732.5 \, kg/m^3$, $\mu_v = 19.11 \times 10^{-6} \, kg/m/s$, $T_{sat} = 253.3°C$, MW = 18,
$k_v = 0.0442 \, W/m/K$, $\lambda = 1697.8 \, kJ/kg$, D = 1.25 in = 0.03175 m

From Equation (2.56),

$$\bar{h}_{fb} = 0.62\left[\frac{g(\rho_l - \rho_v)\lambda k_v^3}{\mu_v(T_w - T_{sat})D}\right]^{1/4} = 0.62\left[\frac{(9.81)(732.5 - 21.14)(1697.8\times10^3)(0.0442)^3}{(19.11\times10^{-6})(400 - 253.3)(0.03175)}\right]^{1/4}$$

$$\bar{h}_{fb} = 203 \, W/m^2/K$$

From Equation (2.58),

$$\bar{h}_{rad} = \frac{\varepsilon\sigma(T_w^4 - T_{sat}^4)}{(T_w - T_{sat})} = \frac{(0.4)(5.6704\times10^{-8})((400 + 273.15)^4 - (253.3 + 273.15)^4)}{(400 - 253.3)}$$

$$\bar{h}_{rad} = 19.9 \, W/m^2/K$$

From Equation (2.57),

$$(\bar{h}_{fb, total})^{4/3} = (203)^{4/3} + 19.9(h_{fb, total})^{1/3}$$

Solving gives

$$\bar{h}_{fb, total} = 218.1 \, W/m^2/K$$

This represents a 7% increase in h_{fb} due to radiation effects.

2.5.3.4   Condensing Heat Transfer

If a vapor is exposed to a cold surface and the liquid does not wet the surface, then the vapor will condense by forming small drops or beads of liquid on the surface. This phenomenon is sometimes seen on window panes and car windshields, when the outside temperature drops and water from the warm humid air inside starts to condense on the cold glass. This phenomenon is known as *dropwise condensation*. The beads that form roll off the surface if it is inclined to the vertical or form a pool on a horizontal surface. Dropwise condensation does not occur, except in rare circumstances, in process heat exchangers, because the surfaces are not smooth and clean but are rough due to fouling and surface scaling. For most practical circumstances, vapors condense on heat-exchanger surfaces by a film condensation mechanism discussed next.

When a vapor close to its saturation temperature, T_{sat}, is exposed to a cold surface at a temperature below T_{sat}, then condensation of the vapor occurs on the surface.

In heat exchangers, this process is often accomplished by exposing a saturated vapor on the shell side of an S-T exchanger to horizontal (or vertical) tubes through which a cooling liquid is passed. At steady state, the liquid forms a film around the circumference of the tube and falls by gravity onto tubes situated below it. This phenomenon is illustrated in Figure 2.30(a).

Shown in Figure 2.30(b) is the condensation of liquid on a flat, vertical plate. As the liquid falls, the liquid layer increases in thickness, and the resistance to heat transfer increases because the path for energy flow through the liquid film becomes longer. In Figure 2.30(c) the condensation on a stack of vertically aligned horizontal tubes is illustrated.

Nusselt's Analysis of Falling-Film Condensation
Using the following assumptions regarding the flow of condensate on a vertical flat plate, Nusselt analyzed the heat transfer process that occurs across the film of flowing liquid condensate:

- Laminar flow of fluid with constant physical properties
- Gas is a pure vapor at T_{sat}. Condensation occurs only at the vapor-liquid interface, that is, no heat transfer resistance in gas

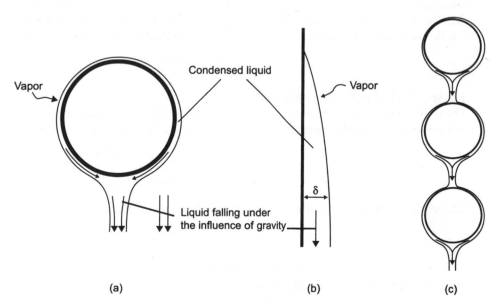

| (a) | (b) | (c) |

Figure 2.30   Liquid condensing on a cold surface: (a) horizontal tube, (b) vertical plate, and (c) a bank of multiple horizontal tubes

- Shear stress at the liquid-vapor boundary is zero, that is, no vapor velocity or vapor thermal boundary layer effects
- Heat transfer across the liquid film is only by conduction

Nusselt's analysis leads to the following expression for the average heat transfer coefficient over the length, L, of the plate:

$$\mathrm{Nu} = \frac{\bar{h}_c L}{k_l} = 0.943 \left[\frac{\rho_l g (\rho_l - \rho_v) \lambda L^3}{\mu_l k_l (T_{sat} - T_w)} \right]^{1/4} \tag{2.59}$$

Alternatively, the average heat transfer coefficient can be written as

$$\bar{h}_c = 0.943 \left[\frac{\rho_l g k_l^3 (\rho_l - \rho_v) \lambda}{\mu_l (T_{sat} - T_w) L} \right]^{1/4} \tag{2.60}$$

where, according to McAdams (1954), all the physical properties of the liquid should be evaluated at the average film temperature, $T_f = T_{sat} - 0.75(T_{sat} - T_w)$.

The total rate at which liquid condenses can then be determined via an energy balance as

$$\dot{m} = \frac{Q}{\lambda} = \frac{\bar{h}_c A (T_{sat} - T_w)}{\lambda} \tag{2.61}$$

The condition for which the expression in Equation (2.60) is valid is determined on the basis of the value of an effective Reynolds number for the falling film, Re_δ, given by

$$\mathrm{Re}_\delta = \frac{4 g \rho_l (\rho_l - \rho_l) \delta^3}{3 \mu_l^3} = \frac{4 \dot{m}}{\mu_l W} \tag{2.62}$$

where δ is the thickness of the falling film (Figure 2.30[b]) and $\dot{m}$ is the mass flow rate of condensate falling down a plate of width W. Equation (2.60) is valid for $\mathrm{Re}_\delta \le 30$. For the region $30 \le \mathrm{Re}_\delta \le 1800$, the flow down the plate is described as wavy laminar and there is a slight enhancement of the heat transfer coefficient such that $\bar{h}_c$ from Equation (2.60) should be multiplied by E, where E is given by

$$E = 0.73 \mathrm{Re}_\delta^{0.0935} \tag{2.63}$$

Falling-Film Condensation on Cylinders
The result from Nusselt for laminar flow on a vertical flat plate is essentially correct, even for **vertically oriented tubes**, except the results from experiments suggest that the coefficient should be increased:

$$\mathrm{Nu} = \frac{\bar{h}_c L}{k_l} = 1.13 \left[\frac{\rho_l g (\rho_l - \rho_v) \lambda' L^3}{\mu_l k_l (T_{sat} - T_w)} \right]^{1/4} \tag{2.64}$$

where the latent heat is modified to include subcooling in the film as

$$\lambda' = \lambda + 0.68 c_{p,l} (T_{sat} - T_w) \tag{2.65}$$

For the case when the flow in the film is turbulent (and $\mathrm{Re}_{\delta,max} > 1800$), McAdams [1954] suggests the use of the following equation:

$$\mathrm{Nu} = \frac{\bar{h}_c L}{k_l} = 0.0077 \left[\frac{\rho_l g (\rho_l - \rho_v) L^3}{\mu_l^2} \right]^{1/3} (\mathrm{Re}_{\delta,max})^{0.4} \tag{2.66}$$

For film condensation on the **outside of horizontally oriented tubes**, the following expression was obtained by Nusselt:

$$\text{Nu} = \frac{\bar{h}_c D_o}{k_l} = 0.728 \left[\frac{\rho_l g (\rho_l - \rho_v) \lambda' D_o^3}{\mu_l k_l (T_{sat} - T_w)} \right]^{1/4} \tag{2.67}$$

Because the path length for the film formation ($\sim \pi D_o / 2$) on tubes used for commercial heat exchangers is relatively small, turbulent flow rarely occurs in horizontal-tube condensers. An example to illustrate use of the correlations for condensing heat transfer is given in Example 2.14.

Example 2.14   (adapted from Bennett and Myers [1982], Problem 25.1)

An S-T condenser contains four rows of four copper tubes per row on a square pitch. The tubes are 1-in, 16 BWG and 6 ft long. Cooling water flows through the tubes such that $h_i = 1000$ BTU/hr/ft^2/°F. The water flow is high so that the temperature on the tube side may be assumed to be constant at 90°F. Pure saturated steam at 5 psig is condensing on the shell side. Determine the capacity of the condenser (Q in BTU/hr) if the condenser tubes are oriented (a) vertically, and (b) horizontally. You should assume that neither water nor steam fouls the heat-exchange surfaces.

Solution

The correct film temperature is unknown and depends on h_o. The solution algorithm is
1. Guess h_o
2. Calculate film temperature and film properties
3. Apply equation for h_o and iterate

Guess $h_o = 1000$ BTU/ hr/ft^2/°F

a. Vertical arrangement

Steam at 227°F, $D_o = 1.0$ in = 0.0833 ft, $D_i = 0.87$ in = 0.0725 ft, $k_{copper} = 220$ BTU/hr/ft/°F
From Equation (2.23) (with no fouling coefficients for either stream),

$$U_o = \left[\frac{1}{h_o} + \cancel{R_{fo}} + \frac{D_o \ln(D_o / D_i)}{2 k_w} + \frac{D_o}{D_i} \cancel{R_{fi}} + \frac{D_o}{D_i} \frac{1}{h_i} \right]^{-1}$$

now $\dfrac{Q}{A_o} = h_o (T_{sat} - T_{wall}) = U_o (T_{sat} - T_{water})$ rearranging

$$\frac{(227 - T_w)}{\dfrac{1}{1000}} = \frac{(227 - 90)}{\dfrac{1}{1000} + \dfrac{(0.0833)\ln(0.0833/0.0725)}{(2)(220)} + \dfrac{(0.0833)}{(0.0725)}\dfrac{1}{1000}}$$

$$(227 - T_w) = \frac{(1 \times 10^{-3})(227 - 90)}{(1 \times 10^{-3} + 2.63 \times 10^{-5} + 1.149 \times 10^{-3})} \Rightarrow T_w = 164.0°\text{F}$$

$$T_f = T_{sat} - 0.75(T_{sat} - T_w) = 227 - 0.75(227 - 164.0) = 179.8°\text{F}$$

For water at 179.8°F → $k_l = 0.3882$ BTU/h/ft/°F, $\rho_l = 60.55$ lb/ft^3, $\lambda = 960$ BTU/lb (at 5 psig), $\mu_l = 0.8161$ lb/ft/h, $c_{p,l} = 0.9720$ BTU/lb/°F, $\rho_v = 0.049$ lb/ft^3

$$\lambda' = \lambda + 0.68 c_{p,l}(T_{sat} - T_w) = 960 + (0.68)(0.9720)(227 - 163.9) = 1002 \text{ BTU/lb}_m$$

Substituting values into Equation (2.64),

$$\text{Nu} = \frac{\bar{h}_c L}{k_l} = 1.13 \left[\frac{\rho_l g (\rho_l - \rho_v) \lambda' L^3}{\mu_l k_l (T_{sat} - T_w)} \right]^{1/4} = 1.13 \left[\frac{(60.55)(32.2 \times 3600^2)(60.55 - 0.049)(1002)(6)^3}{(0.8161)(0.3882)(227 - 163.9)} \right]^{1/4}$$

$$\text{Nu} = \frac{\bar{h}_c L}{k_l} = 12{,}731 \Rightarrow \bar{h}_c = (12{,}731)(0.3882)/(6) = 824 \text{ BTU/hr/ft}^2/°\text{F}$$

Iterating to find the new T_w and T_f gives

$$\frac{(227-T_w)}{\dfrac{1}{824}} = \frac{(227-90)}{\dfrac{1}{824} + \dfrac{(0.0833)\ln(0.0833/0.0725)}{(2)(220)} + \dfrac{(0.0833)}{(0.0725)}\dfrac{1}{1000}}$$

$$(227-T_w) = \frac{(1.214\times10^{-3})(227-90)}{(1.214\times10^{-3} + 2.63\times10^{-5} + 1.149\times10^{-3})} \Rightarrow T_w = 158.1°\text{F}$$

$$T_f = T_{sat} - 0.75(T_{sat} - T_w) = 227 - 0.75(227 - 158.1) = 175.4°\text{F}$$

For water at 175.4°F → k_l = 0.3873 BTU/h/ft/°F, ρ_l = 60.65 lb/ft³, λ = 960 BTU/lb (at 5 psig), μ_l = 0.8405 lb/ft/h, $c_{p,l}$ = 0.9718 BTU/lb/°F, ρ_v = 0.049 lb/ft³

$$\lambda' = \lambda + 0.68 c_{p,l}(T_{sat} - T_w) = 960 + (0.68)(0.9718)(227-158.1) = 1006 \text{ BTU/lb}$$

$$\text{Nu} = 1.13\left[\frac{(60.65)(32.2\times3600^2)(60.65-0.049)(1006)(6)^3}{(0.8405)(0.3873)(227-158.1)}\right]^{1/4} = 12{,}388 \Rightarrow \bar{h}_c = 800 \text{ BTU/hr/°F}$$

One more iteration yields $\bar{h}_c$ = 796 BTU/hr/°F, which is close enough to the previous iteration:

$$\therefore U_o = \left[\frac{1}{h_o} + \cancel{R_{fo}} + \frac{D_o \ln(D_o/D_i)}{2k_w} + \frac{D_o}{D_i}\cancel{R_{fi}} + \frac{D_o}{D_i}\frac{1}{h_i}\right]^{-1}$$

$$U_o = \left[\frac{1}{796} + \frac{(0.0833)\ln(0.0833/0.0725)}{(2)(220)} + \frac{(0.0833)}{(0.0725)}\frac{1}{1000}\right]^{-1} = 411 \text{ BTU/hr/ft}^2/°\text{F}$$

$$Q = U_o A_o \Delta T_{lm} = (411)(16)(\pi)(0.0833)(6)(227-90) = 1.42\times10^6 \text{ BTU/hr}$$

Check for Re_δ:

$$\text{Re}_\delta = \frac{4\dot{m}}{\mu_l W} = \frac{4(Q/\lambda')}{\mu_l(\pi D_o)n_{tubes}} = \frac{4(1.42\times10^6/1006)}{(0.8451)(\pi)(0.0833)(16)} = 1590$$

R_δ < 1800, so the correct equation was used.

b. Horizontal arrangement

Use the same approach as used for vertical tubes.

For iteration 1, use the same temperatures and properties as for vertical tubes (Equation [2.67]):

$$\text{Nu} = \frac{\bar{h}_c D_o}{k_l} = 0.728\left[\frac{\rho_l g(\rho_l-\rho_v)\lambda' D_o^3}{\mu_l k_l(T_{sat}-T_w)}\right]^{1/4}$$

$$\text{Nu} = 0.728\left[\frac{(60.55)(32.2\times3600^2)(60.55-0.049)(1005)(0.0833)^3}{(0.8161)(0.3882)(227-164.0)}\right]^{1/4} = 332$$

$$\bar{h}_c = (332)(0.3882)/(0.0833) = 1546 \text{ BTU/hr/ft}^2/°\text{F}$$

Iterating to find the new T_w and T_f gives

$$\frac{(227-T_w)}{\dfrac{1}{1546}} = \frac{(227-90)}{\dfrac{1}{1546}+\dfrac{(0.0833)\ln(0.0833/0.0725)}{(2)(220)}+\dfrac{(0.0833)}{(0.0725)}\dfrac{1}{1000}}$$

$$(227-T_w)=\frac{(6.468\times10^{-4})(227-90)}{(6.468\times10^{-4}+5.26\times10^{-5}+1.149\times10^{-3})}\Rightarrow T_w=178.4°\text{F}$$

$$T_f = T_{sat} - 0.75(T_{sat} - T_w) = 227 - 0.75(227 - 178.4) = 190.5°\text{F}$$

For water at 190.5°F → k_l = 0.3903 BTU/hr/ft/°F, ρ_l = 60.295 lb/ft^3, λ = 960 BTU/lb (at 5 psig), μ_l = 0.7597 lb/ft/h, $c_{p,l}$ = 0.9728 BTU/lb/°F, ρ_v = 0.049 lb/ft^3

$$\lambda' = \lambda + 0.68c_{p,l}(T_{sat}-T_w) = 960+(0.68)(0.9728)(227-178.4)=992.2\text{ BTU/lb}$$

$$\text{Nu}=0.728\left[\frac{(60.295)(32.2\times3600^2)(60.295-0.049)(992.2)(0.0833)^3}{(0.7597)(0.3903)(227-178.4)}\right]^{1/4}=358\Rightarrow \bar{h}_c=1679\text{ BTU/hr/°F}$$

The third iteration gives $\bar{h}_c$ = 1706 BTU/hr/°F, T_w = 180.9°F, and T_f = 192.4°F.
The fourth iteration gives $\bar{h}_c$ = 1712 BTU/hr/°F—this is close enough to the previous iteration.

$$U_o=\left[\frac{1}{1712}+\frac{(0.0833)\ln(0.0833/0.0725)}{(2)(220)}+\frac{(0.0833)}{(0.0725)}\frac{1}{1000}\right]^{-1}=560\text{ BTU/hr/ft}^2/°\text{F}$$

$$Q=U_oA_o\Delta T_{lm}=(560)(16)(\pi)(0.0833)(6)(227-90)=1.93\times10^6\text{ BTU/hr}$$

Check for Re_δ:

$$\text{Re}_\delta=\frac{4\dot{m}}{\mu_l W}=\frac{4(Q/\lambda')}{\mu_l(2)(L)n_{tubes}}=\frac{4(1.93\times10^6/990)}{(0.7485)(2)(6)(16)}=54.3\text{ laminar flow}$$

The correct equation was used.

2.6 EXTENDED SURFACES

From Examples 2.7 and 2.8, it is clear that the heat transfer coefficients for gases are generally much lower than for liquids or for phase changes. Indeed, the limiting heat transfer coefficient will be the gas film coefficient if a gas is one of the fluids in the exchanger. To increase the effective gas film coefficient, it is often necessary to add some form of extended heat transfer surface or fins to the gas side of the heat exchanger. Some examples of finned tubes were given in Figures 2.22 and 2.23. Some additional arrangements of fins are given in Figure 2.31.

From Figure 2.31, it can be seen that a wide variety of fin geometries is possible, and the equations describing the temperature profile in the fin can be quite complicated. In subsequent sections, the equations for both simple and more complicated geometries are presented without derivation.

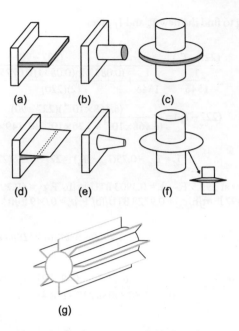

(g)

Figure 2.31   Various fin types: (a) straight fin constant thickness, (b) spine or pin fin constant thickness, (c) annular fin constant thickness, (d) straight fin nonuniform thickness, (e) spine or pin fin nonuniform thickness, (f) annular fin nonuniform thickness, (g) longitudinal fin constant thickness

2.6.1   Rectangular Fin with Constant Thickness

A rectangular fin is shown in Figure 2.31(a). The equations describing the heat conduction in this type of fin, with a constant cross section, are formulated and solved in Appendix 2.B. A key parameter in analyzing the performance of a fin is the fin effectiveness defined as

$$\varepsilon_{fin} = \frac{\text{Heat transferred through fin}}{\text{Heat transferred through fin if fin temperature were } T_0 \text{ throughout}} = \frac{\tanh(mL)}{mL} \quad (2.68)$$

For the rectangular fin, L is the length of the fin and δ is the thickness of the fin; the fin effectiveness is given by Equation (2.68) and is plotted in Figure 2.32. The term m is a dimensionless parameter that is defined in Equation (2.69).

$$m = \sqrt{\frac{2h}{\delta k}} \quad (2.69)$$

The value of ε_{fin} indicates the efficiency by which the additional fin surface area is being utilized. For example, if the thermal conductivity of the fin material is very high or the fin is short, then, all else being equal, the temperature everywhere along the length of the fin is close to the wall temperature, and temperature driving force between the fin and the surroundings will be close to $T_0 - T_{fluid}$. Therefore, the fin surface area is used effectively. If, on the other hand, the fin length, L, is very long compared with its thickness, δ, then the temperature of the fin over a major portion of its length is close to T_{fluid}, and the efficiency of the fin is low, since only a small portion of the fin (close to the wall) has a significant temperature driving force to

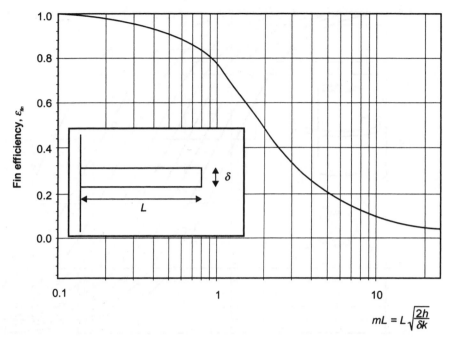

Figure 2.32 Efficiency for rectangular fins with constant cross section

exchange heat with the surrounding fluid. In other words, the good news is that the fin provides more area for heat transfer, but the bad news is that the driving force may decrease down the length of the fin, so all of the benefit of the increased area is not seen. The effectiveness factor quantifies this situation.

In the previous analysis, it was assumed that the tip of the fin did not lose any heat or that an adiabatic boundary condition at the end of the fin was used. This assumption is strictly correct only for fins with a very large value of L/δ. However, the results given in Equation (2.68) and Figure 2.32 can be used with great accuracy if a corrected fin length, L_c, is used, where

$$L_c = L + \frac{1}{2}\delta \tag{2.70}$$

2.6.2   Fin Efficiency for Other Fin Geometries

2.6.2.1   Annular or Circular Fins of Uniform Thickness
These fins are shown in Figure 2.31(c), and the fin efficiency is given by

$$\varepsilon_{fin} = \frac{2a}{b(1-a^2)} \frac{K_1(ab)I_1(b) - K_1(b)I_1(ab)}{K_0(ab)I_1(b) - K_1(b)I_0(ab)} \tag{2.71}$$

where $a = \frac{r_{tube}}{r_{fin}}$, $b = r_{fin}\sqrt{\frac{2h}{k\delta}}$, and I_i and K_i are modified Bessel functions of the first and second kind

of order i, respectively. Equation (2.71) is plotted in Figure 2.33.

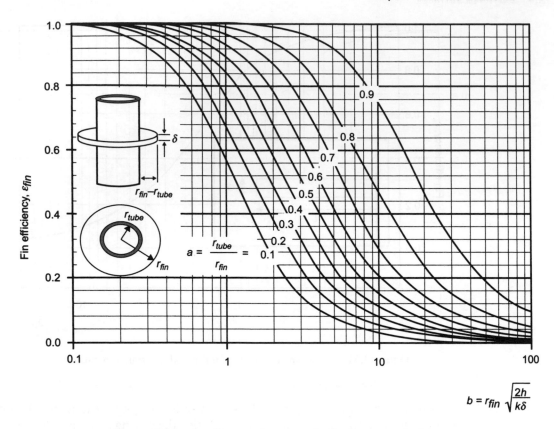

Figure 2.33   Fin efficiency for an annular fin of constant thickness

2.6.2.2   Narrow Triangular Fin

The efficiency for these fins, Figure 2.31(d), is given by

$$\varepsilon_{fin} = \frac{1}{mL}\frac{I_1(2mL)}{I_0(2mL)} \tag{2.72}$$

where $m = \sqrt{\dfrac{2h}{k\delta}}$, and Equation (2.72) is plotted in Figure 2.34.

2.6.3   Total Heat Transfer Surface Effectiveness

Up until this point, the efficiency of single fins of different geometries has been considered. However, multiple fins are always used in real heat exchangers. Consider a set of fins stacked as shown in Figure 2.35. For the case of rectangular straight fins, shown in Figure 2.35(a), consider a section of wall of area $W(\delta + b)$. Without fins, this would be the area exposed to the fluid; with fins, the heat transfer area becomes $Wb + 2LW$. Assuming that the film heat transfer coefficient, h, between the surface and the surrounding fluid does not vary with position, then an energy balance for the surface gives

$$Q = Wbh(T_0 - T_{fluid}) + 2LWh\varepsilon_{fin}(T_0 - T_{fluid}) \tag{2.73}$$

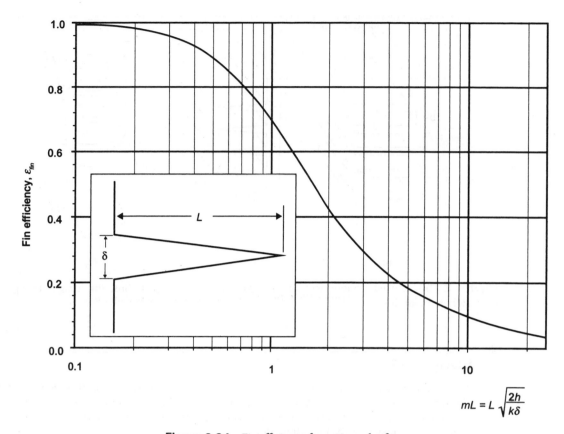

Figure 2.34   Fin efficiency for a triangular fin

If A_{base} is the area of the surface not taken up by fins (equal to Wb for the section of wall considered) and A_{bare} is the total surface area without fins (equal to $W \times (\delta + b)$ for this example) and A_{fin} is the area of the fins (equal to $2WL$ for the single fin considered in this example), then the results from Equation (2.73) can be generalized as follows:

With fins

$$Q = (A_{base} + \varepsilon_{fin} A_{fin})h(T_0 - T_{fluid}) \qquad (2.74)$$

Without fins

$$Q = A_{bare}h(T_0 - T_{fluid}) \qquad (2.75)$$

The enhancement in heat transfer is found by comparing Equations (2.74) and (2.75):

$$\text{Enhancement in heat transfer} = \frac{A_{base} + \varepsilon_{fin} A_{fin}}{A_{bare}} \qquad (2.76)$$

This enhancement of heat transfer depends on the geometry of the fins and the fin efficiency and is illustrated in Example 2.15.

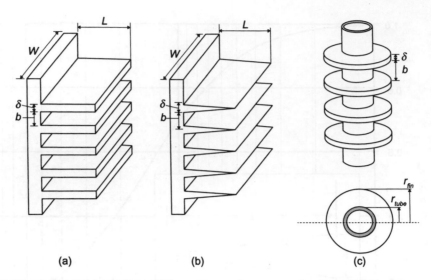

Figure 2.35 Stacked arrangement of fins: (a) straight rectangular fins, (b) straight triangular fins, (c) straight annular fins.

Example 2.15

An aluminum heat transfer surface that is maintained at 100°C is in contact with air at 20°C. The film heat transfer coefficient between the surface and the air is estimated to be 42 W/m²/K. Determine the area of bare surface equivalent required to transfer 4 kW of power and the enhancement in heat transfer for the following cases:

a. A bare flat metal surface

b. A flat metal surface equipped with 50 mm long, 3 mm wide rectangular aluminum fins placed on 10 mm centers

c. A flat metal surface equipped with 50 mm long, 3 mm wide-base triangular aluminum fins placed on 10 mm centers

d. A bare 1-in, 14 BWG tube

e. A 1-in, 14 BWG tube with 2-in diameter annular fins of constant thickness of 3 mm placed on 10 mm centers

f. A 1-in, 14 BWG tube with 3-in diameter annular fins of constant thickness of 3 mm placed on 10 mm centers

Solution

From Table 2.3, for aluminum at 100°C, $k = 206$ W/m/K. From Table 2.4, OD 1-in, 14 BWG tube is 25.4 mm.

a. $Q = 4{,}000 = hA\Delta T = (42)(A)(100-20) \Rightarrow A = \dfrac{4{,}000}{(42)(80)} = 1.19\,\mathrm{m}^2$

b. The arrangement of the fins is shown in Figure E2.15A

$m = (2h/\delta k)^{1/2} = [(2)(42)/(0.003)/(206)]^{1/2} = 11.66$

$L = 0.050\,\mathrm{m}$
$mL = 0.5829$
From Figure 2.32, $\varepsilon_{fin} = 0.90$
From Equation (2.74),

$$Q = (A_{base} + \varepsilon_{fin} A_{fin}) h(T_0 - T_{fluid})$$

$$(A_{base} + \varepsilon_{fin} A_{fin}) = \frac{4000}{(42)(80)} = 1.19 \, \text{m}^2$$

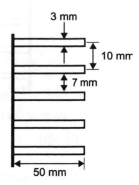

Width of fins into the page = *w* m

Figure E2.15A   Fin arrangement for rectangular fins

Consider an area of bare surface equal to 10 mm by *w* or 0.01*w*. The base area is 0.007*w* and the equivalent fin area is 2(0.05)*w* = 0.1*w*. Normalizing these areas with the bare area gives A_{base} = $A_{bare}(0.007w)/(0.01w) = 0.7 A_{bare}$ and $A_{fin} = A_{base}(0.1w)/(0.01w) = 10 A_{base}$. Substitute these values into the above expression, with ε_{fin} = 0.90, and solve:

$$(A_{base} + \varepsilon_{fin} A_{fin}) = A_{bare}[0.7 + (10)(0.90)] = 9.7 A_{bare} = 1.19 \, \text{m}^2$$

$$\therefore A_{bare} = \frac{1.19}{9.7} = 0.1227 \, \text{m}^2$$

Enhancement in heat transfer = $\dfrac{A_{base} + \varepsilon_{fin} A_{fin}}{A_{bare}} = \dfrac{0.7 + 9.7}{1.0} = 10.4$

It was pointed out previously that the adiabatic tip assumption for the fin may not be valid and that to account for this, an adjustment in the length of the fin can be made. According to Equation (2.70),

$$L_c = L + \frac{1}{2}\delta = 50 + 3.0/2 = 51.5 \, \text{mm}$$

Using this adjusted value of L gives mL_c = (11.66)(0.0515) = 0.600 and ε_{fin} = 0.895 and A_{fin} = 2(0.0515)(w) = 0.103w = 10.3A_{base}.

Substitute these values into the above expression, with ε_{fin} = 0.895, and solve:

$$(A_{base} + \varepsilon_{fin} A_{fin}) = A_{bare}[0.7 + (10.3)(0.895)] = 9.92 A_{bare} = 1.19 \, \text{m}^2$$

$$\therefore A_{bare} = \frac{1.19}{9.92} = 0.120 \, \text{m}^2$$

Enhancement in heat transfer = $\dfrac{A_{base} + \varepsilon_{fin} A_{fin}}{A_{bare}} = \dfrac{0.7 + 9.92}{1.0} = 10.62$ or a 2% increase from the case when an

adiabatic tip is assumed.

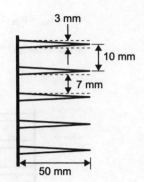

3 mm

10 mm

7 mm

50 mm

Width of fins into the page = w m

Figure E2.15B   Fin arrangement for triangular fins

c. The arrangement of the fins is shown in Figure E2.15B.

$m = (2h/\delta k)^{½} = [(2)(42)/(0.003)/(206)]^{½} = 11.66$

$L = 0.050$ m

$mL = 0.5829$

From Figure 2.34, $\varepsilon_{fin} = 0.86$

From Equation (2.74),

$$Q = (A_{base} + \varepsilon_{fin} A_{fin})h(T_0 - T_{fluid})$$

$$(A_{base} + \varepsilon_{fin} A_{fin}) = \frac{4000}{(42)(80)} = 1.19 \text{ m}^2$$

Consider an area of bare surface equal to 10 mm by w or 0.01w. The base area is 0.007w, and the equivalent fin area is $2(0.05^2 + 0.0015^2)^{1/2} w \cong 0.1w$. Normalizing these areas with the bare area gives $A_{base} = A_{bare}(0.007w)/(0.01w) = 0.7A_{bare}$ and $A_{fin} = A_{base}(0.1w)/(0.01w) = 10 A_{base}$. Substitute these values into the above expression, with $\varepsilon_{fin} = 0.90$, and solve:

$$(A_{base} + \varepsilon_{fin} A_{fin}) = A_{bare}[0.7 + (10)(0.86)] = 9.3 A_{bare} = 1.19 \text{m}^2$$

$$\therefore A_{bare} = \frac{1.19}{9.3} = 0.1280 \text{ m}^2$$

Enhancement in heat transfer = $\dfrac{A_{base} + \varepsilon_{fin} A_{fin}}{A_{bare}} = \dfrac{0.7 + 8.6}{1} = 9.3$

d. From Part (a), $A = 1.19$ m^2— this is equivalent to a length of 1-in tube, L_{tube}:

$$L_{tube} = \frac{A}{\pi D} = \frac{1.19}{(3.142)(0.0254)} = 14.91 \text{ m}$$

e. The arrangement of the fins is shown in Figure E2.15C.

$$a = \frac{r_{tube}}{r_{fin}} = \frac{(0.0127)}{(0.0254)} = 0.5,$$

$$b = r_{fin}\sqrt{\frac{2h}{k\delta}} = (0.0254)\sqrt{\frac{(2)(42)}{(206)(0.003)}} = 0.2961$$

From Figure 2.33, $\varepsilon_{fin} = 0.99$

From Equation (2.74),

$$Q = (A_{base} + \varepsilon_{fin} A_{fin})h(T_0 - T_{fluid})$$

$$(A_{base} + \varepsilon_{fin} A_{fin}) = \frac{4000}{(42)(80)} = 1.19 \text{ m}^2$$

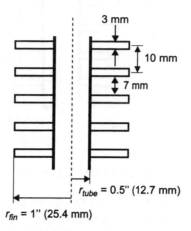

Figure E2.15C   Fin arrangement for 2-in annular fins

Consider an area of bare surface equal to 10 mm of tube length, $A_{bare} = \pi(0.0254)(0.01) = 7.980 \times 10^{-4}$ m. The base area is $\pi(0.0254)(0.007) = 5.586 \times 10^{-4}$ m, and the equivalent fin area is $2\pi(0.0254^2 - 0.0127^2)^{1/2}/4 = 7.6 \times 10^{-4}$ m. Normalizing these areas with the bare area gives $A_{base} = A_{bare}(5.586 \times 10^{-4})/(7.980 \times 10^{-4}) = 0.7A_{bare}$ and $A_{fin} = A_{base}(7.6 \times 10^{-4})/(7.980 \times 10^{-4}) = 0.952 \, A_{base}$. Substitute these values into the above expression, with $\varepsilon_{fin} = 0.99$, and solve:

$$(A_{base} + \varepsilon_{fin} A_{fin}) = A_{bare}[0.7 + (0.952)(0.99)] = 1.642 A_{bare} = 1.19 \text{ m}^2$$

$$\therefore A_{bare} = \frac{1.19}{1.642} = 0.7245 \text{ m}^2$$

$$L_{tube} = \frac{A}{\pi D} = \frac{0.7245}{(3.142)(0.0254)} = 9.08 \text{ m}$$

$$\text{Enhancement in heat transfer} = \frac{A_{base} + \varepsilon_{fin} A_{fin}}{A_{bare}} = \frac{(0.7) + (0.952)(0.99)}{(1)} = 1.64$$

f. The arrangement of the fins is shown in Figure E2.15D.

$$a = \frac{r_{tube}}{r_{fin}} = \frac{(0.0127)}{(0.0381)} = 0.333,$$

$$b = r_{fin}\sqrt{\frac{2h}{k\delta}} = (0.0254)\sqrt{\frac{(2)(42)}{(206)(0.003)}} = 0.2961$$

From Figure 2.33, $\varepsilon_{fin} = 0.978$
From Equation (2.74),

$$Q = (A_{base} + \varepsilon_{fin} A_{fin})h(T_0 - T_{fluid})$$

$$(A_{base} + \varepsilon_{fin} A_{fin}) = \frac{4000}{(42)(80)} = 1.19 \text{ m}^2$$

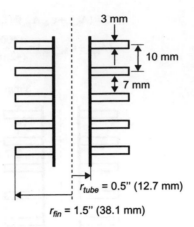

Figure E2.15D   Fin arrangement for 3-in annular fins

Consider an area of bare surface equal to 10 mm of tube length, $A_{bare} = \pi(0.0254)(0.01) = 7.980 \times 10^{-4}$ m. The base area is $\pi(0.0254)(0.007) = 5.586 \times 10^{-4}$ m, and the equivalent fin area is $2\pi(0.0381^2 - 0.0127^2)^{1/2}/4 = 20.27 \times 10^{-4}$ m. Normalizing these areas with the bare area gives $A_{base} = A_{bare}(5.586 \times 10^{-4})/(7.980 \times 10^{-4}) = 0.7A_{bare}$ and $A_{fin} = A_{base}(20.27 \times 10^{-4})/(7.980 \times 10^{-4}) = 2.540\ A_{base}$. Substitute these values into the above expression, with $\varepsilon_{fin} = 0.978$, and solve:

$$(A_{base} + \varepsilon_{fin} A_{fin}) = A_{bare}\left[0.7 + (2.54)(0.978)\right] = 3.184\,A_{bare} = 1.19\ \text{m}^2$$

$$\therefore A_{bare} = \frac{1.19}{3.184} = 0.3737\ \text{m}^2$$

$$L_{tube} = \frac{A}{\pi D} = \frac{0.3737}{(3.142)(0.0254)} = 4.68\ \text{m}$$

$$\text{Enhancement in heat transfer} = \frac{A_{base} + \varepsilon_{fin} A_{fin}}{A_{bare}} = \frac{0.7 + (2.54)(.978)}{1} = 3.184$$

From Example 2.15, it is clear that for all the fins considered, there is a significant enhancement of heat transfer due to the presence of the fins. The enhancement factors for the cases with fins were 10.4, 9.3, 1.64, and 3.184 for cases b, c, e, and f, respectively. The enhancement factor can be considered a measure of the increase in effective film heat transfer coefficient that the fins provide. Thus, the required heat transfer area, based on the base area, is reduced significantly and varies in the range of 40% to 90% reduction for the cases considered in Example 2.15. The choice to use fins is an economic one, since the overall bare heat transfer area will be reduced, but the cost of the heat transfer area increases significantly.

2.7   ALGORITHM AND WORKED EXAMPLES FOR THE DESIGN OF HEAT EXCHANGERS

In this section, the theory needed to design a heat exchanger, which has been presented previously, is brought together along with some of the practical considerations of heat-exchanger design. The approach given here is to provide an algorithm to determine a preliminary design of a heat exchanger for a given service. The final design requires the use of sophisticated software and input from a manufacturer of heat exchangers. However, the algorithm presented here should provide a reasonable design, albeit not optimal, for the service considered.

2.7.1 Pressure Drop Considerations

Up until this point, the equations describing the thermal behavior of the heat exchanger have been covered. However, the design of an exchanger is always a compromise in which higher heat transfer coefficients caused by higher fluid velocities are balanced with the high-pressure drops caused by the higher velocities. It is usual in the design process to specify nominal pressure drops for the shell-side and tube-side fluids. Although these nominal values may be exceeded, they usually act as reasonable upper bounds on the allowable pressure drop. The procedure for estimating the pressure drop for the shell-side fluid was covered in Section 2.5.3.2 using Kern's method. For the tube-side fluid, the standard term for the frictional pressure drop in circular tubes can be used with the addition of four velocity heads per return due to changes in fluid direction for multiple tube passes. The recommended equation is

$$-\Delta P_{f,\,tubes} = 2f\frac{N_{tp}L_{tube}}{D_i}\rho u_i^2 + 2(N_{tp}-1)\rho u_i^2 = \left[2f\frac{N_{tp}L_{tube}}{D_i}+2(N_{tp}-1)\right]\rho u_i^2 \qquad (2.77)$$

where N_{tp} is the number of tube passes, u_i is the velocity inside the tubes, and D_i is the inside tube diameter. In Equation (2.77), the first term on the right-hand side accounts for the frictional pressure loss in the tubes, and the second term accounts for losses occurring at the ends of the exchanger where the fluid changes direction.

Once a given arrangement of tubes, shells, baffles, and so on, has been made, more sophisticated CFD (computational fluid dynamics) models can be constructed to determine more accurate estimates of the expected pressure drops through the exchanger.

2.7.2 Design Algorithm

The design process for S-T heat exchangers is an iterative one, because the value of U must be known, but it cannot be calculated until values for the shell diameter, baffle spacing, number of tubes, and so on, have been determined, which in turn cannot be calculated unless U is known. Once the calculations have been completed, the assumptions about the construction parameters may be checked and modified and calculations reworked. Following such a process, a final acceptable design can be found. Although not unique, the following algorithm for the design of an S-T heat exchanger is recommended:

1. Establish the energy balance around the heat exchanger:

$$Q = \dot{m}_{tube}\Delta h_{tube} = \dot{m}_{tube}c_{p,\,tube}\Delta t_{tube} = \dot{M}_{shell}\Delta h_{shell} = \dot{M}_{shell}C_{p,\,shell}\Delta T_{shell}$$

2. Using the heuristics in Section 2.2.1.7, determine which fluid should go on the shell side and which on the tube side.
3. For the tube side, determine the mass flowrate, inlet temperature, outlet temperature, and enthalpy change.

4. For the shell side, determine the mass flowrate, inlet temperature, outlet temperature, and enthalpy change.

5. Determine all the fluid properties, μ, ρ, k, c_p, at the appropriate average temperatures.

6. Determine the maximum allowable pressure drops for the tube side and shell side.

7. Determine the number of shell passes, N_{shells}, using Equation (2.16) and the number of tube passes, N_{tp} = 2(or 4 or 6...)N_{shells}, and determine the LMTD correction factor, F_{N-2N} and LMTD.

8. Choose the shell and tube arrangement. Determine the following:

   a. Tube length, L_{tube}: Standard lengths are 8, 12, 16, and 20 ft
   b. Baffle cut, BC: 15% to 45% of D_s with 20% to 35% being "standard"
   c. Baffle spacing: L_b ranges from a minimum of $0.2D_s$ or 2 in to a maximum of D_s
   d. Tube diameter, D_o: Standard sizes are ¾ and 1 in
   e. Tube thickness: Ranges from 12 to 18 BWG (see Table 2.4) and actual value will depend on the tube-side fluid pressure and temperature
   f. Tube arrangement (square or triangular) and pitch, p: Typical values are $p = 1.25D_o$ for triangular and $p = D_o + 0.25$ in for square

9. Choose a tube-side velocity between 1 and 3 m/s (this should not exceed 3 m/s), and from this value estimate the number of tubes per pass, the total number of tubes in the exchanger, N_{tubes}, and the inside heat transfer coefficient, h_i.

10. Using the tube count information from Tables 2.6 and 2.7, determine the shell diameter.

11. With the shell diameter from Step 10 and the information in Step 8, calculate the shell-side heat transfer coefficient using Equation (2.43).

12. Combine the heat transfer coefficients and fouling and tube metal resistance to give the overall outside heat transfer coefficient, U_o, Equation (2.23).

13. Determine the total heat transfer area required, $A_{o,new} = Q/(U_o \Delta T_{lm} F_{N-2N})$ and compare with the total area assumed from $A_o = \pi D_o L_{tube} N_{tubes}$.

14. Determine the tube-side and shell-side pressure drops. If either of the pressure drops exceeds the maximum allowable values, adjust the shell/tube configuration to accommodate. Recommended actions include the following:
    Tube side: Reduce the tube-side velocity by increasing the number of tubes.
    Note that $\Delta P_{tube} \propto (1/N_{tubes})^2$
    Shell side: Increase the baffle spacing or tube pitch.
    Note that $\Delta P_{shell} \propto (1/L_b)^3$
    If pressure drop is too low, consider increasing the number of tube passes (for increasing ΔP_{tube}) or decreasing the baffle spacing (for increasing ΔP_{shell}). These actions increase the overall heat transfer coefficient, which reduces heat transfer area and the cost of the heat exchanger.

15. Adjust the number of tubes based on the new estimated area and tube-side pressure drop and, if needed, adjust the baffle spacing based on shell-side pressure drop. Iterate from Step 8 until the solution converges—namely, $A_{o,new}$ in Step 13 is between A_o and 1.2 A_o.

Example 2.16   (Modified from Kern [1950], Example 7.3)

Determine a preliminary rating for a heat exchanger with the following service:
19,900 kg/h of kerosene leaves a distillation column at 200°C and will be cooled to 93°C by 67,880 kg/h of crude oil from storage that is to be heated from 38°C to 76°C. A maximum pressure drop of 70 kPa is permissible for both streams. A scale factor (R_f) of 0.000053m²K/W should be used for this

Table 2.6 Tube Counts for Shell-and-Tube Heat Exchangers—Fixed Tubesheet (Equilateral Triangular Pitch)

Shell Diameter, D_s (in)	1-pass		2-pass		4-pass		6-pass		8-pass	
D_t–p(in)	¾–1	1–1¼	¾–1	1–1¼	¾–1	1–1¼	¾–1	1–1¼	¾–1	1–1¼
8	33	15	28	16	—	—	—	—	—	—
10	57	33	56	32	44	24	—	—	—	—
12	91	57	90	52	72	44	—	—	—	—
13¼	117	73	110	62	96	60	66	34	—	—
15¼	157	103	154	92	134	78	104	56	82	—
17¼	217	133	208	126	180	104	156	82	124	66
19¼	277	163	264	162	232	138	202	112	170	90
21¼	343	205	326	204	292	176	258	150	224	120
23¼	423	247	398	244	360	212	322	182	286	154
25	493	307	468	292	424	258	388	226	344	190
27	577	361	556	346	508	308	464	274	422	240
29	667	427	646	410	596	368	548	338	496	298
31	765	481	746	462	692	422	640	382	588	342
33	889	551	858	530	802	486	744	442	694	400
35	1007	633	972	608	912	560	852	514	798	466
37	1127	699	1088	688	1024	638	964	586	902	542

Source: Adapted with permission from Couper et al. (2012); also see www.red-bag.com/tubesheet-layout.html.

service, which is mainly caused by the crude oil. Based on previous experience, 1-in diameter, 16 ft long, 12 BWG tubes should be used (these may be changed if needed), and the crude oil should flow inside the tubes.

Solution

The steps in the design algorithm are as follows.

1. Energy balance (using average C_p values):

$$\text{Kerosene (shell side): } (19{,}900)(2486)(200 - 93) = 5.29 \times 10^3 \text{ MJ/h}$$

$$\text{Crude oil (tube side): } (67{,}880)(2052)(76 - 38) = 5.29 \times 10^3 \text{ MJ/h}$$

$$Q = 5.29 \times 10^3 \text{ MJ/h} = 1.4694 \text{ MW}$$

2. Choice of shell- and tube-side fluids:

   Set crude in tubes and kerosene in shell since crude is more fouling than kerosene.

3. Tube side—crude oil:

$$\dot{m}_{tube} = 67{,}880 \text{ kg/h}$$

$$t_{in} = 38°C \text{ and } t_{out} = 76°C$$

Table 2.7   Tube Counts for Shell-and-Tube Heat Exchangers—Fixed Tubesheet (Square Pitch)

Shell Diameter, D_s (in)	1-pass		2-pass		4-pass		6-pass		8-pass	
D_t–p(in)	¾–1	1–1¼	¾–1	1–1¼	¾–1	1–1¼	¾–1	1–1¼	¾–1	1–1¼
8	33	17	26	12	—	—	—	—	—	—
10	53	33	48	26	48	24	—	—	—	—
12	85	45	78	40	72	40	—	—	—	—
13¼	101	65	94	56	88	48	54	—	—	—
15¼	139	83	126	76	126	74	78	44	—	—
17¼	183	111	172	106	142	84	116	66	94	—
19¼	235	139	222	136	192	110	158	88	132	74
21¼	287	179	280	172	242	142	212	116	174	94
23¼	355	215	346	218	308	188	266	154	228	128
25	419	255	408	248	366	214	324	184	286	150
27	495	303	486	298	440	260	394	226	352	192
29	587	359	560	348	510	310	460	268	414	230
31	665	413	644	402	590	360	536	318	490	280
33	765	477	746	460	688	414	634	368	576	334
35	865	545	840	522	778	476	724	430	662	388
37	965	595	946	584	880	534	818	484	760	438

Source: Adapted from Couper et al. (2012); also see http://www.red-bag.com/tubesheet-layout.html.

4. Shell side—kerosene:

$$\dot{m}_{shell} = 19{,}900 \text{ kg/h}$$

$$T_{in} = 200°C \text{ and } T_{out} = 93°C$$

5. Fluid properties:

Crude oil (properties at $T_{bulk, ave} = (38 + 76)/2 = 57°C$)

$$\rho = 855 \text{ kg/m}^3$$

$$\mu = 0.003595 \text{ kg/m/s}$$

$$k = 0.1332 \text{ W/m/K}$$

$$c_p = 2052 \text{ J/kg/K}$$

$$\text{Pr} = c_p \mu / k = 55.4$$

Kerosene (properties at $T_{bulk, ave} = (200 + 93)/2 = 146.5°C$)

$$\rho = 815 \text{ kg/m}^3$$

$$\mu = 0.000401 \text{ kg/m/s}$$

$$k = 0.1324 \, \text{W/m/K}$$

$$C_p = 2486 \, \text{J/kg/K}$$

$$\text{Pr} = c_p \mu / k = 7.53$$

6. Set maximum pressure drops:

$$\Delta P_{tube, \, max} = 70 \, \text{kPa and } \Delta P_{shell, \, max} = 70 \, \text{kPa}$$

7. Determine the number of shell (N_{shells}) and tube passes (N_{tp}):

$$(2N), LMTD, \text{ and } F_{N-2N}$$

$$P = \frac{(t_2 - t_1)}{(T_1 - t_1)} = \frac{(76 - 38)}{(200 - 38)} = 0.2346$$

$$R = \frac{(T_1 - T_2)}{(t_2 - t_1)} = \frac{(200 - 93)}{(76 - 38)} = 2.8158$$

From Equation (2.16),

$$N_{shells} = \frac{\ln\left[\dfrac{1 - PR}{1 - P}\right]}{\ln\left[\dfrac{1}{R}\right]} = \frac{\ln\left[\dfrac{1 - (0.2346)(2.8158)}{1 - 0.2346}\right]}{\ln\left[\dfrac{1}{2.8158}\right]} = \frac{\ln(0.4434)}{\ln(0.3551)} = 0.7855$$

$$\text{Use } N_{shells} = 1 \text{ and } N_{tp} = 2 \times N_{shells} = 2$$

The T-Q diagram for this exchanger is shown in Figure E2.16.

$$LMTD = \frac{\Delta T_1 - \Delta T_2}{\ln\left(\dfrac{\Delta T_1}{\Delta T_2}\right)} = \frac{124 - 55}{\ln\left(\dfrac{124}{55}\right)} = 84.88^\circ \, \text{C}$$

Use Equation (2.13) to determine F_{1-2}:

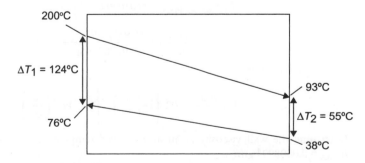

Figure E2.16   T-Q diagram for crude oil heat exchanger

$$F_{1-2} = \frac{\sqrt{(R^2+1)}}{(R-1)} \frac{\ln\left[\dfrac{1-P}{1-PR}\right]}{\ln\left[\dfrac{2-P\left(R+1-\sqrt{(R^2+1)}\right)}{2-P\left(R+1+\sqrt{(R^2+1)}\right)}\right]}$$

$$= \frac{\sqrt{(2.8158)^2+1}}{(2.8158-1)} \frac{\ln\left[\dfrac{1-0.2346}{1-(0.2346)(2.8158)}\right]}{\ln\left[\dfrac{2-0.2346\left(2.8158+1-\sqrt{(2.8158^2+1)}\right)}{2-0.2346\left(2.8158+1+\sqrt{(2.8158^2+1)}\right)}\right]} = 0.8935$$

8. Choose shell-and-tube arrangement:

   a. Tube length: L_{tube} = 16 ft = (16)(0.3048) = 4.877 m

   b. Baffle cut: BC = 25% of shell diameter, D_s

   c. Baffle spacing: $L_b = 0.20D_s$

   d. Tube diameter: D_o = 1-in = 0.0254 m

   e. Tube thickness: Δx_{wall} for12 BWG (from Table 2.4), tube thickness (Δx_{wall}) = 0.109 in = 0.002769 m

   $$D_i = D_o - 2\,\Delta x_{wall} = 0.0254 - 2(0.002769) = 0.01986 \text{ m}$$

   f. Tube arrangement (square or triangular) and pitch, p: Use a square arrangement to allow for cleaning on the outside of the tubes and a pitch of 1.25 in

9. Choose a tube-side velocity and determine number of tubes and tube-side heat transfer coefficient, h_i:

   u_i = 2 m/s. A relatively high velocity is chosen, as the crude is quite viscous and turbulent flow is desired. This can be modified later if the ΔP on the tube side is too high.

   A mass balance gives

   $$n_{tube/pass} \frac{\pi}{4} D_i^2 u_i \rho_{crude} = \dot{m}_{crude}$$

   $$\Rightarrow n_{tube/pass} = \frac{4\dot{m}_{crude}}{\pi D_i^2 u_i \rho_{crude}} = \frac{4(67880/3600)}{\pi(0.01986)^2(2)(855)} = 35.6 \text{ so use } 36$$

   $$N_{tubes} = N_{tp}n_{tube/pass} = (2)(36) = 72$$

   Actual velocity in tubes,

   $$u_i = \frac{4\dot{m}_{crude}}{\pi D_i^2 n_{tube/pass} \rho_{crude}} = \frac{4(67880/3600)}{\pi(0.01986)^2(36)(855)} = 1.977 \text{ m/s}$$

   $$\text{Re}_{crude} = \frac{D_i u_i \rho}{\mu} = \frac{(0.01986)(1.977)(855)}{(0.003595)} = 9338$$

   From Equation (2.26),

   $$\text{Nu} = \frac{h_i D_i}{k} = 0.023\left[1+\left(\frac{D_i}{L}\right)^{0.7}\right]\text{Re}^{0.8}\,\text{Pr}^{1/3}\left(\frac{\mu}{\mu_w}\right)^{0.14}$$

   Determine the wall viscosity at an average wall temperature = (146.5 + 57)/2 = 102°C $\mu_{w,crude}$ = 0.000899 kg/m/s:

   $$\text{Nu} = 0.023\left[1+\left(\frac{0.01986}{4.8768}\right)^{0.7}\right]9338^{0.8}55.4^{1/3}\left(\frac{0.003595}{0.000899}\right)^{0.14} = 163.1$$

$$h_i = \mathrm{Nu}\frac{k}{D_i} = (163.1)\frac{0.1332}{0.01986} = 1094 \text{ W/m}^2/\text{K}$$

10. Determine shell diameter:

    For 1-in tubes in a square arrangement with $p = 1.25$ in, Table 2.7 indicates that a 15¼ in shell diameter will accommodate 76 tubes in a two-pass arrangement.
11. Determine shell-side heat transfer coefficient, h_o:

    Diameter of shell: $D_s = 15.25$ in $= 0.3874$ m
    Baffle spacing: $L_b = 0.2D_s = (0.2)(0.3874) = 0.07747$ m
    From Equation (2.37), determine the hydraulic diameter of the shell, $D_{H,s}$

$$D_{H,s} = \frac{1.273p^2 - D_o^2}{D_o} = \frac{(1.273)(1.25)^2 - (1)^2}{(1)} = 0.9891 \text{ in} = 0.02512 \text{ m}$$

From Equation (2.41), determine the flow area on the shell side:

$$A_s = D_s L_b \frac{p - D_o}{p} = (0.3874)(0.07747)\frac{(1.25-1)}{(1.25)} = 0.00600 \text{ m}$$

Shell-side velocity,

$$u_s = \frac{\dot{m}_s}{\rho A_s} = \frac{(19,900/3600)}{(815)(0.00600)} = 1.130 \text{ m/s}$$

Shell-side Reynolds number,

$$\mathrm{Re}_s = \frac{D_{H,s}u_s\rho}{\mu} = \frac{(815)(1.13)(0.02512)}{(0.000401)} = 57,703$$

From Equation (2.43),

$$\mathrm{Nu} = \frac{h_s D_{H,s}}{k_f} = j_h \mathrm{Re}\,\mathrm{Pr}^{1/3}\left(\frac{\mu}{\mu_w}\right)^{0.14}$$

with

$$j_h = 1.249(BC)^{-0.329}\mathrm{Re}^{-0.470} = (1.249)(25)^{-0.329}(57,703)^{-0.470} = 0.002505$$

The value of μ_w corresponding to a wall temperature of 102°C is 0.00055 kg/m/s.

$$\mathrm{Nu} = j_h \mathrm{Re}\,\mathrm{Pr}^{1/3}\left(\frac{\mu}{\mu_w}\right)^{0.14} = (0.002505)(57,703)(7.53)^{1/3}\left(\frac{0.0004}{0.00055}\right)^{0.14} = 271.0$$

$$h_s = \mathrm{Nu}\frac{k_f}{D_{H,s}} = (271.0)\frac{(0.1324)}{(0.02512)} = 1428 \text{ W/m}^2/\text{K}$$

12. Combine heat transfer resistance to give U_o:

    From Equation (2.23),
    Assume the exchanger will be made of carbon steel; from Table 2.3, $k_w = 45.8$ W/m/K (@102°C), and from the problem statement, $R_{fi} = 0.000053$ m²K/W:

$$U_o = \left[\frac{1}{1428} + 0 + \frac{(0.0254)\ln(0.0254/0.01986)}{(2)(45.8)} + \frac{(0.0254)}{(0.01986)}(0.000053) + \frac{(0.0254)}{(0.01986)}\frac{1}{(1094)}\right]^{-1}$$

$$U_o = 498.7 \text{ W/m}^2/\text{K}$$

13. Determine the total heat transfer area required:

$$A_o = \frac{Q}{U_o \Delta T_{lm} F_{1-2}} = \frac{(1.4694 \times 10^6)}{(498.7)(84.88)(0.8935)} = 38.9 \text{ m}^2$$

14. Determine the tube-side and shell-side pressure drops:

Tube-side ΔP:

From Equation (2.77), the tube-side pressure loss is given by

$$-\Delta P_{f,tubes} = \left[2f \frac{N_{tp} L_{tube}}{D_i} + 2(N_{tp} - 1) \right] \rho u_i^2$$

where $N_{tp} = 2$, $u_i = 1.977$ m/s and f is found from the Pavlov equation (Equation [1.16]),

$$\frac{1}{\sqrt{f}} = -4\log \left[\frac{1}{3.7} \frac{e}{D} + \left(\frac{6.81}{\text{Re}} \right)^{0.9} \right]$$

for drawn tubing, $e = 1.5 \times 10^{-6}$ m and $D = D_i = 0.01986$ and $\text{Re} = 9338$. Substituting these values into the equation for f gives

$$\frac{1}{\sqrt{f}} = -4\log \left[\frac{1}{3.7} \frac{e}{D} + \left(\frac{6.81}{\text{Re}} \right)^{0.9} \right] = -4\log \left[\frac{1}{3.7} \frac{1.5 \times 10^{-6}}{0.01986} + \left(\frac{6.81}{9338} \right)^{0.9} \right] = 11.270$$

$$f = 0.007872$$

$$-\Delta P_{f,tubes} = \left[2(0.007872) \frac{(2)(4.8768)}{(0.01986)} + 2(2-1) \right] (855)(1.977)^2 = 32.5 \text{ kPa}$$

This is less than the maximum allowable value of 70 kPa.

Shell-side ΔP:

The shell-side ΔP is obtained from Equation (2.44):

$$-\Delta P_f = \frac{4 j_f G_s^2 D_s (N_B + 1)}{2\rho D_{H,s}} \left[\frac{\mu}{\mu_w} \right]^{-0.14}$$

And for Re > 300,

$$j_f = 5.46(BC)^{-0.674} \text{Re}^{-0.189} = (5.46)(25)^{-0.674}(57,703)^{-0.189} = 0.07854$$

$$(N_B + 1) = L_{tube}/L_b = (4.877)/(0.07747) = 62.95 = 63$$

$$G_s = \frac{\dot{m}_s}{A_s} = \frac{(19,900)}{(3600)(0.0060)} = 921.3$$

$$-\Delta P_f = \frac{4(0.07854)(921.3)^2(15.25)(0.0254)(63)}{2(815)(0.02512)} \left[\frac{0.00040}{0.00055} \right]^{-0.14} = 166.0 \text{ kPa}$$

This is greater than the maximum allowable value of 70 kPa.

15. Adjust the number of tubes based on the new estimated area and tube-side pressure drop and, if needed, adjust the baffle spacing based on shell-side pressure drop. Iterate from Step 9 until converged.

Number of tubes:

$$A_o = N_{tubes} \pi D_o L_{tube}$$

$$\therefore N_{tubes} = \frac{A_o}{\pi D_o L_{tube}} = \frac{(38.9)}{\pi (0.0254)(4.8768)} = 100$$

Because the number of tubes has increased, if the 1–2 configuration is kept, then ΔP_{tube} and h_i will both decrease and a solution will not be found. Change the configuration to 1–4.

Shell-side pressure drop:
Using an $L_b = 0.2 D_s$, the value of ΔP_{shell} was 166.0 kPa, which is greater than the maximum allowable value of 70 kPa. Therefore, adjust L_b by a factor $= (166.0/70)^{1/3} = 1.33$ so that $L_b = (1.33)(0.2 D_s) = 0.266 D_s$, and round up to give $L_b = 0.3 D_s$.

Iterate from Step 9 with $N_{tubes} = 104$ (corresponding to 19.25 in shell diameter from Table 2.7) and $L_b = 0.3 D_s$.

The results for the next and subsequent iterations are given in Table E2.16.

Table E2.16   Results for Worked Example 2.16

Step in Design Algorithm	Iteration 1	Iteration 2	Iteration 3	Iteration 4
	$N_{tubes} = 76$ $L_b = 0.2 D_s$	$N_{tubes} = 104$ (4 passes) $L_b = 0.3 D_s$	$N_{tubes} = 140$ (4 passes) $L_b = 0.3 D_s$	$N_{tubes} = 140$ (4 passes) $L_b = 0.2 D_s$ Use 14 ft tubes
9	$h_i = 1094 \text{ W/m}^2\text{/K}$	$h_i = 1419 \text{ W/m}^2\text{/K}$	$h_i = 1119 \text{ W/m}^2\text{/K}$	$h_i = 1121 \text{ W/m}^2\text{/K}$
10	$D_s = 15.25$ in	$D_s = 19.25$ in	$D_s = 21.25$ in	$D_s = 21.25$ in
11	$h_s = h_o = 1435 \text{ W/m}^2\text{/K}$	$h_s = h_o = 904 \text{ W/m}^2\text{/K}$	$h_s = h_o = 814 \text{ W/m}^2\text{/K}$	$h_s = h_o = 1009 \text{ W/m}^2\text{/K}$
12	$U_o = 499.5 \text{ W/m}^2\text{/K}$	$U_o = 466.6 \text{ W/m}^2\text{/K}$	$U_o = 398.8 \text{ W/m}^2\text{/K}$	$U_o = 441.0 \text{ W/m}^2\text{/K}$
13	$A_o = 38.8 \text{ m}^2$ Requires 100 tubes	$A_o = 41.56 \text{ m}^2$ Requires 107 tubes	$A_o = 48.6 \text{ m}^2$ Requires 125 tubes	$A_o = 44.0 \text{ m}^2$ Requires 130 tubes
14	$\Delta P_{tube} = 32.5$ kPa $\Delta P_{shell} = \textbf{165.9 kPa}$	$\Delta P_{tube} = \textbf{129.3 kPa}$ $\Delta P_{shell} = 23.3$ kPa	$\Delta P_{tube} = \textbf{75.5 kPa}$ $\Delta P_{shell} = 16.4$ kPa	$\Delta P_{tube} = 68.7$ kPa $\Delta P_{shell} = 44.2$ kPa
15	Adjust $N_{tubes} = 104$ (Table 2.7 with $D_s = $ 19.25 in), use 4 tube passes, and adjust baffle spacing $= 0.30 D_s$	$\Delta P_{tube} > 70$ kPa Increase number of tubes per pass by $(129.3/70)^{1/2} = 1.31$ $\therefore N_{tubes} = 142$ (Table 2.7 with $D_s = 21.25$ in) Use 4 tube passes with 35 tubes per pass = 140 tubes	Area required = 48.6 m^2 Area = 54.5 m^2 Overdesign = 1.121 or 12.1% Note: ΔP_{tube} is still a bit high and ΔP_{shell} is low Reduce the length of tubes to 14 ft and set $L_b = 0.2 D_s$	Area required = 43.96 m^2 Area = 48.35 m^2 Overdesign = 1.10 or 10% $\Delta P_{tube} < 70$ kPa $\Delta P_{shell} < 70$ kPa **Acceptable design**

The preliminary design for this equipment gives an exchanger with the following specifications:

Number of tubes: 142
Length of Tubes: 14 ft
Tube type: 1-in diameter 12 BWG
Tube arrangement: 1.25-in square pitch
Shell passes: 1; tube passes = 4
Shell diameter: 21.25 in
Baffle cut: 25% of the shell diameter (5.31 in)
Baffle spacing: 20% of the shell diameter (4.25 in)
Overdesign: 10%

2.8   PERFORMANCE PROBLEMS

The main focus of this chapter has been developing the ideas and relationships that describe the heat transfer process in process heat exchangers and then bringing all these ideas together to design a heat exchanger. It has been shown that many, sometimes complex, relationships must be considered in order to design a heat exchanger successfully. However, once the heat exchanger has been designed, built, and installed into a process, further analysis of the heat exchanger is still necessary. As any process engineer will attest, a heat exchanger (or any other piece of process equipment) will hardly ever (never) run under the exact conditions for which it was designed. This does not mean that the design algorithm or any of the calculations were incorrect but rather that the heat exchanger is now part of a complex chemical process with the main objective to make profitable chemicals from less-expensive feed stocks. Therefore, in the day-to-day operations of the plant, things will change almost continuously, and the heat exchanger will have to operate over a spectrum of conditions in order for the process to produce the desired amount of product chemicals.

An example of off-design conditions will occur when there is high market demand for the chemical(s) produced from a process. Under these conditions, the process may have to be run to maximize the products (and profits) from the plant. Thus, the process might be pushed to 110% or 120% of its design capacity. Obviously, every heat exchanger in the plant will be operating quite far from its design conditions. The opposite might occur when there is weak market demand. The question addressed in this section is, for a given existing heat exchanger, how is the performance of that heat exchanger predicted?

2.8.1   What Variables to Specify in Performance Problems

When determining the performance of an existing heat exchanger, the same equations that were used in the design must be solved, specifically, the enthalpy balances for both streams and the design equation $Q = UA\Delta T_{lm}F$. However, unlike in the design calculation, the conditions of the streams leaving the exchanger are usually unknown, but the conditions entering are known. In addition, the size (heat transfer area) is known and fixed. For example, consider the exchanger designed in Section 2.7.2. This exchanger was built as specified and has been operating in the plant. The current operating conditions are

Tube Side: Crude Oil $\dot{m}_{tube}$ = 67,880 kg/h, t_{in} = 38°C and t_{out} = 77.8°C
Shell Side: Kerosene $\dot{m}_{shell}$ = 19,900 kg/h, T_{in} = 200°C and T_{out} = 88.0°C

It should be noted that these conditions are not the design conditions because the exchanger was built with approximately 10% additional area required to perform the design. Thus, more heat

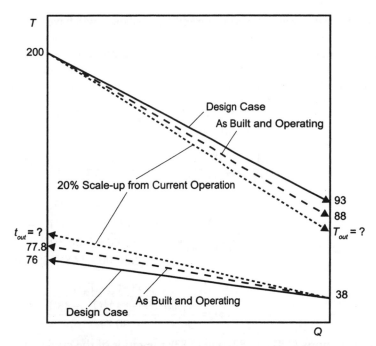

Figure 2.36 T-Q diagram for performance problem

is transferred between the streams, and the crude leaves hotter and the kerosene leaves cooler than the design specification, which is to be expected because of the overdesign.

Now consider the case when the plant is to be scaled up. It is desired that the flows of crude and kerosene be increased by 20%. If the temperatures into the exchanger remain the same as in the design case, the exit temperatures of both streams must be determined. This problem is illustrated in Figure 2.36. For performance problems (where the equipment has been designed, installed, and is operating), the easiest way to determine the new operating conditions is via a ratio analysis, shown in the next section. The terms *rating* and *performance* are synonymous when considering how existing equipment performs. However, the term *performance problem* is preferred in this text.

2.8.2   Using Ratios to Determine Heat-Exchanger Performance

From the problem given in Example 2.16, it is clear that the design of the exchanger is quite involved. Rather than revisit all the equations used in this example, a simpler approach is adopted for evaluating the performance problem. As was explained in Chapter 1, the idea is to take the ratio of the important equations describing the heat transfer process; that is, the two enthalpy balances and the design equation for the new case (subscript 2) and the base case (subscript 1). Thus,

$$\frac{Q_2}{Q_1} = \frac{\dot{m}_2 c_{p2}(t_2 - t_1)_2}{\dot{m}_1 c_{p1}(t_2 - t_1)_1} \tag{2.78}$$

$$\frac{Q_2}{Q_1} = \frac{\dot{M}_2 C_{p2}(T_1 - T_2)_2}{\dot{M}_1 C_{p1}(T_1 - T_2)_1} \tag{2.79}$$

$$\frac{Q_2}{Q_1} = \frac{U_2 A_2 \Delta T_{lm,2} F_2}{U_1 A_1 \Delta T_{lm,1} F_1} \tag{2.80}$$

If there are changes in phase, the enthalpy balances will have heats of vaporization instead of sensible heats, but the approach is similar. The assumption is then made that, for relatively small changes in flows, the physical properties of the fluids remain constant and also that the *LMTD* correction factors remain constant. The validity of these assumptions can be checked later. For the problem considered here, the three equations can be rewritten:

$$\frac{Q_2}{Q_1}=\frac{\dot{m}_2 \ell_{p2}(t_{out}-38)}{\dot{m}_1 \ell_{p1}(77.8-38)} \Rightarrow Q=m\frac{(t_{out}-38)}{39.8}=1.2\frac{(t_{out}-38)}{39.8} \tag{2.81}$$

$$\frac{Q_2}{Q_1}=\frac{\dot{M}_2 \ell_{p2}(200-T_{out})}{\dot{M}_1 \ell_{p1}(200-88)} \Rightarrow Q=M\frac{(200-T_{out})}{112}=1.2\frac{(200-T_{out})}{112} \tag{2.82}$$

$$\frac{Q_2}{Q_1}=\frac{U_2 A_2 \Delta T_{lm,2} F_2}{U_1 A_1 \Delta T_{lm,1} F_1}=\frac{U_2}{U_1}\frac{\frac{(200-t_{out})-(T_{out}-38)}{\ln[(200-t_{out})/(T_{out}-38)]}}{\frac{(200-77.8)-(88-38)}{\ln[(200-77.8)/(88-38)]}}$$

$$\Rightarrow Q=\frac{U_2}{U_1}\frac{\frac{(238-t_{out}-T_{out})}{\ln[(200-t_{out})/(T_{out}-38)]}}{80.79} \tag{2.83}$$

In Equations (2.81), (2.82), and (2.83), the ratios of the crude oil and kerosene flowrates for cases 1 and 2 are given as *m* and *M*, respectively. For the case of 20% scale-up, these ratios are 1.2 each. The ratio of the heat transferred is designated as Q, but it should be noted that Q is not 1.2 and must be determined.

In Equation (2.83), the ratio of the overall heat transfer coefficients must be determined. For the base case, from the last iteration in Table E2.16, $h_o=1009\ \text{W/m}^2/\text{K}$, $h_i=1121\ \text{W/m}^2/\text{K}$, and

$$U_o=U_1=\left[\frac{1}{h_o}+0+\frac{(0.0254)\ln(0.0254/0.01986)}{(2)(45.8)}+\frac{(0.0254)}{(0.01986)}(0.000053)+\frac{(0.0254)}{(0.01986)}\frac{1}{h_i}\right]^{-1} \tag{2.84}$$

$$U_1=\left[\frac{1}{1009}+1.36\times10^{-4}+\frac{1.2789}{1121}\right]^{-1}=441\ \text{W/m}^2/\text{K} \tag{2.85}$$

In Equation (2.84), the inside (tube) and outside (shell) heat transfer coefficients are dependent on the flows of the tube- and shell-side fluids. From Equations (2.27) and (2.36) for turbulent flow in the tubes and the shell, the heat transfer coefficients are related to the mass flow of fluids through the Reynolds number as follows:

$$h_{tube}\propto \text{Re}^{0.8} \quad \text{and} \quad h_{shell}\propto \text{Re}^{0.6}$$

It should be noted that in Kern's method, the exponent on the Reynolds number is a little below 0.6, but this difference is small, and 0.6 can be used for shell-side flow. Using these relationships and noting that for constant physical properties Re is proportional to the mass flowrate, Equation (2.84) may be rewritten for the new case as

$$U_2=\left[\frac{1}{1009M^{0.6}}+1.36\times10^{-4}+\frac{1.2789}{1121m^{0.8}}\right]^{-1}$$

$$U_2=\left[\frac{1}{(1009)(1.2)^{0.6}}+1.36\times10^{-4}+\frac{1.2789}{(1121)(1.2)^{0.8}}\right]^{-1}=497.4 \tag{2.86}$$

Table 2.8   Changes in h as a Function of Re and ΔT

Flow Regime	Tube Side, h_i	Shell Side, h_o	Flow in Annulus*, h_o
Turbulent	$\propto \mathrm{Re}^{0.8}$ or $\propto m^{0.8}$	$\propto \mathrm{Re}^{0.6}$ or $\propto m^{0.6}$	$\propto \mathrm{Re}^{0.8}$ or $\propto m^{0.8}$
Laminar	$\propto \mathrm{Re}^{1/3}$ or $\propto m^{1/3}$	$\propto \mathrm{Re}^{0.45}$ or $\propto m^{0.45}$	$\propto \mathrm{Re}^{0.45}$ or $\propto m^{0.45}$
Phase Change			
Boiling		$\propto (T_W - T_{sat})^3$ pool boiling	
		$\propto (T_W - T_{sat})^2$ convective boiling	
		$\propto (T_W - T_{sat})^{-1/4}$ film boiling	
Condensation		$\propto (T_{sat} - T_W)^{-1/4}$	

*Useful in analyzing double-pipe heat exchangers.

Substituting Equations (2.85) and (2.86) in (2.83) gives

$$\Rightarrow Q = \frac{497.4}{441} \frac{(238 - t_{out} - T_{out})}{\dfrac{\ln[(200 - t_{out})/(T_{out} - 38)]}{80.79}} \tag{2.87}$$

Equations (2.81), (2.82), and (2.87) may now be solved simultaneously to give

$$t_{out} = 76.64°C, T_{out} = 91.26°C, Q = 1.165$$

Thus, the crude temperature decreases from 77.8°C to 76.4°C, and the kerosene temperature increases from 88°C to 91.3°C. Clearly, the average fluid properties would not change significantly for these temperature changes; similarly, the value of the *LMTD* correction factor hardly changes from 0.894 to 0.885, a change of ~1%. It should also be noted that the amount of heat transferred between the streams increases by only 16.5%, because the heat transfer coefficients are not linearly dependent on the mass flow rates.

Table 2.8 gives the dependence of individual heat transfer coefficients on the mass flow rate (effectively the Reynolds number, Re). Also shown in Table 2.8 are the relationships to be used when phase changes are considered. For phase changes, the mass flow rates do not affect the heat transfer coefficients, but the ΔT between the wall and fluid may be important.

2.8.3   Worked Examples for Performance Problems

To reinforce the concepts introduced in the preceding section, some worked examples are provided.

Example 2.17

An S-T exchanger is designed to exchange heat between two process streams. The *T-Q* diagram is shown in Figure E2.17. The inside and outside heat transfer coefficients are equal ($h_i = h_o = 1000 \ \mathrm{W/m^2/K}$). If the shell-side fluid flow rate is increased by 25% and the tube-side flow and the temperatures of both streams entering the heat exchanger remain unchanged, calculate

  a. The outlet temperatures of both streams assuming that the service is non-fouling and that the tube wall is thin and offers negligible heat transfer resistance.

  b. The outlet temperatures of both streams assuming that the service is non-fouling and the tubes are 1-in, 14-BWG and made of copper.

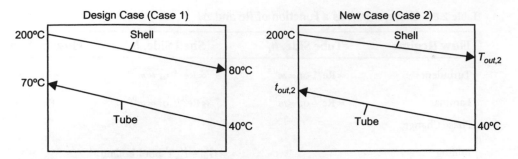

Figure E2.17   T-Q diagrams for the design and new cases

Solution

a. For the design case (Case 1),

$$U_{o,1} = \left[\frac{1}{h_o} + \cancel{R_{fo}} + \frac{D_o \ln(\cancel{D_o}/\cancel{D_i})}{2k_w} + \frac{D_o}{D_i}\cancel{R_{fi}} + \frac{\cancel{D_o}}{\cancel{D_i}}\frac{1}{h_i} \right]^{-1} = \left[\frac{1}{(1,000)} + \frac{1}{(1,000)} \right]^{-1}$$

$$U_{o,1} = 500 \ \text{W/m}^2/\text{K} \qquad\qquad\qquad (\text{E2.17a})$$

$$\Delta T_{lm,1} = \frac{(200-70)-(80-40)}{\ln\left[\dfrac{(200-70)}{(80-40)}\right]} = 76.36°\text{C} \qquad\qquad (\text{E2.17b})$$

The ratio form of the energy balance and design equations are

$$\frac{Q_2}{Q_1} = \frac{\dot{M}_2 \mathscr{E}_{p2}(T_{in}-T_{out})_2}{\dot{M}_1 \mathscr{E}_{p1}(T_{in}-T_{out})_1} \Rightarrow Q = M\frac{(200-T_{out,2})}{(200-80)} = 1.25\frac{(200-T_{out,2})}{120} \qquad (\text{E2.17c})$$

$$\frac{Q_2}{Q_1} = \frac{\dot{m}_2 \mathscr{E}_{p2}(t_{out}-t_{in})_2}{\dot{m}_1 \mathscr{E}_{p1}(t_{out}-t_{in})_1} \Rightarrow Q = m\frac{(t_{out,2}-40)}{(70-40)} = \frac{(t_{out,2}-40)}{30} \qquad (\text{E2.17d})$$

$$Q = \frac{Q_2}{Q_1} = \frac{U_{o,2}\cancel{A}_{o,2}\Delta T_{lm,2}\cancel{F}_2}{U_{o,1}\cancel{A}_{o,1}\Delta T_{lm,1}\cancel{F}_1} = \frac{U_2 \Delta T_{lm,2}}{U_1 \Delta T_{lm,1}} \qquad\qquad (\text{E2.17e})$$

where

$$\Delta T_{lm,2} = \frac{(200-t_{out,2})-(T_{out,2}-40)}{\ln\left[\dfrac{(200-t_{out,2})}{(T_{out,2}-40)}\right]} \qquad\qquad (\text{E2.17f})$$

and using the relationships from Table 2.8,

$$U_{o,2} = \left[\frac{1}{(1,000)M^{0.6}} + \frac{1}{(1,000)m^{0.8}} \right]^{-1} = \left[\frac{1}{(1,000)(1.25)^{0.6}} + \frac{1}{(1,000)(1)^{0.8}} \right]^{-1}$$

$$U_{o,2} = 533.4 \ \text{W/m}^2/\text{K} \qquad\qquad\qquad (\text{E2.17g})$$

Substituting Equations (E2.17f) and (E2.17g) into (E2.17e) gives

$$Q = \frac{U_2 \Delta T_{lm,2}}{U_1 \Delta T_{lm,1}} = \frac{533.4}{500} \frac{\dfrac{(200-t_{out,2})-(T_{out,2}-40)}{\ln\left[\dfrac{(200-t_{out,2})}{(T_{out,2}-40)}\right]}}{76.36} \qquad (\text{E2.17h})$$

Solving Equations (E2.17c), (E2.17d), and (E2.17h) simultaneously gives

$$T_{out,2} = 90.0°C, t_{out,2} = 74.4°C, Q = 1.146$$

b. $D_i = 0.02118$ m, $D_o = 0.0254$, $k_w = k_{copper} = 380$ W/m/K

For the design case (Case 1),

$$U_{o,1} = \left[\frac{1}{h_o} + \cancel{R}_{fo} + \frac{D_o \ln(D_o/D_i)}{2k_w} + \frac{D_o}{D_i}\cancel{R}_{fi} + \frac{D_o}{D_i}\frac{1}{h_i} \right]^{-1}$$

$$U_{o,1} = \left[\frac{1}{(1,000)} + \frac{(0.0254)\ln(0.0254/0.02118)}{(2)(380)} + \frac{(0.0254)}{(0.02118)}\frac{1}{(1,000)} \right]^{-1}$$

$$U_{o,1} = 453.5 \ \text{W/m}^2/\text{K} \tag{E2.17i}$$

$$\Delta T_{lm,1} = 76.36°C \tag{E2.17j}$$

The ratio form of the energy balance and design equations are

$$\frac{Q_2}{Q_1} = \frac{\dot{M}_2 \mathscr{C}_{p2}(T_{in}-T_{out})_2}{\dot{M}_1 \mathscr{C}_{p1}(T_{in}-T_{out})_1} \Rightarrow Q = 1.25\frac{(200-T_{out,2})}{120} \tag{E2.17k}$$

$$\frac{Q_2}{Q_1} = \frac{\dot{m}_2 \mathscr{C}_{p2}(t_{out}-t_{in})_2}{\dot{m}_1 \mathscr{C}_{p1}(t_{out}-t_{in})_1} \Rightarrow Q = \frac{(t_{out,2}-40)}{30} \tag{E2.17l}$$

$$Q = \frac{Q_2}{Q_1} = \frac{U_2 \Delta T_{lm,2}}{U_1 \Delta T_{lm,1}} \tag{E2.17m}$$

where

$$\Delta T_{lm,2} = \frac{(200-t_{out,2})-(T_{out,2}-40)}{\ln\left[\dfrac{(200-t_{out,2})}{(T_{out,2}-40)}\right]} \tag{E2.17n}$$

$$U_{o,2} = \left[\frac{1}{(1,000)M^{0.6}} + \frac{D_o \ln(D_o/D_i)}{2k_w} + \frac{D_o}{D_i}\frac{1}{(1,000)m^{0.8}} \right]^{-1}$$

$$U_{o,2} = \left[\frac{1}{(1,000)(1.25)^{0.6}} + \frac{(0.0254)\ln(0.0254/0.02118)}{(2)(380)} + \frac{(0.0254)}{(0.02118)}\frac{1}{(1,000)(1)^{0.8}} \right]^{-1}$$

$$U_{o,2} = 480.8 \ \text{W/m}^2/\text{K} \tag{E2.17o}$$

Substituting Equations (E2.17i), (E2.17j), (E2.17n), and (E2.17o) into (E2.17m) and solving (E2.17k), (E2.17l), and (E2.17m) gives

$$T_{out,2} = 90.3°C, t_{out,2} = 74.3°C, Q = 1.143$$

These results are virtually the same as for Part (a), which is not surprising, since the resistance due to the conduction through the wall can often be neglected.

Example 2.18

A hot, viscous process stream is cooled using cooling water (cw). The exchanger is a double-pipe design in which the cooling water flows in the annulus and the viscous process fluid flows in the inner pipe. The T-Q diagram for the exchanger at design conditions is shown in Figure E2.18 along with an image of a double-pipe exchanger. At design conditions, the cw is in the turbulent regime, but the process fluid is in the laminar flow regime. The process is to be scaled down such

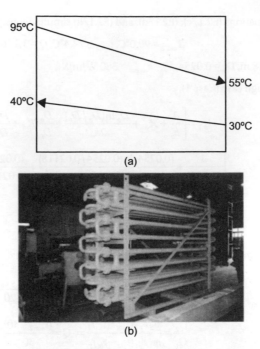

Figure E2.18   *T-Q* diagram for design case and photo of a double-pipe heat exchanger (Courtesy of Protherm Systems [Pty] Ltd., www.protherm.co.za/product/protherm-shell-tube-heat-exchangers)

that the process flow is reduced to 72% of its current value, but the inlet temperature is maintained at 95°C. It is desired to keep the exit temperature of the process at the current value of 55°C by reducing the cooling water flow rate. However, the maximum temperature of the cooling water is set at 45°C because, if it gets hotter than this value, then excessive scaling of the heat exchanger will result.

Data:

$h_i = 125$ W/m^2/K
$h_o = 3500$ W/m^2/K
$R_o = 1.76 \times 10^{-4}$ m^2/K/W, $R_i = 0$
Inner tube = 0.75-in, 16-BWG copper ($D_{i,i} = 0.01576$ m $D_{i,o} = 0.01905$ m)
Outer tube = 1.25-in, 16-BWG copper ($D_{o,i} = 0.02845$ m $D_{o,o} = 0.03175$)
$k_{copper} = k_w = 380$ W/m/K

a. Can this operation be accomplished? Determine the cw flow rate and exit temperature of cw.

b. If the desired reduction in process flow rate cannot be achieved, determine the minimum process flow rate and the cw flow rate to accomplish this.

Solution

a. For the design (base case or Case 1) condition,

$$U_{o,1} = \left[\frac{1}{h_o} + R_{fo} + \frac{D_o \ln(D_o/D_i)}{2k_w} + \frac{D_o}{D_i} \cancel{R_{fi}} + \frac{D_o}{D_i} \frac{1}{h_i} \right]^{-1}$$

$$= \left[\frac{1}{(3,500)} + 1.76 \times 10^{-4} + \frac{(0.01905)\ln(0.01905/0.01576)}{(2)(380)} + \frac{(0.01905)}{(0.01576)} \frac{1}{(125)} \right]^{-1}$$

$$= \left[\frac{1}{(3,500)} + 1.81 \times 10^{-4} + (1.2088) \frac{1}{(125)} \right]^{-1}$$

$$U_{o,1} = 98.65 \text{ W/m}^2\text{/K} \tag{E2.18a}$$

and

$$\Delta T_{lm,1} = \frac{(95-40)-(55-30)}{\ln\left[\dfrac{(95-40)}{(55-30)}\right]} = 38.05°\text{C} \tag{E2.18b}$$

For the scaled down case (Case 2), the unknowns are the new cooling water flowrate, $\dot{m}_2$, and the cooling water outlet temperature, t_2. Using the relationships in Table 2.8, the value of U_2 is given by

$$U_{o,2} = \left[\frac{1}{(3,500)\left(\dfrac{m_2}{m_1}\right)^{0.8}} + 1.81\times10^{-4} + (1.2088)\frac{1}{(125)\left(\dfrac{M_2}{M_1}\right)^{1/3}}\right]^{-1}$$

$$= \left[\frac{1}{(3,500)m^{0.8}} + 1.81\times10^{-4} + (1.2088)\frac{1}{(125)(0.72)^{1/3}}\right]^{-1}$$

$$U_{o,2} = \left[\frac{1}{(3,500)m^{0.8}} + 1.0970\times10^{-2}\right]^{-1} \tag{E2.18c}$$

The energy balance and design equations (noting that for a double-pipe heat exchanger the flow is pure countercurrent and $F = 1$) are

$$\frac{Q_2}{Q_1} = \frac{\dot{m}_2 \mathscr{C}_{p2}(t_{out}-t_{in})_2}{\dot{m}_1 \mathscr{C}_{p1}(t_{out}-t_{in})_1} \Rightarrow Q = m\frac{(t_{out}-30)}{(40-30)} = m\frac{(t_{out}-30)}{10} \tag{E2.18d}$$

$$\frac{Q_2}{Q_1} = \frac{\dot{M}_2 \mathscr{C}_{p2}(T_{in}-T_{out})_2}{\dot{M}_1 \mathscr{C}_{p1}(T_{in}-T_{out})_1} \Rightarrow Q = M\frac{(95-55)}{(95-55)} = 0.72 \tag{E2.18e}$$

$$Q = \frac{Q_2}{Q_1} = \frac{U_{o,2}\cancel{A}_{o,2}\Delta T_{lm,2}\cancel{F}_2}{U_{o,1}\cancel{A}_{o,1}\Delta T_{lm,1}\cancel{F}_1} = \frac{U_2 \Delta T_{lm,2}}{U_1 \Delta T_{lm,1}} \tag{E2.18f}$$

where

$$\Delta T_{lm,2} = \frac{(95-t_{out,2})-(55-30)}{\ln\left[\dfrac{(95-t_{out,2})}{(55-30)}\right]} = \frac{(70-t_{out,2})}{\ln\left[\dfrac{(95-t_{out,2})}{(25)}\right]} \tag{E2.18g}$$

Substituting Equations (E2.18a), (E2.18b), (E2.18c), and (E2.18g) into (E2.18f) and solving Equations (E2.18d), (E2.18e), and (E2.18f) simultaneously gives

$$m = 0.2840 \text{ or } \dot{m}_2 = 0.284\dot{m}_1 \text{ and } t_{out,2} = 55.4°\text{C}$$

This solution is not practical, since reducing the cooling water flow to 28.4% of the design case increases the exit temperature to well above 45°C, and excessive fouling of the exchanger would result. This fouling is due to the reverse solubility of many of the salts found in water, which start to precipitate out at temperatures greater than 45°C.

b. The unknown variables now become the process flowrate and the cooling water flowrate (i.e., m and M). In addition, the exit temperature of the cooling water is set to 45°C, which is the maximum allowable temperature.

The ratio form of the energy balance and design equations are

$$\frac{Q_2}{Q_1} = \frac{\dot{m}_2 \mathscr{C}_{p2}(t_{out}-t_{in})_2}{\dot{m}_1 \mathscr{C}_{p1}(t_{out}-t_{in})_1} \Rightarrow Q = m\frac{(45-30)}{(40-30)} = 1.5m \tag{E2.18h}$$

$$\frac{Q_2}{Q_1}=\frac{\dot{M}_2\mathscr{C}_{p2}(T_{in}-T_{out})_2}{\dot{M}_1\mathscr{C}_{p1}(T_{in}-T_{out})_1}\Rightarrow Q=M\frac{(95-55)}{(95-55)}=M \qquad (E2.18i)$$

$$Q=\frac{Q_2}{Q_1}=\frac{U_{o,2}A_{o,2}\Delta T_{lm,2}F_2}{U_{o,1}A_{o,1}\Delta T_{lm,1}F_1}=\frac{U_2\Delta T_{lm,2}}{U_1\Delta T_{lm,1}} \qquad (E2.18j)$$

where

$$\Delta T_{lm,2}=\frac{(95-45)-(55-30)}{\ln\left[\dfrac{(95-45)}{(55-30)}\right]}=36.07°C \qquad (E2.18k)$$

and

$$U_{o,2}=\left[\frac{1}{(3,500)\left(\dfrac{\dot{m}_2}{\dot{m}_1}\right)^{0.8}}+1.81\times10^{-4}+(1.2088)\frac{1}{(125)\left(\dfrac{\dot{M}_2}{\dot{M}_1}\right)^{1/3}}\right]^{-1}$$

$$U_{o,2}=\left[\frac{1}{(3,500)m^{0.8}}+1.81\times10^{-4}+(1.2088)\frac{1}{(125)M^{1/3}}\right]^{-1} \qquad (E2.18l)$$

Substituting Equations (E2.18k) and (E2.18l) into (E2.18j) and solving (E2.18h), (E2.18i), and (E2.18j) simultaneously gives

$$M = 0.9065, m = 0.6043, Q = 0.9064$$

Thus, the maximum scale-down is a little less than 10%, and the cw flowrate must be reduced to 60.43% of the design value.

It is important to note the implications of this result on the control and flexibility of operation for this process. If scale-down below 10% is required, then the only feasible way to achieve this is for the process temperature to drop below 55°C. This may not be a problem, but if, for example, the viscosity of the process fluid became too large at temperatures less than 55°C, then this design would need to be changed.

Example 2.19

A reboiler for a column is supplied with saturated steam (in shell) at a temperature of 145°C and reboils the bottom product (in tubes) that leaves the column at 122°C. It is desired to increase the boil-up rate by 15% from the design case. How can this change be accomplished?

Solution

The *T-Q* diagram for the current operation of the reboiler is shown in Figure E2.19.

In looking at the relevant relationships, Equations (2.78), (2.79), and (2.80), and noting that the heats of vaporization will not change, that the area of the exchanger is constant, and the *LMTD* correction factor, *F* = 1:

$$\frac{Q_2}{Q_1}=\frac{\dot{M}_2\lambda_{s2}}{\dot{M}_1\lambda_{s1}}=M \qquad (E2.19a)$$

$$\frac{Q_2}{Q_1}=\frac{\dot{m}_2\lambda_{p2}}{\dot{m}_1\lambda_{p1}}=m \qquad (E2.19b)$$

$$\frac{Q_2}{Q_1}=\frac{U_2A_2\Delta T_{lm,2}F_2}{U_1A_1\Delta T_{lm,1}F_1}=\frac{U_2\Delta T_2}{U_1\Delta T_1}=\frac{U_2}{U_1}\frac{\Delta T_2}{(23)} \qquad (E2.19c)$$

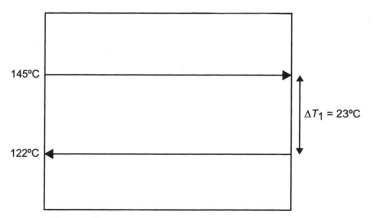

Figure E2.19   *T-Q* diagram for current operation

From Table 2.8, it can be seen that neither the condensing- nor the boiling-side coefficients are a function of flowrate. Thus, ignoring the effects of the changing wall temperature, $U_2/U_1 = 1$. This reduces Equation (E2.19c) to

$$\frac{Q_2}{Q_1} = \frac{\cancel{U}_2}{\cancel{U}_1} \frac{\Delta T_2}{(23)} = \frac{\Delta T_2}{(23)} \tag{E2.19d}$$

It must be concluded that the only way to increase the boil-up rate by 15% (i.e., $Q_2/Q_1 = 1.15$) is to increase ΔT_2 to $(1.15)(23) = \textbf{26.45°C}$. This would most likely be done by increasing the steam temperature (and pressure) to **148.45°C**. Alternatively, the column pressure could be lowered to bring the temperature of the boiling process liquid in the reboiler down to $(122 - 3.45) = \textbf{118.55°C}$. However, that would involve decreasing the pressure at which the column operates, which might not be possible or desirable.

This solution is not quite correct, since the temperature driving forces for boiling and condensation will change slightly, that is, the wall temperature will change. In order to find the change in the wall temperature, the relative magnitudes of the condensing and boiling heat transfer coefficients must be known. This problem is considered in the next example.

Example 2.20

Revisiting Example 2.19, by how much should the steam temperature be increased in order to achieve the 15% increase in boil-up rate for the case when the process- and steam-side heat transfer coefficients are approximately equal?

Assume that this is a non-fouling service, the conduction resistance of the tube wall may be assumed negligible, and $D_i \cong D_o$. In addition, the process fluid may be assumed to be in the convective boiling regime.

Solution

For the original design case, Equation (2.23), reduces to

$$U_{o,1} = \left[\frac{1}{h_o} + \cancel{R}_{fo} + \frac{D_o \ln(\cancel{D}_o / \cancel{D}_i)}{2k_w} + \frac{D_o}{D_i} \cancel{R}_{fi} + \frac{\cancel{D}_o}{\cancel{D}_i} \frac{1}{h_i} \right]^{-1} = \left[\frac{1}{h_{o,1}} + \frac{1}{h_{i,1}} \right]^{-1}$$

and

$$h_{o,1} = h_{i,1} = h$$

Writing an energy balance across the inside and outside films gives

$$\cancel{h}_{o,1}(145 - T_{w,1}) = \cancel{h}_{i,1}(T_{w,1} - 122) \Rightarrow T_{w,1} = 133.5°C \tag{E2.20a}$$

For the new case (15% scale-up), the energy balance becomes

$$h_{o,2}(T_{steam,2} - T_{w,2}) = h_{i,2}(T_{w,2} - 122) \tag{E2.20b}$$

The design relationship for this case is

$$\frac{Q_2}{Q_1} = \frac{U_2 \cancel{A_2} \Delta T_{lm,2} \cancel{F_2}}{U_1 \cancel{A_1} \Delta T_{lm,1} \cancel{F_1}} = \frac{U_2 \Delta T_2}{U_1 \Delta T_1} = \frac{U_2}{U_1} \frac{\Delta T_2}{(23)} = \frac{U_2}{U_1} \frac{(T_{steam,2} - 122)}{(23)} = 1.15 \tag{E2.20c}$$

and from Table 2.8, U_2 can be written in terms of the temperature driving forces as

$$U_2 = \left[\frac{1}{h_{o,2}} + \frac{1}{h_{i,2}} \right]^{-1} = \left[\frac{1}{h_{o,1}\left[\dfrac{T_{steam,2} - T_{w,2}}{T_{steam,1} - T_{w,1}} \right]^{-1/4}} + \frac{1}{h_{i,2}\left[\dfrac{T_{w,2} - T_{process,2}}{T_{w,1} - T_{process,1}} \right]^{2}} \right]^{-1}$$

$$U_2 = \left[\frac{1}{h\left[\dfrac{T_{steam,2} - T_{w,2}}{145 - 133.5} \right]^{-1/4}} + \frac{1}{h\left[\dfrac{T_{w,2} - 122}{133.5 - 122} \right]^{2}} \right]$$

$$U_2 = h\left[\frac{1}{\left[\dfrac{T_{steam,2} - T_{w,2}}{11.5} \right]^{-1/4}} + \frac{1}{\left[\dfrac{T_{w,2} - 122}{11.5} \right]^{2}} \right] \quad \text{and} \quad U_1 = \frac{h}{2} \tag{E2.20d}$$

Substituting the relationships for $h_{o,2}$ and $h_{i,2}$ from Table 2.8 in Equation (E2.20b) gives

$$\cancel{h}_{o,1}\left[\frac{(T_{steam,2} - T_{w,2})}{(145 - 133.5)} \right]^{-1/4}(T_{steam,2} - T_{w,2}) = \cancel{h}_{i,1}\left[\frac{(T_{w,2} - 122)}{(133.5 - 122)} \right]^{2}(T_{w,2} - 122)$$

$$(T_{steam,2} - T_{w,2})^{3/4} = (0.004106)(T_{w,2} - 122)^{3} \tag{E2.20e}$$

Substituting Equation (E2.20d) into (E2.20c) and solving the resulting equation simultaneously with Equation (E2.20e) gives the two unknowns, $T_{steam,2}$ and $T_{w,2}$:

$$T_{steam,2} = 147.9°C, T_{w,2} = 134.0°C$$

Note that the answer differs only a small amount from that obtained in Example 2.19 where $T_{steam,2} = 148.45°C$.

It should be noted that in Example 2.20, an additional relationship was used, namely, a flux balance across the wall given by Equations (E2.20a) and (E2.20b), in order to determine the unknown wall temperature T_w. The final problem considers a situation where the heat transfer coefficients and exchanger area are unknown, but one heat transfer coefficient is known to be limiting.

Example 2.21

A condenser is used to condense saturated steam to saturated liquid at 150°C (shell) using a process stream entering at 30°C and exiting at 50°C (tubes). Assume that the condensation resistance is negligible compared to that of the process stream. The condensation throughput must be decreased by 50%. By what factor must the flowrate of the process stream be reduced?

The T-Q diagram for this problem is shown in Figure E2.21.

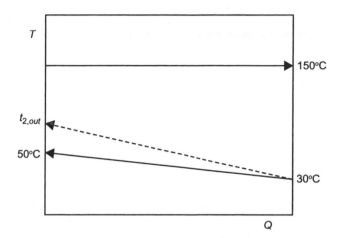

Figure E2.21 T-Q diagram for design case and new case

Solution

In this problem, the area and heat transfer coefficients are not known; nevertheless, there is sufficient information to solve the problem.

$$\frac{Q_2}{Q_1} = \frac{\dot{m}_2 \ell_{p2}(t_{out}-t_{in})_2}{\dot{m}_1 \ell_{p1}(t_{out}-t_{in})_1} \Rightarrow Q = m\frac{(t_{out,2}-30)}{(50-30)} = 0.05 m(t_{out,2}-30) \qquad \text{(E2.21a)}$$

$$\frac{Q_2}{Q_1} = \frac{\dot{M}_2 \lambda_2}{\dot{M}_1 \lambda_1} \Rightarrow Q = M = 0.5 \qquad \text{(E2.21b)}$$

$$\frac{Q_2}{Q_1} = \frac{U_{o,2}A_{o,2}\Delta T_{lm,2}F_2}{U_{o,1}A_{o,1}\Delta T_{lm,1}F_1} = \frac{U_2\Delta T_{lm,2}}{U_1\Delta T_{lm,1}} \qquad \text{(E2.21c)}$$

$$U_1 = h_{i,1} \quad \text{and} \quad U_2 = h_{i,2} = h_{i,1}m^{0.8}$$

$$\therefore \frac{U_2}{U_1} = m^{0.8} \qquad \text{(E2.21d)}$$

$$\Delta T_{lm,1} = \frac{(150-30)-(150-50)}{\ln\left[\dfrac{150-30}{150-50}\right]} = 109.7°C \qquad \text{(E2.21e)}$$

$$\Delta T_{lm,2} = \frac{(150-30)-(150-t_{out,2})}{\ln\left[\dfrac{150-30}{150-t_{out,2}}\right]} = \frac{t_{out,2}-30}{\ln\left[\dfrac{120}{150-t_{out,2}}\right]} \qquad \text{(E2.21f)}$$

Substituting Equations (E2.21d), (E2.21e), and (E2.21f) into Equation (E2.21c) gives

$$\frac{Q_2}{Q_1} = \frac{U_2 \Delta T_{lm,2}}{U_1 \Delta T_{lm,1}} = \frac{m^{0.8}}{109.7} \frac{t_{out,2} - 30}{\ln\left[\dfrac{120}{150 - t_{out,2}}\right]} \tag{E2.21g}$$

Solving Equations (E2.21a), (E2.21b), and (E2.21g) simultaneously gives
$t_{out,2} = 53.3°C$, $m = 0.4291$ or reduce the process stream to 42.91% of design flowrate.

WHAT YOU SHOULD HAVE LEARNED

- The meaning of *LMTD* (ΔT_{lm}) and the ability to calculate it along with the *LMTD* correction factor (F_{N-2N}) for the common heat-exchanger types
- How to calculate the individual heat transfer coefficients for any type of flow inside tubes (h_i) or on the shell side (h_o) of a heat exchanger
- When to use fins to improve the heat transfer coefficient and how to calculate the fin effectiveness (ε_{fin})
- How to compute the overall heat transfer coefficient (U) and apply the design procedure to obtain a preliminary design of a heat exchanger, including the number of tubes and shell diameter and other details of the design such as tube pitch (p), baffle spacing (L_B) and baffle cut (BC), number of shell (N_{shell}) and tube passes (N_{tube}), and tube length (L_{tube})
- Given a heat-exchanger design, how to determine the performance of the exchanger given a change in tube-side or shell-side conditions using the concept of ratios

NOTATION

Symbol	Definition	SI Units
A	area	m^2
A_c	cross-sectional area	m^2
b	fin spacing	m
BC	baffle cut (% of shell diameter)	
B_o	boiling number	
C_f	material constant for surfaces used in boiling heat transfer	
C_p, c_p	heat capacity	kJ/kg/°C or kJ/kmol/°C
D	diameter	m

Symbol	Definition	SI Units
f	factor used in convective boiling correlation	
F	correction for multipass heat exchangers	
f	friction factor	
g	acceleration due to gravity	m/s^2
G	superficial mass velocity	$kg/m^2/s$
h	individual heat transfer coefficient	$W/m^2/K$
H	enthalpy or specific enthalpy	kJ or kJ/kg
k	thermal conductivity	W/m/K
L	length or spacing (for baffles)	m
m	dimensionless parameter used in fin effectivenes	
m	ratio of tube side flows in performance problem analysis	
M	molecular weight	g/mol
M	ratio of shell side flows in performance problem analysis	
$\dot{M}, \dot{m}$	flowrate	kg/s
N	number	
Nu	Nusselt number	
p	tube pitch (distance between centers of adjacent tubes)	m
P	dimensionless temperature approach used in log-mean temperature correction factor	
P	pressure	bar or kPa
Pr	Prandtl number	
Q	rate of heat transfer	W or MJ/h
R	ratio of heat capacities in shell and tube exchanger used in log-mean temperature correction factor	
R	heat transfer resistance	m^2K/W
Re	Reynolds number	
s	suppression factor used in convective boiling correlation	

Symbol	Definition	SI Units
t	time	s, min, h, or yr
T, t	temperature	K, R, °C, or °F
u	flow velocity	m/s
U	overall heat transfer coefficient	w/m²/K
$\dot{v}$	volumetric flowrate	m³/s
W	width of fin	m
x	wall or film thickness	m
x	mass faction of vapor in stream	
X_{tp}	Martinelli's two-phase flow parameter	
z	distance	m

GREEK SYMBOLS

δ	condensing film thickness or fin thickness	m
Δx	difference in variable x $(x_2 - x_1)$	
ε	void fraction	
ε	emissivity	
ε	effectiveness (for fins)	
λ, λ'	heat of vaporization, adjusted heat of vaporization	kJ/kg
μ	viscosity	kg/m/s
σ	surface tension	N/m
σ	Stefan-Boltzmann constant	W/m⁻²/ K⁻⁴
ρ	density	kg/m³

SUBSCRIPTS

1	base or design case or input
2	new case or output
b	baffle
$bare$	bare fin
$base$	fin base

c	cold
c	corrected
c	critical
cb	convective boiling
cocurrent	designating a cocurrent arrangement for an S-T heat exchanger
countercurrent	designating a counter-current arrangement for an S-T heat exchanger
fb	film boiling
fin	fin
$film$	film
h	hot
H	hydraulic
i	index
i	inside
in	inlet
l	liquid
max	maximum
o	outside
out	outlet
pb	pool boiling
r	reduced (pressure)
rad	radiation
s	surface
sat	saturated
$s, shell$	shell (side) of heat exchanger
$t, tube$	tube (side) of heat exchanger
tp	tube passes
v	vapor
w	wall
y	designation for exchanger type in effectiveness factor, y = 1–2, 2–4, 3–6, etc.

REFERENCES

Bell, K. J., and A. C. Mueller. 2014. *Wolverine Engineering Data Book II*. Ulm, Germany: Wieland-Werke AG. http://ichemengineer-heattransfer.blogspot.com/2014/11/wolverine-heat-transfer-data-book-ii-by.html.

Bennett, C. O., and J. E. Myers. 1983. *Momentum, Heat, and Mass Transfer*, 3rd ed. New York: McGraw-Hill.

Bowman, R. A. 1936. "Mean Temperature Difference Correction in Multipass Exchangers." *Ind Eng Chem* 28: 541–544.

Bowman, R. A., A. C. Mueller, and W. M. Nagle. 1940. "Mean Temperature Difference in Design." *Trans ASME* 62: 283–294.

Bromley, L. A. 1950. "Heat Transfer in Stable Film Boiling." *Chem Eng Progr* 46: 221–227.

Chen, C. Y., G. A. Hawkins, and H. L. Solberg. 1946. "Heat Transfer in Annuli." *Trans ASME* 68: 99–106.

Churchill, S. W., and M. Bernstein. 1977. "A Correlating Equation for Forced Convection from Gases and Liquids to a Circular Cylinder in Crossflow." *J Heat Transfer, Trans ASME* 99: 300–306.

Cichelli, M. T., and C. F. Bonilla. 1945. "Heat Transfer to Liquids Boiling Under Pressure." *Trans Am Inst Chem Engrs* 41: 755–787.

Clark, S. H., and W. M. Kays. 1953. "Laminar Flow Forced Convection in Rectangular Ducts." *Trans ASME* 75: 859–866.

Cooper, M. G. 1984. "Saturated Nucleate Pool Boiling—A Simple Correlation." First UK National Heat Transfer Conf., *IChemE Symp Ser*, No. 86, 2, 785–793.

Couper, J. R., W. R. Penney, J. R. Fair, and S. M. Walas. 2012. *Chemical Process Equipment—Selection and Design*, 3rd ed. Oxford, UK: Butterworth-Heinmann.

Dittus, F. W., and L. M. K. Boelter. 1985. "Heat Transfer in Automobile Radiators of Tubular Type." *Int Comm Heat Mass Transfer* 12: 3–22; modified from *Univ. Calif. Publ. in Engineering* 2: 443–461 (1930).

Geankoplis, C. J. 2003. *Transport Processes and Separation Process Principles*, 4th ed. Upper Saddle River, NJ: Prentice Hall.

Green, D. W., and R. H. Perry. 2008. *Perry's Chemical Engineers' Handbook*, 8th ed. New York: McGraw-Hill.

Gungor, K. E., and R. H. S. Winterton. 1986. "A General Correlation for Flow Boiling in Tubes and Annuli." *Int J Heat Mass Transfer* 29: 351–358.

Hausen, Z. 1943. *Ver Deut Beth Verfahrenstech* 4: 91.

Kern, D. Q. 1950. *Process Heat Transfer*. New York: McGraw-Hill.

McAdams, W. H. 1954. *Heat Transmission*, 3rd ed. New York: McGraw-Hill.

Mukherjee, R. 1998. "Effectively Design Shell-and-Tube Heat Exchangers." *Chem Eng Prog* 2: 21–37.

Rohsenow, W. M. (Ed.). 1964. *Developments of Heat Transfer.* Cambridge, MA: MIT Press.

Shelton, S. M. "Thermal Conductivity of Some Irons and Steels over the Temperature Range 100 to 500°C." *J Res Nat Stand* 12 (April 1934).

Sieder, E. N., and G. E. Tate. 1936. "Heat Transfer and Pressure Drop of Liquids in Tubes." *Ind Eng Chem* 28: 1429–1435.

Tinker, T. 1958. "Shellside Characteristics of Shell-Tube Heat Exchangers—A Simplified Rating System for Commercial Heat Exchangers." *Trans ASME* 80: 36–52.

Tubular Exchanger Manufacturers Association. 2013. *Standards of the Tubular Exchanger Manufacturers Association,* 9th ed. Tarrytown, NY: Tubular Exchanger Manufacturers Association.

Žukauskas, A. 1972. "Heat Transfer from Tubes in Crossflow." *Advances in Heat Transfer* 8: 93–160.

Zuber, N. 1958. "On the Stability of Boiling Heat Transfer." *Trans ASME* 80: 711.

APPENDIX 2.A   HEAT–EXCHANGER EFFECTIVENESS CHARTS

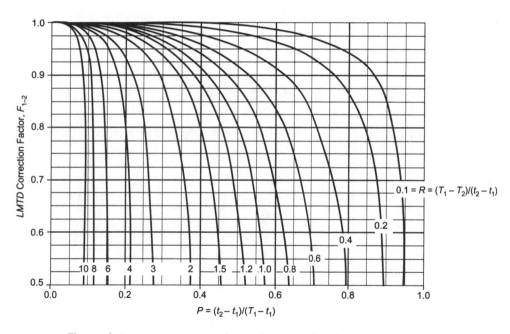

Figure A.1   *LMTD* correction factors for 1–2 shell-and-tube heat exchanger

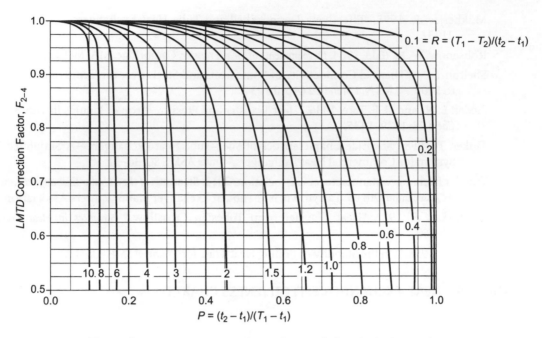

Figure A.2   *LMTD* correction factors for 2–4 shell-and tube-heat exchanger

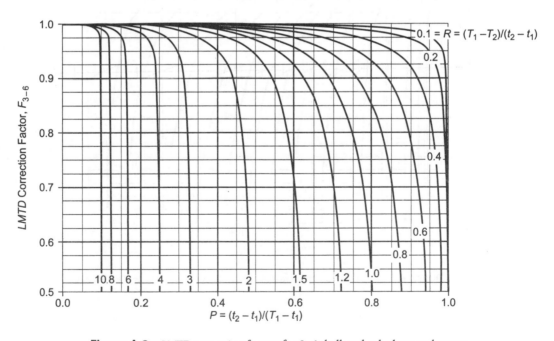

Figure A.3   *LMTD* correction factors for 3–6 shell-and-tube heat exchanger

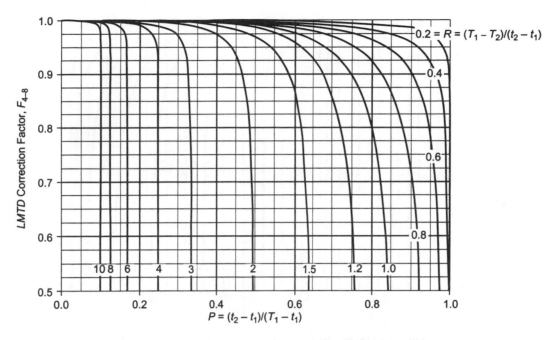

Figure A.4   *LMTD* correction factors for 4–8 shell-and-tube heat exchanger

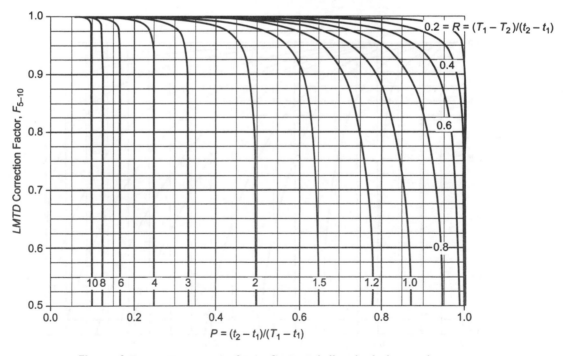

Figure A.5   *LMTD* correction factors for 5–10 shell-and-tube heat exchanger

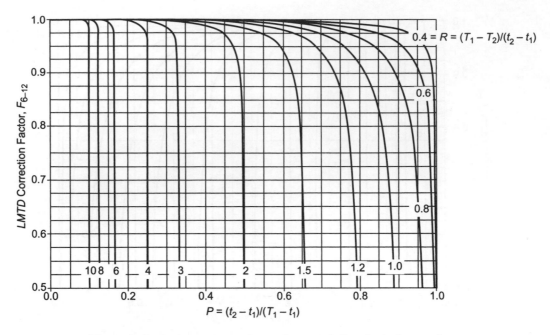

Figure A.6   *LMTD* correction factors for 6–12 shell-and-tube heat exchanger

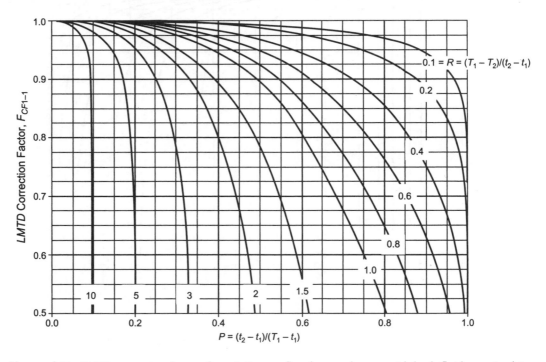

Figure A.7   *LMTD* correction factors for a 1–1 cross-flow heat exchanger with both fluids unmixed (use this for air coolers with a single tube pass)

For *LMTD* correction factors for other configurations, see CheCalc's *LMTD* Correction Factor Charts at http://checalc.com/solved/LMTD_Chart.html.

APPENDIX 2.B   DERIVATION OF FIN EFFECTIVENESS FOR A RECTANGULAR FIN

Rectangular Fin with Constant Thickness

The heat transfer process occurring in the fin consists of conduction through the fin and convectional loss from the side of the fin. The geometry and notation used in this derivation are shown in Figure B.1. A differential flux balance on a slice through the fin yields

$$\delta W \, q\big|_x = \delta W \, q\big|_{x+\Delta x} + 2W \, \Delta x h (T_x - T_{fluid}) \tag{B.2.1}$$

where W is the width of the fin, δ is the fin thickness, h is the heat transfer coefficient between the fin and the surrounding fluid, and T_{fluid} is the temperature of the surrounding fluid. Equation (B.2.1) means that the heat flow into a differential section of the fin by conduction equals the heat flow out of the differential section of the fin by conduction down the length of the fin plus the heat exchanged with the surroundings in that differential section of the fin by convection. It is important to remember that heat is exchanged with the surroundings on both the top and bottom faces of the fin. Dividing through by $\delta W \Delta x$ and taking the limit as $\Delta x \to 0$ yields the following differential equation:

$$\frac{dq}{dx} + \frac{2h}{\delta}(T_x - T_{fluid}) = 0 \tag{B.2.2}$$

Substituting Fourier's law of heat conduction $q = -k\dfrac{dT}{dx}$ in Equation (B.2.2) gives

$$\frac{d^2T}{dx^2} = \frac{2h}{\delta k}(T_x - T_{fluid}) \tag{B.2.3}$$

Introducing the new variable $T' = T_x - T_{fluid}$ gives

$$\frac{d^2T'}{dx^2} = \frac{2h}{\delta k}T' \tag{B.2.4}$$

The boundary condition at $x = 0$ is $T = T_0$ or $T' = T_0 - T_{fluid}$. Several possible boundary conditions exist for $x = L$, the simplest being that $dT'/dx = 0$, which means that there is no heat lost from

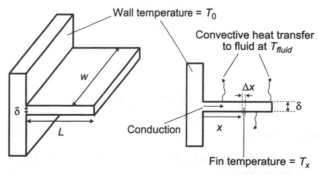

Figure B.1   Energy balance on a section of rectangular fin

the tip of the fin (adiabatic fin tip). This is a good approximation for the case where the tip of the fin has a very small surface area compared to the faces of the fin. The second boundary condition is reasonable and accurate for long, thin fins, but for short thick fins, the amount of heat lost from the fin tip may be significant. Nevertheless, the adiabatic fin tip assumption may be used for all types of fin with an adjustment that is discussed in Section 2.6. With these boundary conditions, Equation (B.2.4) can be integrated as follows:

$$\frac{d^2T'}{dx^2} = \frac{2h}{\delta k}T' \Rightarrow \frac{d^2T'}{dx^2} - m^2T' = 0$$

where $m = (2h/\delta k)^{1/2}$

$$T' = A\sinh(mx) + B\cosh(mx)$$

$$x = 0 \quad T' = T_0 - T_{fluid} = A\sinh(0) + B\cosh(0) \Rightarrow B = T_0 - T_{fluid}$$

$$x = L \quad \frac{dT'}{dx} = 0 = Am\cosh(mL) + Bm\sinh(mL) \Rightarrow A = -B\tanh(mL)$$

$$\therefore T' = A\sinh(mx) + B\cosh(mx) \Rightarrow T' = -B\tanh(mL)\sinh(mx) + B\cosh(mx)$$

$$\frac{T_x - T_{fluid}}{T_0 - T_{fluid}} = \cosh(mx) - \tanh(mL)\sinh(mx) = \frac{\cosh(mx)\cosh(mL) - \sinh(mL)\sinh(mx)}{\cosh(mL)}$$

$$\frac{T_x - T_{fluid}}{T_0 - T_{fluid}} = \frac{\cosh[m(L-x)]}{\cosh(mL)} \tag{B.2.5}$$

The total heat transferred into the fin is $(\delta W)q_{x=0}$ or

$$\delta W\, q\big|_{x=0} = -\delta Wk\frac{dT}{dx}\bigg|_{x=0} = -\delta Wk\left[Am\cosh(0) + Bm\sinh(0)\right] = \delta Wkm\tanh(mL)(T_0 - T_{fluid}) \tag{B.2.6}$$

At this point, the efficiency of the fin is introduced and is defined as

$$\varepsilon_{fin} = \frac{\text{Heat transferred through fin}}{\text{Heat transferred through fin if fin temperature were } T_0 \text{ throughout}} \tag{B.2.7}$$

Substituting the result from Equation (B.2.6) into Equation (B.2.7) gives

$$\varepsilon_{fin} = \frac{\delta Wkm\tanh(mL)(T_0 - T_{fluid})}{2hWL(T_0 - T_{fluid})} = \frac{\delta km\tanh(mL)}{2hL} = \frac{\tanh(mL)}{mL} \tag{B.2.8}$$

This result is plotted in Figure 2.32.

PROBLEMS

Short Answer Problems

1. Why is it necessary to include an *LMTD* correction factor (*F*) into the design equation for most practical heat exchangers?

2. Explain why you agree or disagree with this statement: Because the heat transfer coefficients for gases are always much lower than for liquids, the limiting resistance for condensers (where a gas is turned into a liquid) is always on the condensing side of the heat exchanger.

3. For shell-and-tube exchangers in which the flow is turbulent and there is no phase change taking place for either stream, how does the heat transfer coefficient change with a 50% increase in the mass flowrate for the

   a. Shell-side fluid?

   b. Tube-side fluid?

4. Repeat Problem 2.3 assuming laminar flow for both fluids.

5. Would you expect the heat transfer coefficient for a given fluid to be higher if the fluid were in laminar flow or turbulent flow? Explain your answer.

6. Give three reasons for placing a specific fluid on the tube side of an S-T heat exchanger.

7. Give three reasons for placing a specific fluid on the shell side of an S-T heat exchanger.

8. Why are fins (extended surfaces) used in some heat exchangers?

9. Give explanations for the following terms for S-T heat exchangers:

   a. Baffle cut

   b. Number of tube passes

   c. Tube pitch

   d. Tube arrangement

10. State the basic design equation for an S-T heat exchanger in which there are no phase changes, and draw a sketch of the T-Q diagram with the relevant terms shown.

11. Draw the T-Q diagram for the following cases:

    a. Condensing (pure) vapor using cooling water

    b. Distillation reboiler using condensing steam as the heating media

    c. Process liquid stream cooled by a another process stream

Problems to Solve

12. A process fluid (Stream 1) (C_p = 2100 J/kg/K) enters a heat exchanger at a rate of 3.4 kg/s and at a temperature of 135°C. This stream is to be cooled with another process stream (Stream 2) (c_p = 2450 J/kg/K) flowing at a rate of 2.65 kg/s and entering the heat exchanger at a temperature of 55°C. Determine the following (you may assume that the heat capacities of both streams are constant):

    a. The exit temperature of Stream 1 if pure countercurrent flow occurs in the exchanger and the minimum approach temperature between the streams anywhere in the heat exchanger is 10°C

    b. The exit temperature of Stream 1 if pure cocurrent flow occurs in the exchanger and the minimum approach temperature between the streams anywhere in the heat exchanger is 10°C

13. Repeat Problem 2.12 for the case when the specific heat capacities of both streams vary linearly with temperature with the following values:

    a. Stream 1: C_p = 2000 + 3(T − 100) J/kg/K

    b. Stream 2: c_p = 2425 + 5(T − 50) J/kg/K

14. The T-Q diagrams for three S-T heat-exchanger designs are given in Figure P2.14(a–c). Determine the number of S-T passes required to accomplish these designs.

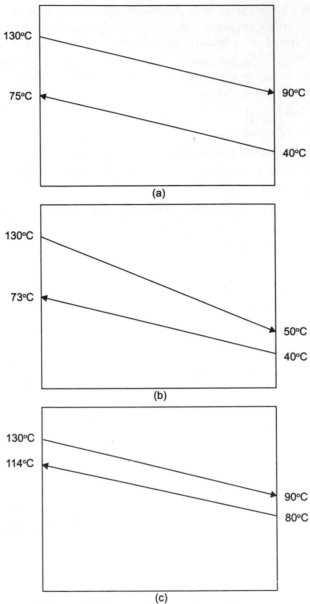

Figure P2.14 *T*-*Q* diagrams for Problem 2.14

15. At a given point in a thin-walled heat exchanger, the shell-side fluid is at 100°C and the tube side fluid is at 20°C. Ignoring any fouling resistances and the resistance of the wall, determine the wall temperature for the following cases:

   a. The inside and outside heat transfer coefficients are equal.

   b. The inside coefficient has a value 3 times that of the outside coefficient.

   c. The inside coefficient has a value ⅓ that of the outside coefficient.

   d. The inside coefficient is limiting.

16. In a heat exchanger, water (c_p = 4200 J/kg/K) flows at a rate of 1.5 kg/s, and toluene (c_p = 1953 J/kg/K) flows at a rate of 2.2 kg/s. The water enters at 50°C and leaves at 90°C, and the toluene enters at 190°C. For this situation, do the following:

    a. Sketch the T–Q diagram if the flows are countercurrent.

    b. Sketch the T–Q diagram if the flows are concurrent.

17. At a given location in a double-pipe heat exchanger, the bulk temperature of the fluid in the annulus is 100°C, and the bulk temperature of the fluid in the inner pipe is 20°C. The tube wall is very thin, so the resistance due to the metal wall may be ignored. Likewise, fluid fouling resistances may also be ignored. The individual heat transfer coefficients at this point in the heat exchanger are

$$h_i = 100 \text{ W/m}^2/\text{K and } h_o = 500 \text{ W/m}^2/\text{K}$$

    a. What is the temperature of the wall at this location?

    b. What is the heat flux across the wall at this location?

    c. If there was a fouling resistance of 0.001 m²K/W on the inside surface of the inner pipe, what would the temperature of the wall be at this location?

18. A single phase fluid ($\dot{m}$ = 2.5 kg/s and c_p = 1800 J/kg/K) is to be cooled from 175°C to 75°C by exchanging heat with another single phase fluid ($\dot{m}$ = 2.0 kg/s and c_p = 2250 J/kg/K), which is to be heated from 50°C to 150°C. You have the choice of one of the following five heat exchangers:

	Shell Passes	Tube Passes
Exchanger 1	1	2
Exchanger 2	2	4
Exchanger 3	3	6
Exchanger 4	4	8
Exchanger 5	5	10

    Which exchanger do you recommend using for this service? Carefully explain your answer.

(For the following problems and where information on organic materials is not given in the problem, the physical properties may be obtained from a process simulator using the SRK thermodynamics package or from using suitable handbooks.)

19. What is the critical nucleate boiling flux for water at 20 atm?

20. Water flows inside a long ¾-in, 16 BWG tube at an average temperature of 110°F. Determine the inside heat transfer coefficient for the following cases:

    a. Velocity = 0.1 ft/s

    b. Velocity = 3 ft/s

21. Use Equations (2.47) and (2.48) to estimate the heat flux and heat transfer coefficient for boiling acetone at 1 atm pressure for a temperature driving force of 10°C. You may assume a C_f value of 0.01 for this problem.

22. Use the Sieder-Tate equation to determine the inside heat transfer coefficient for a fluid flowing inside a 16 ft long, 1-in tube (14 BWG) at a velocity of 2.5 ft/s that is being cooled and has an average temperature of 176°F and an average wall temperature of 104°F. Consider the following fluids:

    a. Acetone (liquid)

    b. Isopropanol (liquid)

    c. Methane at 1 atm pressure—use a velocity of 50 ft/s

    d. Compare the results from Parts (a), (b), and (c) using the Dittus-Boelter equation

23. Water at 5 atm pressure and 30°C flows at an inlet velocity of 1 m/s into a square duct 2.5 m long that has a cross section of 25 mm by 25 mm. The wall of the duct is maintained at a temperature of 120°C by condensing steam. What will the temperature of the water be when it exits the duct?

24. An S-T condenser contains six rows of five copper tubes per row on a square pitch. The tubes are ¾-in, 14 BWG, and 3 m long. Cooling water flows through the tubes such that h_i = 2000 W/m²/K. The water flow is high so that the temperature on the tube side may be assumed to be constant at 35°C. Pure, saturated steam at 2 bar is condensing on the shell side. Determine the capacity of the condenser (Q in kW) if the condenser tubes are oriented (a) vertically, and (b) horizontally.

25. Liquid dimethyl ether (DME) flows across the outside of a bank of tubes. It enters at 100°C and leaves at 50°C. The DME enters at a flowrate of 20 kg/s, the shell diameter is 15 in, the baffle spacing is 6 in, the baffle cut (BC) is 15%, and 1-in OD tubes on a 1.25-in pitch are used. The fluid (cooling water) inside the tubes may be assumed to be at a constant temperature of 35°C and the inside coefficient is expected to be much higher than the shell-side coefficient, and thus the wall temperature may be taken as 35°C.

    Use Kern's method to determine the average heat transfer coefficient for the shell side for the following arrangements:

    a. Square pitch

    b. Equilateral triangular pitch

26. A double-pipe heat exchanger consists of a length of 1-in, schedule-40 pipe inside an equal length of 3 in, schedule-40 pipe. Water flows at a velocity of 1.393 m/s in the annular region between the pipes and enters the heat exchanger at 30°C. Oil flows in the inner pipe at an average velocity of 1 m/s and enters at 100°C. The water and oil flow countercurrently. Following are the properties of the water and oil:

Water @30°C	Oil @100°C	Oil @50°C
ρ = 998 kg/m³	ρ = 690 kg/m³	ρ = 727 kg/m³
c_p = 4216 J/kg/K	c_p = 2421 J/kg/K	c_p = 2201 J/kg/K
k = 0.60 W/m/K	k = 0.1179 W/m/K	k = 0.1295 W/m/K
μ = 700 × 10⁻⁶ kg/m/s	μ = 511 × 10⁻⁶ kg/m/s	μ = 945 × 10⁻⁶ kg/m/s

Determine how long the pipe lengths must be in order for the oil to leave the exchanger at 50°C. You may assume that both fluids are clean and that there is no fouling or tube wall resistances.

27. A 3 m long, 1.25-in, BWG 14 copper tube is used to condense ethanol at 3 bar pressure. Cooling water at 30°C flows through the inside of the tube at a high rate such that the wall temperature may be assumed to be 30°C and the inside heat transfer coefficient may be assumed to be much greater than the outside coefficient. Determine how much vapor will condense (kg/h) assuming no fouling resistance for the following cases:

    a. The tube is oriented vertically.

    b. The tube is oriented horizontally.

28. Air (1 atm and 30°C) flows crosswise over a bare copper tube (1-in, BWG 16). The approach velocity of the air is 20 m/s. Water enters the tube at 140°C and leaves the tube at an average temperature of 80°C. The average velocity of the water in the tube is 1 m/s. Determine the length of tube required to cool the water to the desired 80°C.

29. Oil flows inside a thin-walled copper tube of diameter D_i = 30 mm. Steam condenses on the outside of the tube, and the tube wall temperature may be assumed to be constant at the temperature of the steam (150°C). The oil enters the tube at 30°C and a flow rate of 1.6 kg/s. The properties of the oil are as follows:

	30°C	50°C	150°C
Density, ρ (kg/m³)	886	882	864
Thermal conductivity, k (W/m/K)	0.2	0.2	0.18
Specific heat capacity, c_p (J/kg/K)	2000	2000	1950
Viscosity, μ (kg/m/s)	5×10^{-3}	4×10^{-3}	4×10^{-4}

    a. Calculate the inside heat transfer coefficient, h_i, using the appropriate correlation. You should assume that the bulk oil temperature changes from 30°C to 50°C along the tube.

    b. Using the result from Part (a), calculate the length of tube required to heat the oil from 30°C to 50°C.

30. Use the approach given in Example 2.12 to determine the length of tubes necessary to vaporize an organic liquid (acetic acid at 1 bar) flowing inside a set of vertical ¾-in, BWG 16 tubes using condensing steam on the outside of the tubes to provide the energy for vaporization. The major resistance to heat transfer is expected to be on the inside of the tubes, and the wall temperature, as a first approximation, may be assumed to be at the temperature of the condensing steam, which for this case is 125°C. It may be assumed that the value of the vapor quality, x, varies from 0.05 to 0.95 in the tube. The physical parameters for acetic acid are

    ρ_v = 1.893 kg/m³, ρ_l = 939.7 kg/m³, μ_v = 11.32 × 10⁻⁶ kg/m/s, μ_l = 390.1 × 10⁻⁶ kg/m/s, T_{sat} = 117.6°C, P_c = 57.9 bar, M = 60, k_l = 0.1423 W/m/K, λ = 405 kJ/kg, $c_{p,l}$ = 2.434 kJ/kg/K, $c_{p,v}$ = 1.319 kJ/kg/K, λ = 0.04 kg/s/tube

31. Repeat Problem 2.28 to find the length of a tube fitted with 2-in diameter, 1/16-in thick annular fins spaced a distance of 3/16-in apart.

32. An air heater consists of a shell-and-tube heat exchanger with 24 longitudinal fins on the outside of the tubes. The tubes are 1.5-in, 14 BWG, and the fins are 0.75 mm thick and are 15 mm long. A total of 8 fins are spaced uniformly around the circumference of each tube. The tubes

are made from carbon steel, and their length is 3 m. Air at 0.8 kg/s is being heated from 30°C to 200°C in the shell, and the heat transfer coefficient may be taken as h_o = 25 W/m²/K (without fins). Steam is condensing at 254°C in the tubes, and at that temperature λ = 1700 kJ/kg. The tube-side heat transfer coefficient may be taken as h_i = 6000 W/m²/K, and no fouling occurs for this service.

a. Calculate the overall heat transfer coefficient with and without fins.

b. Calculate the heat transfer area and the number of tubes needed with and without fins.

33. Your assignment is to design a replacement condenser for an existing distillation column. Space constraints dictate a vertical-tube condenser with a maximum height (equals tube length) of 3 m. An organic is condensed at a rate of 12,000 kg/h at a temperature of 75°C, and cooling water is used, entering at 30°C and exiting at 40°C. The person you are replacing has done some preliminary calculations suggesting that a 1–4 exchanger (water in tubes) using 1-in 16 BWG copper tubes on 1.25-in equilateral triangular pitch with a 37-in shell diameter would be suitable. However, there are only partial calculations to support this claim, and the person who performed the original design is unavailable for consultation. Complete the detailed heat transfer calculations to evaluate the suitability of this heat-exchanger design.

Data for condensing organic:

ρ_f = 800 kg/m³

ρ_g = 5.3 kg/m³

λ = 800 kJ/kg

c_{pl} = 2600 J/kg/K

k_f = 0.15 W/m/K

μ_f = 0.4 × 10⁻³ kg/m/s

Heat transfer coefficient for tube side: 6000 W/m²/K

Typical fouling coefficient for plant cooling water: 1200 W/m²/K (=1/R_{fi})

Assume no fouling on condensing side

34. A 1–2 shell-and-tube heat exchanger has the following dimensions:

Tube length: 20 ft

Tube diameter: 1-in, BWG 14 carbon steel (k_{cs} = 45 W/m/K)

Number of tubes in shell: 608

Shell diameter: 35 in

Tube arrangement: Triangular pitch, center-to-center = 1.25 in

Number of baffles: 19

Baffle spacing: 1 ft

Baffle: Horizontal baffle with baffle cut = 25%

The fluids in the shell and tube sides of the exchanger have the following properties:

	Shell Side	Tube Side
Inlet temperature (°C)	120	30
Mass flowrate, $\dot{m}$(kg/s)	120	180

	Shell Side	Tube Side
Specific heat capacity, c_p (kJ/kg/K)	2.0	4.2
Thermal conductivity, k (W/m/K)	0.2	0.61
Density, ρ (kg/m³)	850	1000
Viscosity, μ (kg/m/s)	5.0×10^{-4}	0.72×10^{-3}

It may be assumed that neither fluid changes phase, and the viscosity correction factor at the wall may be ignored for both fluids. Determine the following:

a. The inside heat transfer coefficient

b. The shell-side heat transfer coefficient

c. The overall heat transfer coefficient (assuming that fouling may be ignored)

d. The exit temperatures of both fluids

35. A 1–2 S-T heat exchanger is used to cool oil in the tubes from 91°C to 51°C using cooling water at 30°C. In the design case, the water exits at 40°C. The resistances on the water and oil sides are equal. What are the new cooling water outlet temperature and the required cooling water flowrate if the oil throughput must be increased by 25% but the outlet temperature must be maintained at 51°C?

36. Repeat Problem 2.35 if the oil side provides 80% of the total resistance to heat transfer.

37. Repeat Problem 2.35 for the case when the oil throughput must be increased by 25% but the cooling water flow rate remains unchanged from the base case. Determine the new outlet temperatures for both the process and cooling water streams?

38. For the situation in Problem 2.35, suppose that the flowrate of the process stream must be reduced while keeping the process exit temperature at 51°C. Therefore, it will be necessary to reduce the flow of the cooling water stream. Determine the maximum scale-down of the process fluid that can occur without the exit cooling water temperature exceeding the limit of 45°C (when excessive fouling is known to occur). Plot the results as the ratio of the process stream flowrate to the base case value (x-axis) versus the cooling water exit temperature.

39. A reboiler is a heat exchanger used to add heat to a distillation column. In a typical reboiler, an almost pure material is vaporized at constant temperature, with the energy supplied by condensing steam at constant temperature. Suppose that steam is condensing at 254°C to vaporize an organic at 234°C. It is desired to scale up the throughput of the distillation column by 25%, meaning that 25% more organic must be vaporized. What will be the new operating conditions in the reboiler (numerical value for temperature, qualitative answer for pressure)? Suggest at least two possible answers.

40. In an S-T heat exchanger, initially $h_o = 500$ W/m²/K and $h_i = 1500$ W/m²/K. If the mass flowrate of the tube-side stream is increased by 30%, what change in the mass flowrate of the shell-side stream is required to keep the overall heat transfer coefficient constant?

41. A reaction occurs in an S-T reactor. One type of S-T reactor is essentially a heat exchanger with catalyst packed in the tubes. For an exothermic reaction, heat is removed by circulating a heat transfer fluid through the shell. In this situation, the reaction occurs isothermally at 510°C. The Dowtherm™ always enters the shell at 350°C. In the design (base) case, it exits at 400°C. In the base case, the heat transfer resistance on the reactor side is equal to that on the Dowtherm side. If it is required to increase throughput in the reactor by 25%, what is the required Dowtherm flowrate and the new Dowtherm exit temperature? You should assume that the reaction temperature remains at 510°C.

42. In Problem 2.41, the reactor is now a fluidized bed where Dowtherm circulates through tubes in the reactor with the reaction in the shell. The Dowtherm then flows to a heat exchanger in which boiler feed water is vaporized on the shell side to high-pressure steam at 254°C. This removes the heat absorbed by the Dowtherm stream in the reactor so the Dowtherm can be recirculated to the reactor. So, the Dowtherm is in a closed loop. The resistance in the steam boiler is all on the Dowtherm side. In the base case for the reactor, the resistance on the reactor side is four times that on the Dowtherm side. The desired increase in production can be accomplished by adding 25% more catalyst to the bed and operating at the same temperature, which is what is assumed to happen in this problem.

   a. Write the six equations needed to model the performance of this system.

   b. How many unknowns are there? Solve for all the unknowns.

   c. If the temperature of the reactor is to be maintained at 510°C, determine the amount of process scale-up and all other unknowns for the following cases:
      i.   10% increase in Dowtherm flowrate
      ii. 25% increase in Dowtherm flowrate
      iii. 50% increase in Dowtherm flowrate

   d. Plot the results from Parts (b) and (c) in the form of a performance curve in which the amount of process scale-up (Q_2/Q_1 on the y-axis) is plotted as a function of the increase in Dowtherm flowrate ($M_{DT,2}/M_{DT,1}$ on the x-axis).

43. It is necessary to decrease the capacity of an existing distillation column by 30%. As a consequence, the amount of liquid condensed in the shell of the overhead condenser must decrease by the same amount (30%). In this condenser, cooling water (in tubes) is available at 30°C, and, under present operating conditions, exits the condenser at 40°C. The maximum allowable return temperature without a financial penalty assessed to your process is 45°C. Condensation takes place at 85°C.

   a. If the limiting resistance is on the cooling water side, what is the maximum scale-down possible based on the condenser conditions without incurring a financial penalty? What is the new outlet temperature of cooling water? By what factor must the cooling water flow change?

   b. Repeat Part (a) if the resistances are such that the cooling water heat transfer coefficient is three times the condensing heat transfer coefficient. You may assume that the value of the condensing heat transfer coefficient does not change appreciably from the design case. Does your solution exceed the maximum cooling water return temperature of 45°C? If so, can you suggest a way to decrease the condenser duty by 30% that would not violate the cooling water return temperature constraint?

44. A reaction occurs in a well-mixed fluidized bed reactor maintained at 450°C. Heat is removed by Dowtherm A™ circulating through a coil in the fluidized bed. In the design or base case, the Dowtherm enters and exits the reactor at 320°C and 390°C, respectively. In the base case, the heat transfer resistance on the reactor side is two times that on the Dowtherm side. It may be assumed that for the fluidized bed, the heat transfer coefficient on the fluidized side is essentially constant and independent of the throughput.

   The Dowtherm is cooled in an external exchanger that produces steam at 900 psig (T_{sat} = 279°C). The Dowtherm pump limits the maximum increase in Dowtherm flowrate, so the pump limits the heat removal rate based on its pump and system curves. For the current situation, the maximum increase in Dowtherm flowrate through the reactor and boiler is estimated to be 34%. By how much can the process be scaled-up? You may assume that the limiting heat transfer coefficient in the steam boiler is Dowtherm that flows through the tube side of the exchanger.

CHAPTER

3

Separation Equipment

> **WHAT YOU WILL LEARN**
> - The separation basis and separating agent for the most common chemical engineering separations
> - How to determine the size of tray columns and packed columns
> - The key design parameters affecting tray columns and packed columns
> - The internals of tray and packed columns
> - The impact of the reboiler and condenser on the design and performance of distillation columns
> - The economic trade-offs for tray and packed columns
> - The performance of existing tray and packed columns
> - The types of equipment used in extraction, their advantages and their disadvantages
> - The type of equipment used for gas-permeation membrane separations

3.0 INTRODUCTION

The purpose of this chapter is to introduce the fundamental relationships needed to design separation devices. Then, the design and performance of equipment used for the most common chemical process separations is discussed. The details of the typical graphical methods taught in basic separations courses are presented. However, the use of these graphs to provide a conceptual understanding of the behavior of separation equipment is emphasized. This chapter is not designed to replace a complete text on separation processes (Wankat, 2017; Seader, Henley, and Roper, 2011). It is meant to provide a conceptual summary of typical chemical engineering separations and provide equipment information that complements existing separation processes textbooks. The focus is on binary distillation, binary gas permeation, and absorption and stripping involving one solute and two solvents. The overriding concepts affecting these separations can be learned from these simple cases and are generally applicable to more complex systems.

Separations require a basis and a separating agent. The separation basis is the physical property being exploited. For example, when drying hair with a hair dryer, the difference in boiling points (or volatility) between water and hair is the separation basis. Distillation also exploits the difference in boiling points between components, as most students have seen in organic chemistry lab.

Table 3.1   Typical Chemical Engineering Separations

Separation	Basis	Separating Agent
Distillation	Volatility difference	Energy
Absorption, stripping, extraction, leaching	Solubility difference	Mass (additional phase)
Crystallization (from melt)	Melting point difference	Energy
Crystallization (from solution)	Solubility in solution	Energy or mass (whatever changes solubility)
Adsorption, ion exchange	Difference in surface equilibrium	Solid adsorbent
Gas permeation membranes	Different rate of mass transfer through membrane	Membrane

This illustrates the subtle differences in the nomenclature used for separations. Distillation refers to the separation where both components can vaporize at typical operating conditions. Evaporation refers to the separation where one component does not vaporize at typical operating conditions. In a chemical engineering context, a solid can be separated from a solvent by evaporating the solvent. Two components that have boiling points that differ by 20°C, for example, can be separated by preferentially boiling the component with the lower boiling point. Another difference between these two separations is that the solvent obtained through evaporation will be essentially pure; however, in distillation the lower boiling component will contain some of the higher boiling component.

The separating agent is employed to exploit the separation basis to effect the separation. In distillation and evaporation, the separating agent is energy. Most students are also familiar with extraction from organic chemistry lab, where a solute is transferred from one phase to another, immiscible, phase. In this case, the destination solvent is the separating agent, and the general category is called *mass separating agents*. Another familiar mass separating agent is involved in the brewing of real (not instant) coffee, in which hot water removes the flavor ingredients from the solid coffee bean, but the coffee bean does not dissolve in the water. This solid-liquid separation is called *leaching*.

Table 3.1 shows some typical separations used in chemical engineering, their basis, and the separating agent.

3.1   BASIC RELATIONSHIPS IN SEPARATIONS

Most separation processes require simultaneous solution of three fundamental relationships. As with typical chemical engineering equipment, the fundamental equations involved in separations start with the material balance and the energy balance. Then, depending on the specific separation process and/or specific equipment being used, the third relationship could be an equilibrium relationship, a mass transfer relationship, or a rate expression.

3.1.1   Mass Balances

The exact form of the mass balance depends on the separation basis. For a separation basis not involving a mass separating agent, such as energy or a membrane, for example, as illustrated in Figure 3.1(a), the overall mass balance is of the form

$$F = L + V \tag{3.1}$$

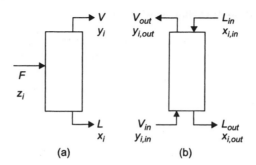

Figure 3.1   Input and output structure of separation involving (a) energy separating agent or membrane and (b) mass separating agent

and the component mass balance for species i is

$$Fz_i = Lx_i + Vy_i \qquad (3.2)$$

where the stream flowrates (F, L, V) are either in mass or mole units, and the fractions (x_i, y_i, z_i) are either mass fractions or mole fractions, respectively.

For a mass separating agent, such as extraction, as illustrated in Figure 3.1(b), wherein solute i is transferred from the liquid stream L to the vapor stream V or vice versa, the mass balances are

$$L_{in} + V_{in} = L_{out} + V_{out} \qquad (3.3)$$

$$L_{in}x_{i,in} + V_{in}y_{i,in} = L_{out}x_{i,out} + V_{out}y_{i,out} \qquad (3.4)$$

where, once again, if the flowrates are in mass units, the fractions are mass fractions, and if the flowrates are in mole units, the fractions are mole fractions.

The nomenclature used for these mass balances is in anticipation of liquid-vapor separations, hence the use of L and V. However, the mass balances are applicable to any separation modeled by either case in Figure 3.1.

3.1.2   Energy Balances

For the system illustrated in Figure 3.1(a), the overall energy balance is

$$Fh_F + Q = Lh_L + VH_v \qquad (3.5)$$

where Q is the heat duty, in energy/mass or energy/mole, H is a vapor enthalpy, and h is a liquid enthalpy, both in energy/mass or energy/mole.

For the system illustrated in Figure 3.1(b), the energy balance is

$$L_{in}h_{in} + V_{in}H_{in} + Q = L_{out}h_{out} + V_{out}H_{out} \qquad (3.6)$$

For liquid-liquid separations such as extraction, all enthalpies are for liquids, and the energy balance is not usually needed, since little or no energy of solution is involved in transferring a solute between liquid phases, so the process is essentially isothermal. For some vapor-liquid separations, such as absorption and stripping, the energy balance may be involved, since dissolving a gas in a liquid is often accompanied by a heat of solution, which makes the process nonisothermal.

3.1.3   Equilibrium Relationships

For an equilibrium separation, it is assumed that the outlet streams are in equilibrium, or if the actual behavior is modeled as an approach to equilibrium, an equilibrium expression for each component must be solved along with the mass and energy balances. In effect, it is assumed that the phases are well mixed for a sufficient residence time to reach equilibrium. In general, the equilibrium relationship is of the form

$$y_i = m_i x_i \tag{3.7}$$

where, for vapor-liquid separations

$$m_i = \frac{P_i^* \varphi_i^* \gamma_i \exp\left[\dfrac{V_i^L (P_i - P_i^*)}{RT}\right]}{\varphi_i^V P} \tag{3.8}$$

where P_i is the partial pressure of component i; P_i^* is the vapor pressure of component i; φ_i^* is the fugacity coefficient for component i at saturation; γ_i is the activity coefficient for component i; the exponential is known as the Poynting correction factor, with V_i^L being the liquid molar volume; and φ_i^V is the fugacity coefficient for component i in the vapor phase. The Poynting correction factor only deviates from unity at very high pressures. The fugacity coefficients only deviate from unity at high pressures, and the activity coefficient only deviates from unity for nonideal systems. Therefore, for ideal systems at low pressures, Equation (3.8) reduces to

$$\frac{y_i}{x_i} = m_i = \frac{P_i^*}{P} \tag{3.9}$$

which is Raoult's law. For liquid-liquid separations,

$$m_i = \frac{P_i^{*I} \varphi_i^{*I} \gamma_i^I \exp\left[\dfrac{V_i^{L,I} (P_i^I - P_i^{*I})}{RT}\right]}{P_i^{*II} \varphi_i^{*II} \gamma_i^{II} \exp\left[\dfrac{V_i^{L,II} (P_i^{II} - P_i^{*II})}{RT}\right]} \tag{3.10}$$

where the superscripts I and II refer to the two liquid phases. Usually, for liquid-liquid systems, experimental data are used, and if the data appear to be linear, a constant value for m can be determined. For gases dissolving, but not condensing, in liquids, m can be related to Henry's law. In Henry's law, the partial pressure in the vapor phase is related to the liquid mole fraction by $p_A = H_A x_A$, where H_A has pressure units, so m in Equation (3.7) becomes H_A/P, where P is the total pressure.

3.1.4   Mass Transfer Relationships

3.1.4.1   Continuous Differential Model

If the two phases being contacted in a separation are not well mixed, equilibrium is not reached between the phases. A model for this type of separation is similar to that for a countercurrent heat exchanger and is illustrated in Figure 3.2. In this development, it is assumed that there is one solute being transferred between phases, from the V phase to the L phase. The overall mass balances are Equations (3.3) and (3.4). Paralleling the heat transfer development in Chapter 2, the differential mass balance between two points S and $S + \Delta S$ is

$$(Vy)_{S+\Delta S} - (Vy)_S - N_y \Delta S = 0 \tag{3.11}$$

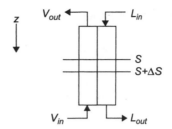

Figure 3.2 Model for continuous differential separation

where N_y is the flux of solute in mass or moles/interface area/time and S is the interfacial (mass transfer) area/flow cross-sectional area, which is assumed to be zero at $z = 0$ and increase proportionally with the coordinate z. Equation (3.11) reduces to

$$\frac{d(Vy)}{dS} = N_y \tag{3.12}$$

A similar development under the assumption that transfer is from the L phase to the V phase gives

$$\frac{d(Lx)}{dS} = N_x \tag{3.13}$$

3.1.4.2 Two-Film Model
In order to apply Equation (3.12) or (3.13), a model for mass transfer between phases is needed. For transfer between immiscible phases (liquid-liquid or gas-liquid), the two-film model can be used. In this model, which is illustrated in Figure 3.3, there is a mass transfer resistance on each side of the interface described by a mass transfer coefficient, which is similar to a heat transfer coefficient. However, for mass transfer, there is a discontinuity in concentration at the film/film interface due to the difference in solubilities in the two phases, while for heat transfer the temperature is continuous across the films.

The flux on the V side for this case is (Treybal, 1980)

$$N_y = \frac{c_y D_{ABy}}{\delta_y} \ln\left(\frac{1 - y_{Ai}}{1 - y_A}\right) \tag{3.14}$$

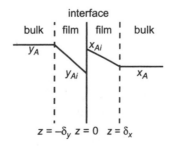

Figure 3.3 Two-film model for mass transfer between immiscible phases

where c_y is the total molar concentration, D_{ABy} is the diffusion coefficient of solute (A) in the V phase, δ_y is the film thickness in the V phase, y_A is the mole fraction in the V phase, and y_{Ai} is the mole fraction in the V phase at the interface. Similarly, for the L phase,

$$N_x = \frac{c_x D_{ABx}}{\delta_x} \ln\left(\frac{1-x_A}{1-x_{Ai}}\right) \tag{3.15}$$

In this illustration, it is assumed that the direction of mass transfer is from the V phase to the L phase; however, the result is the same for transport in the opposite direction. At steady state, the flux from the V phase must be equal to the flux into the L phase. Equating Equations (3.14) and (3.15) yields

$$\frac{1-y_{Ai}}{1-y_A} = \left(\frac{1-x_A}{1-x_{Ai}}\right)^{F_x/F_y} \tag{3.16}$$

where $F_y = (c_y D_{ABy}/\delta_y)$ and $F_x = (c_x D_{ABx}/\delta_x)$. It is observed that F_x and F_y have the same units as N, moles/interfacial area/time. Since this model assumes a stagnant film, and in a real application there will also be convection, the parameters F_x and F_y are effectively mass transfer coefficients. The relationships for these mass transfer coefficients parallel those for heat transfer and are dependent on the actual flow geometry involved. Equation (3.16) can be used to calculate the values of y_{Ai} and x_{Ai} from the values far from the interface, based on the assumption that the interface is at equilibrium, so that $y_{Ai} = m_A x_{Ai}$. It is important to understand that, when this model is used to describe a separation, equilibrium occurs only at the interface, whereas, in the well-mixed, equilibrium model described in Section 3.1.3, the phases are in equilibrium at all points in both phases, meaning that the outlet streams are in equilibrium. Equations (3.14), (3.15), and (3.16) can be simplified for special cases, such as dilute solutions, and these relationships are available (Treybal, 1980).

In the two-film model, the two mass transfer coefficients, F_x and F_y, can be combined into an overall mass transfer coefficient, just as in heat transfer. For an overall mass transfer coefficient based on the V phase, defined as K_y, this result is

$$\frac{1}{K_y} = \frac{1}{F_y} + \frac{m}{F_x} \tag{3.17}$$

and the overall mass transfer coefficient based on the L phase, defined as K_x, is

$$\frac{1}{K_x} = \frac{1}{F_x} + \frac{1}{mF_y} \tag{3.18}$$

3.1.4.3   Transfer Units

To determine the interfacial area required for a given separation, the differential equation obtained by using Equation (3.14) in Equation (3.12) or by using Equation (3.15) in Equation (3.13) must be solved. The former case is illustrated. The differential equation to be solved is

$$\frac{d(Vy)}{dS} = F_y \ln\left(\frac{1-y}{1-y_i}\right) \tag{3.19}$$

where the subscript A has been dropped. The result, presented here without derivation, is

$$S = \int_{y_{A,in}}^{y_{A,out}} \frac{V dy}{F_y(1-y)\ln\left(\frac{1-y_i}{1-y}\right)} \tag{3.20}$$

In turbulent flow, the mass transfer coefficient is proportional to $Re^{0.8}$ or a power close to 0.8 (just as in heat transfer), so the approximation that V/F_y is constant is often made. In this case, Equation (3.20) becomes

$$S = \frac{V}{F_y} \int_{y_{A,in}}^{y_{A,out}} \frac{dy}{(1-y)\ln\left(\frac{1-y_i}{1-y}\right)} \tag{3.21}$$

This form has the advantage of having one term (V/F_y) that is dependent on the flow geometry and one term (the integral) that is dependent only on the compositions of the phases for any flow geometry. These two terms are usually given the definitions "height of a transfer unit" (H) and "number of transfer units" (N), respectively. The use of the word *height* is based on a typical application to vertical, packed columns, although, the word *length* might be a better term. The height of a transfer unit is a measure of 1/separation efficiency, and the number of transfer units is a measure of the difficulty of the separation. The larger the height of a transfer unit, the less efficient the separation, and the larger the number of transfer units, the more difficult the separation. Therefore,

$$H_V = \frac{V}{F_y} \tag{3.22}$$

$$N_V = \int_{y_{A,in}}^{y_{A,out}} \frac{dy}{(1-y)\ln\left(\frac{1-y_i}{1-y}\right)} \tag{3.23}$$

and

$$S = H_V N_V \tag{3.24}$$

A parallel derivation is possible for the L phase, and the results are

$$H_L = \frac{L}{F_x} \tag{3.25}$$

$$N_L = \int_{x_{A,in}}^{x_{A,out}} \frac{dx}{(1-x)\ln\left(\frac{1-x}{1-x_i}\right)} \tag{3.26}$$

and

$$S = H_L N_L \tag{3.27}$$

In principle, the size (interfacial area) of a continuous differential separation unit calculated from either Equation (3.24) or Equation (3.27) is identical. This is similar to the area of a heat exchanger being identical for the two cases of the overall heat transfer coefficient based on the internal surface area (U_i) and the overall heat transfer coefficient based on the external surface area (U_o). The decision on which method to use is based on computational issues; however, it is

Table 3.2   **Summary of Equations for Calculating Transfer Units**

z (Units of Length)	Height of Transfer Unit (Length)	Number of Transfer Units	Comments
$z = N_V H_V$	$H_V = \dfrac{V}{F_y}$	$N_V = \displaystyle\int_{y_{A,in}}^{y_{A,out}} \dfrac{dy}{(1-y)\ln\left(\dfrac{1-y_i}{1-y}\right)}$	Usually used when gas phase has limiting mass transfer resistance
$z = N_L H_L$	$H_L = \dfrac{L}{F_x}$	$N_L = \displaystyle\int_{x_{A,in}}^{x_{A,out}} \dfrac{dx}{(1-x)\ln\left(\dfrac{1-x}{1-x_i}\right)}$	Usually used when liquid phase has limiting mass transfer resistance
$z = N_{oV} H_{oV}$	$H_{oV} = \dfrac{V}{K_y}$ $H_{oV} = H_V + \left(\dfrac{mV}{L}\right)H_L$	$N_{oV} = \displaystyle\int_{y_{A,in}}^{y_{A,out}} \dfrac{dy}{(1-y)\ln\left(\dfrac{1-y^*}{1-y}\right)}$	$y^* = mx$, where x corresponds to y value on operating line; usually used when gas phase has limiting mass transfer resistance
$z = N_{oL} H_{oL}$	$H_{oL} = \dfrac{L}{K_x}$ $H_{oL} = H_L + \left(\dfrac{L}{mV}\right)H_V$	$N_{oL} = \displaystyle\int_{x_{A,in}}^{x_{A,out}} \dfrac{dx}{(1-x)\ln\left(\dfrac{1-x}{1-x^*}\right)}$	$x^* = y/m$, where y corresponds to x value on operating line; usually used when liquid phase has limiting mass transfer resistance

generally true that the transfer unit expression for the phase with the limiting resistance is the better method to use.

It is also possible to define transfer units based on the overall mass transfer coefficients in Equations (3.17) and (3.18). All cases are presented in Table 3.2.

3.1.5   Rate Expressions

Some separations are based on differential rates of transport between components, so the rate of transport, not equilibrium, is used in conjunction with the mass and energy balances. The application that is treated here is membrane separations. The model for membrane transport is shown in Figure 3.4. An external resistance to mass transfer exists on either side of the membrane, which is characterized by a mass transfer coefficient (F_i). This model resembles heat transfer across a solid with external resistance, which was discussed for cylindrical coordinates in Chapter 2. The major difference is the concentration "jump" at the interface, while for the heat transfer case, the temperature is continuous across the interface. This is due to the different solute solubility in the membrane from that in the external phase. The steady-state flux (moles/interface area/time) of solute A across the membrane is

$$N_A = F_1(C_{A1} - C_{A1i}) = \frac{D_{AB}}{t_m}\left(C_{A1m} - C_{A2m}\right) = F_2\left(C_{A2i} - C_{A2}\right) \tag{3.28}$$

Assuming that the interfaces are at equilibrium, and using the equilibrium expressions $C_{A1m} = m_1 C_{A1i}$ and $C_{A2m} = m_2 C_{A2i}$, Equation (3.28) can be rearranged into a series resistance form:

$$N_A = \frac{m_1 C_{A1} - m_2 C_{A2}}{\dfrac{m_1}{F_1} + \dfrac{t_m}{D_{AB}} + \dfrac{m_2}{F_2}} \tag{3.29}$$

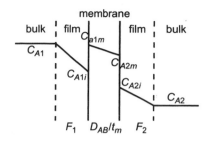

Figure 3.4 Model for mass transfer across membrane

It is observed that the denominator of Equation (3.29) resembles the series resistance form in Equation (2.24). The difference, other than geometry, is the different solubilities of the solute in the membrane and in the two external phases. Analogous parameters are not present in the heat transfer form, because the temperature criterion for equilibrium at an interface is equal temperatures. For mass transfer, the criterion for equilibrium is the ratio of the solubilities in the two phases, m_i, which is often called the *partition coefficient*.

In some cases, it is assumed that the external resistance is negligible, in which case Equation (3.29) reduces to

$$N_A = \frac{m_1 C_{A1} - m_2 C_{A2}}{\dfrac{t_m}{D_{AB}}} = \frac{D_{AB}}{t_m}\left(m_1 C_{A1} - m_2 C_{A2}\right) \qquad (3.30)$$

In the particular application discussed later in this chapter, the two external phases are gases, so it can be assumed that $m_1 = m_2 = m$, so Equation (3.30) further reduces to

$$N_A = \frac{m D_{AB}}{t_m}\left(C_{A1} - C_{A2}\right) = \frac{P}{t_m}\left(C_{A1} - C_{A2}\right) \qquad (3.31)$$

where P is defined as the membrane permeability. (Note that some references define the permeability as P/t_m.) In addition, for applications involving gas permeation, the concentrations are often expressed as partial pressures using the ideal gas law.

3.2 ILLUSTRATIVE DIAGRAMS

3.2.1 *TP-xy* Diagrams

In traditional chemical processes, vapor-liquid separations are by far the most common. The simplest vapor-liquid separation is a partial condensation or partial vaporization. The discussion here is limited to binary mixtures with illustrations for ideal mixtures; however, the concepts apply to mixtures of any number of components and nonideal mixtures. In a partial condensation, a vapor mixture is brought into the two-phase region, and the vapor and liquid phases in equilibrium are at different mole fractions. Partial condensation can occur by either cooling and/or compressing a vapor mixture. Figure 3.5 denotes the equipment involved, and the separation is shown on a T-xy diagram. It is important to understand that either a heat exchanger or a compressor is needed to change the temperature or pressure, respectively. Quite often student problems and process simulators lump both pieces of equipment into one unit. Calculations can be performed this way, but the correspondence to actual equipment is lost. In Figure 3.5, the use of compression to form a two-phase mixture is shown only for illustrative purposes, since liquid droplets damage

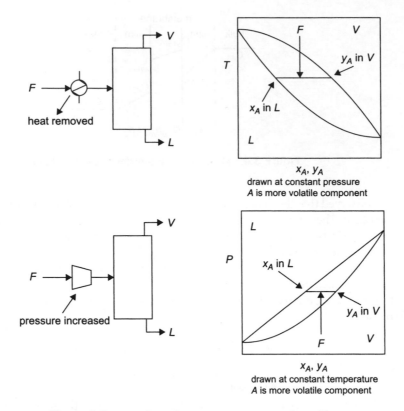

Figure 3.5   Partial condensation equipment and equilibrium

compressor rotors, meaning that this operation is not used. If compression was to be used, there would need to be cooling to facilitate condensation, since compression increases the temperature of the vapor. Figure 3.6 illustrates partial vaporizations, either by heating a liquid mixture into the two-phase region or by reducing the pressure of a liquid to form a vapor-liquid mixture. Once again, all relevant equipment is shown.

Quite often, all four of these operations are called *flash* separations. Technically, only the reduction in pressure is a "flash" separation, specifically a flash vaporization.

The equations used to solve any of these flash separations are the material balances, Equations (3.1) and (3.2); the energy balance, Equation (3.5); and the equilibrium expression, Equation (3.8). If Equations (3.1), (3.2), and (3.8) are combined, Equations (3.32) and (3.33) can be obtained.

$$\sum_{i=1}^{C} x_i = 1 = \sum_{i=1}^{C} \frac{z_i}{1 + \dfrac{V}{F}(m_i - 1)} \tag{3.32}$$

$$\sum_{i=1}^{C} y_i = 1 = \sum_{i=1}^{C} \frac{m_i z_i}{1 + \dfrac{V}{F}(m_i - 1)} \tag{3.33}$$

where the subscript i represents each component, and there are C components. In principle, either of these equations can be used to solve for one unknown, either V, temperature, or pressure (inside m_i) if the other two are specified. Then, the mole fractions can be found. It is also possible, in principle, to solve for two of V, T, or P, with any two outlet parameters, including component mole fractions and fractional recoveries. However, since Equations (3.32) and (3.33) are not always

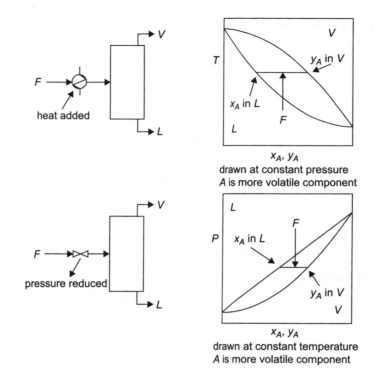

Figure 3.6 Partial vaporization equipment and equilibrium

monotonic, numerical methods can fail. Additionally, the difference between Equations (3.32) and (3.33) is always monotonic, so it is a better choice for a numerical solution.

$$\sum_{i=1}^{C} y_i - \sum_{i=1}^{C} x_i = 0 = \sum_{i=1}^{C} \frac{(m_i - 1)z_i}{1 + \frac{V}{F}(m_i - 1)} \tag{3.34}$$

Equation (3.34) is often called the Rachford-Rice equation (Wankat, 2017). The energy balance, Equation (3.5), is only used to calculate the heat duty on the heat exchanger used for the partial condensation.

Example 3.1

In the production of cumene from propylene and benzene, the feed propylene contains propane, which does not react with the benzene. The reaction occurs in the vapor phase. The separation section of the process starts with a partial condensation of the unreacted benzene (present in excess to enhance selectivity) and cumene to remove the propane. For this example, it is assumed that all propylene in the feed reacts and that there are no unwanted by-products in the reactor effluent. The feed to the partial condenser contains 51 mol% benzene, 44 mol% cumene, and the remainder propane. Ideal behavior is assumed, and the Antoine coefficients are shown in Table E3.1 and are assumed to be valid over the temperature range of this problem.

If a partial condensation of 200 kmol/h occurs at 90°C and 1.75 bar, what are the flowrates and mole fractions of the streams leaving the flash drum?

Table E3.1   Antoine Coefficients for Example 3.1

	$\log_{10} P^{\bullet}$ (mm Hg) $= A - \dfrac{B}{T(°C) + C}$		
	A	B	C
Propane	6.80398	803.810	246.990
Benzene	6.90565	1211.033	220.790
Cumene	6.93666	1460.793	207.777

Solution

Using Raoult's law, the values of m_i can be calculated from $m_i = P_i^{\bullet}/P$, where $P_i^{\bullet}$ is obtained from the Antoine coefficients in Table E3.1. Since z_i are known, the only unknown in Equation (3.34) is V/F. Since there are only three terms in Equation (3.34), the result is a quadratic in V/F, although, in general, Equation (3.34) would be solved using an equation solver. The result is $V/F = 0.0393$, so $V = 7.86$ kmol/h, and $L = 192.14$ kmol/h. The mole fractions are obtained from

$$x_i = \frac{z_i}{1 + \dfrac{V}{F}(m_i - 1)} \qquad \text{(E3.1a)}$$

$$y_i = \frac{m_i z_i}{1 + \dfrac{V}{F}(m_i - 1)} \qquad \text{(E3.1b)}$$

which are the individual terms in Equations (3.32) and (3.33), respectively, and m is given by Equation (3.9). The results are

	x_i	y_i
Propane	0.029	0.568
Benzene	0.515	0.395
Cumene	0.457	0.037

If V/F were known, either the temperature or pressure could be obtained using the same method, solving for one unknown.

Example 3.2

In the process in Example 3.1, the vapor stream contains too much benzene, a valuable reactant. The feed flowrate of benzene is 102 kmol/h, while the flowrate of benzene in the liquid phase is (192.14 kmol/h)(0.515) = 98.95 kmol/h, which is about 97% recovery of benzene in the liquid and a loss of about 3 kmol/h of benzene, which has a value of several million dollars/ year. Under what operating conditions would 99% recovery of benzene in the liquid stream be possible?

Solution

In this case, V/F, T, and P are initially unknown. As in Example 3.1, two specifications are needed. One is the desired fractional recovery. The other is either T or P. V/F cannot be specified, because the material balance is determined by the fractional recovery specification. Therefore, if T is specified, the required pressure can be calculated, and vice versa.

The fractional recovery specification must be used to determine V/F. This specification is

$$\frac{Lx_B}{Fz_B} = \frac{Lx_B}{200(0.51)} = 0.99 \tag{E3.2a}$$

which means that $Lx_B = 100.98$, so $Vy_B = 1.02$, since there are 102 kmol/h of benzene in the feed. Taking the ratio of these terms,

$$\frac{Vy_B}{Lx_B} = \frac{Vm_B}{L} = \frac{1.02}{100.98} = 99 \tag{E3.2b}$$

so

$$\frac{Vm_B}{F-V} = \frac{\dfrac{V}{F}m_B}{1-\dfrac{V}{F}} = 99 \tag{E3.2c}$$

and rearrangement yields

$$\frac{V}{F} = \frac{1}{1+99m_B} \tag{E3.3d}$$

If Equation (E3.3d) is inserted into Equation (3.34), then if the pressure is known, temperature is the only unknown, and vice versa. Therefore, there are actually an infinite number of temperature-pressure combinations that solve the problem. For this exercise, the pressure will be determined at 90°C, the original temperature in Example 3.1, and the temperature will be determined at 1.75 atm, the original pressure in Example 3.1. An equation solver is used, and at 90°C, the pressure is 2.14 atm, and at 1.75 atm, the temperature is 81.2°C. The mole fractions in each phase could then be calculated just as in Example 3.1.

Examples 3.1 and 3.2 illustrate that, without consideration of energy requirements, Equation (3.34) can be used to solve for a single unknown with two specifications. The energy balance, Equation (3.5), can be used the get the heat duty. Problems that couple the energy balance with Equation (3.33) are also possible.

It is observed from the T-xy diagrams that the vapor phase is enriched in the more volatile component, while the liquid phase is enriched in the less volatile component. The horizontal line connecting the vapor and liquid phases in equilibrium is called a *tie line*. Any mixture brought to a point in the two-phase region separates into two phases connected by the tie line. It is further observed that neither phase is very pure in the enriched component. Now, suppose that the feed is vapor that is partially condensed, and the vapor phase, V_1, is partially condensed again, as illustrated in Figure 3.7, and the process is repeated several times. It is observed that the more volatile component can asymptotically approach purity. Similarly, Figure 3.7 also illustrates that the less volatile component can asymptotically approach purity by partially vaporizing the liquid phase. Each heat exchanger/drum combination is called a *stage*, and these are called *equilibrium stages*, because it is assumed that the two phases leaving the stage are in equilibrium. This sequence appears promising; however, as shown Figure 3.7, there are a significant number of waste streams, since only the top or bottom stream is the desired product. Furthermore, a significant number of heat exchangers is required. Suppose the top "waste" stream is recycled to the second-from-the-top stage. It can provide the necessary heat sink in place of one partial condenser. This is illustrated in Figure 3.8. Similarly, if the bottom waste stream is recycled from the second-from-the bottom stage, it can provide the energy for one partial vaporization. This is also illustrated in Figure 3.8. Every waste stream can be returned to the adjacent stage, and Figure 3.9 illustrates

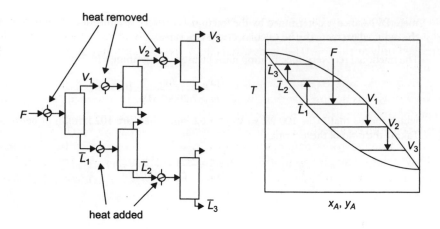

Figure 3.7   Effect of adding multiple stages to partial condensation/partial vaporization process

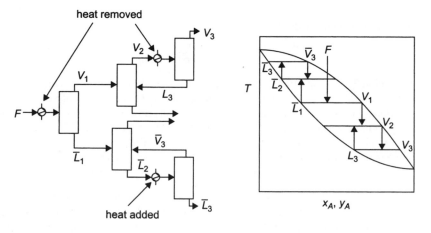

Figure 3.8   Effect of using impure streams as heat source and heat sink

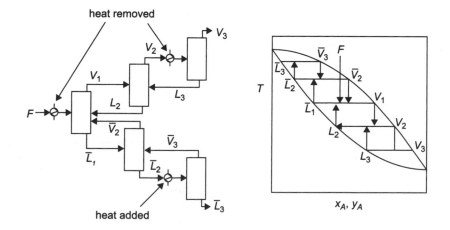

Figure 3.9   The final arrangement with heat only added at bottom and heat only removed at top

the entire process, and it is observed that there is one feed stream and two exit streams, both of which can be very pure in one of the components. Heat is added only at the bottom and energy is removed only at the top. This is how distillation works, although as will be seen later, the actual equipment is more compact.

It is important to realize that while the T-xy diagram provides a mental picture of how distillation works, it is not used for calculations, though it was before high-speed computing. The calculations are done exactly how the diagrams suggest, using a series of mass and energy balances combined with equilibrium expressions from stage to stage, as shown in Section 3.1.

If it is understood that energy is required to "unmix" components, which lowers the system entropy (since mixing is spontaneous), it is no surprise that energy must also be rejected to the surroundings. Overall, energy input is required to unmix the components, but energy must also be rejected to the surroundings, just as in a power cycle.

3.2.2 McCabe-Thiele Diagram

The McCabe-Thiele diagram is a graphical representation of distillation, and it can also be used to represent separations using mass separating agents. It is valid only for certain systems subject to certain assumptions. While current computational power makes the McCabe-Thiele diagram somewhat obsolete, an understanding of the diagram provides conceptual insights that apply to all types of distillation operations and to all types of separations using mass separating agents.

3.2.2.1 Distillation

The McCabe-Thiele diagram can be used to represent distillation using either tray columns or packed columns.

Tray Columns

A schematic of a tray distillation column is shown in Figure 3.10(a), along with a detailed sketch of the internals of a distillation column in Figure 3.10(b). It is a tower with trays containing holes, sometimes with caps or similar devices on top of the holes. A level of liquid is present on each tray, and gas bubbles up through the tray from the tray below. Each tray behaves like a flash operation. The feeds to each tray, one from the top and the other from the bottom, are at different temperatures so that the liquid and vapor on the tray are assumed to come to equilibrium. Since the vapor bubbles through the holes in the tray, the bubble motion is assumed to create a well-mixed condition, so that the exit streams from the tray are at the same conditions as the vapor and liquid on the tray.

The material and energy balances, as described by Equations (3.3), (3.4), and (3.6), are written for every tray. An alternative method is to write these balances from the top to each tray above the feed and the bottom to each tray below the feed. One of these sets of balances, along with balances on the condenser, reboiler, and feed tray, are solved simultaneously. Therefore, the number of trays must be known, so that all balances can be written. This means that the outlet conditions can be predicted for a fixed number of trays, feed location, reflux ratio (L_0/D in Figure 3.10[a]), and feed conditions, which is a simulation, not a design. To design a column this way requires iterations, until the number of trays, reflux ratio, and feed location provide the desired outlet conditions. This is why process simulators require that the number of trays and feed location be provided for a rigorous distillation calculation.

There is a simplification that allows a graphical method to design a distillation column. It only works for binary distillations under certain circumstances. However, a complete understanding of this method provides a complete understanding of the operation of a distillation column. This is the approach taken here.

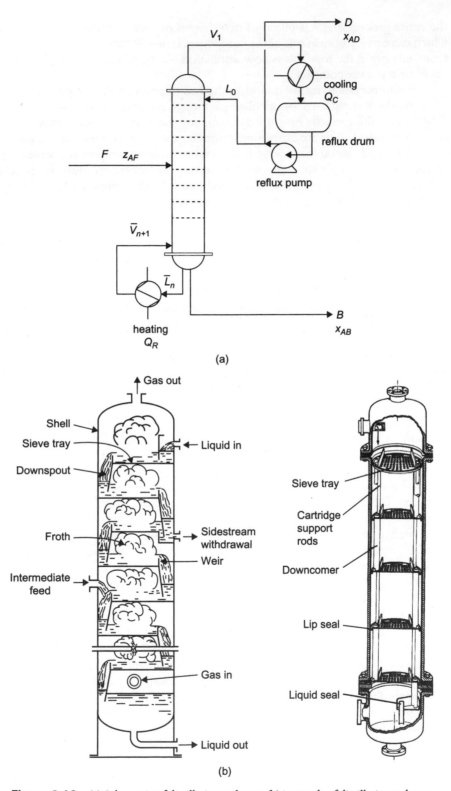

Figure 3.10   (a) Schematic of distillation column, (b) internals of distillation column
(b from Couper et al. [2012])

The assumption that simplifies binary distillation calculations is called *constant molar overflow* (also called *constant molal overflow*). The assumptions of constant molar overflow are

- Molar heat of vaporization (λ) is the same for each component.
- Column is adiabatic.
- Sensible heat is small compared to latent heat ($C_p \Delta T << \lambda$).

As a consequence, the moles of vapor condensing on a tray equal the moles of liquid vaporizing on the tray. Therefore, the molar flowrates of liquid and vapor do not change from tray to tray in a given section of the column (a section is either above the feed or below the feed). Additionally, there is no need for an energy balance on any tray, because the energy need for vaporization equals the energy given up by condensation, and since no heat is lost, the heats of vaporization are identical, and they are much larger than any temperature changes between adjacent trays.

Material balances written from the top of the column to a tray (j) above the feed, as illustrated in the top portion of Figure 3.11, are

$$V = L + D \tag{3.35}$$

$$V y_{A,j+1} = L x_{A,j} + D x_{AD} \tag{3.36}$$

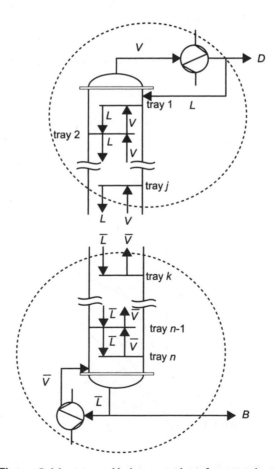

Figure 3.11   Material balance envelope for tray column

where the assumption of constant molar flowrates on every tray is used, so that L and V are constant and not indexed by tray number. (Note that trays are numbered from top to bottom.) Equation (3.36) is written on the more volatile component, A. Equations (3.35) and (3.36) can be rearranged to

$$y_{A,j+1} = \frac{L}{V}x_{A,j} + \left(1 - \frac{L}{V}\right)x_{AD} = \frac{L}{V}x_{A,j} + \frac{x_{AD}}{R+1} \tag{3.37}$$

where $R = L/D$. Equation (3.37) is now written as

$$y_A = \frac{L}{V}x_A + \frac{x_{AD}}{R+1} \tag{3.38}$$

where the subscript for the tray number has been dropped, since the equation is the same for a balance written from the top of the column to any tray. Equation (3.38) is the equation of a straight line with slope L/V and intercept $x_{AD}/(R + 1)$. While it is possible to plot this line if L/V is known, an easier method is to observe that when $y_A = x_A$, $x_A = x_{AD}$. Therefore, two points are known, the intercept and (x_{AD}, x_{AD}). When this line is plotted on an equilibrium diagram, the line labeled 1 in Figure 3.12 is obtained. On Figure 3.12, the curve represents the vapor-liquid equilibrium, which can be obtained from the end-points of the tie lines on a T-xy diagram such as shown in Figure 3.9. These data can be predicted from Raoult's law or, in the most general case, Equation (3.8). The diagonal line is just a plot of $y_A = x_A$.

A similar analysis can be done by writing a material balance from the bottom of the column to any tray below the feed. The situation is illustrated in the lower portion of Figure 3.11. The equations are

$$\bar{L} = B + \bar{V} \tag{3.39}$$

$$\bar{L}x_{A,k} = \bar{V}y_{A,k+1} + Bx_{AB} \tag{3.40}$$

where the overbar indicates molar flowrates below the feed. Equations (3.39) and (3.40) can be rearranged to yield, with removal of the index subscript

$$y_A = \frac{\bar{L}}{\bar{V}}x_A + \left(1 - \frac{\bar{L}}{\bar{V}}\right)x_{AB} \tag{3.41}$$

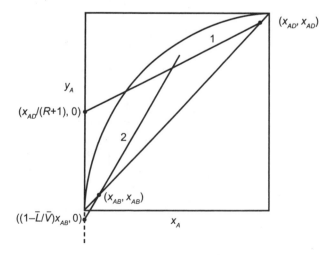

Figure 3.12 Material balance (operating) lines for distillation column

Equation (3.41) is the equation of a straight line with slope $\bar{L}/\bar{V}$ and intercept $\left(1-\bar{L}/\bar{V}\right)x_{AB}$. If $y_A = x_A$, $x_A = x_{AB}$. Therefore, two points are known, the intercept and (x_{AB}, x_{AB}). This line is plotted on Figure 3.12 and labeled 2. It is observed that the intercept is negative because the slope $\bar{L}/\bar{V} > 1$.

The upper section of a distillation column, above the feed, is called the *enriching* section or the *rectification* section. The lower section, below the feed, is called the *stripping* section. The feed tray is the boundary between these two sections. In general, liquid in the feed goes down and vapor goes up; however, what exactly happens on the feed tray depends on whether the feed is saturated liquid, saturated vapor, a vapor-liquid mixture, superheated vapor, or subcooled liquid. It is important to understand that these terms are defined relative to the conditions on the feed tray. For example, if the feed is saturated liquid at 30°C, but the tray temperature is 50°C, then the feed is considered subcooled.

To complete the column model, the top and bottom sections must match at the feed. Figure 3.13 illustrates the feed tray.

The material and energy balances on the feed tray are

$$F + \bar{V} + L = V + \bar{L} \tag{3.42}$$

$$Fh_F + \bar{V}H_{f+1} + Lh_{f-1} = VH_f + \bar{L}h_f \tag{3.43}$$

where the subscript f indicates the feed tray number, uppercase H indicates vapor enthalpy, lowercase h indicates liquid enthalpy, and h_F is the enthalpy of the feed regardless of its phase. If the feed stage is assumed to be at equilibrium, then the vapor and liquid enthalpies are for saturated conditions. From the constant molar overflow assumption, the enthalpies of each phase are constant across the trays, so, rearranging Equations (3.42) and (3.43), while dropping the subscripts involving f, yields an equation for the quality of the feed, q, defined as the fraction of the feed in the saturated liquid state (as opposed to the quality of steam, which is defined as the fraction of vapor in the steam):

$$q = \frac{\bar{L}-L}{F} = 1 - \frac{V-\bar{V}}{F} = \frac{H-h_F}{H-h} \tag{3.44}$$

If the feed is saturated liquid, $q = 1$, so the liquid flowrate below the feed is equal to the liquid flowrate above the feed plus the feed flowrate. The vapor flowrates above and below the feed are identical.

If the feed is saturated vapor, $q = 0$, so the vapor flowrate above the feed is equal to the vapor flowrate below the feed plus the feed flowrate. The liquid flowrates above and below the feed are identical.

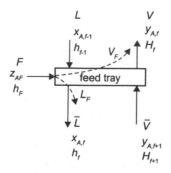

Figure 3.13   Vapor and liquid flows on feed tray

If the feed is a vapor-liquid mixture, $0 < q < 1$, so the liquid flowrate below the feed is equal to the liquid flowrate above the feed plus the liquid flowrate in the feed, L_F. The vapor flowrate above the feed is the vapor flowrate below the feed plus the vapor flowrate in the feed, V_F. Simply put, saturated liquid goes down, and saturated vapor goes up.

Subcooled liquid and superheated vapor are not as simple. For a subcooled feed, the feed enthalpy, h_F, is less than the saturated liquid enthalpy, h. Therefore, $q > 1$, which means that the liquid flowrate below the feed is greater than the liquid flowrate above the feed plus the feed flowrate. How is this possible? Since the denominator of the enthalpy expression in Equation (3.44) is the heat of vaporization (also called *latent heat*), more energy is required to bring the feed to equilibrium on the feed tray than can be supplied by the condensing vapor. In order for the subcooled liquid in the feed to come to the equilibrium conditions on the feed tray, enthalpy is required. This enthalpy comes from condensing some of the saturated vapor. Therefore, the liquid flowrate below the feed increases by more than the feed flowrate. Additionally, the vapor flowrate above the feed is lower than the vapor flowrate below the feed, since some vapor condenses on the feed tray.

A similar explanation exists for superheated vapor. Since $h_F > H$, $q < 0$. The superheated vapor must give up enthalpy to become saturated on the tray, which is obtained by vaporizing some liquid on the tray. Therefore, vapor flowrate increases above the feed tray and the liquid flowrate decreases below the feed tray.

It can now be seen that

$$F = L_F + V_F, \quad V - \bar{V} = V_F, \quad \bar{L} - L = L_F, \quad q = \frac{L_F}{F} \tag{3.45}$$

where the subscript F indicates feed. Adding Equations (3.36) and (3.40) with Equation (3.45), and then applying the overall more volatile component balance

$$Fz_{AF} = Dx_{AD} + Bx_{AB} \tag{3.46}$$

yields

$$y_A = \frac{q}{q-1}x_A + \frac{1}{1-q}z_{AF} \tag{3.47}$$

Equation (3.47) is the equation of a line with slope $q/(q-1)$ and is called the q-line. If $y_A = x_A$, it can be seen that one point on the line is (z_{AF}, z_{AF}). Since the balances from the top and bottom sections were combined to give Equation (3.47), the top section, the bottom section, and the q-line must all intersect. This is illustrated in Figure 3.14 for a vapor-liquid mixed feed. Since the slope of the q-line is $q/(q-1)$, it can be seen that there are five possible q-lines, representing the different possible feeds, which are illustrated in Figure 3.15. Table 3.3 summarizes these results.

The ideal feed location is where the feed stream and the tray conditions match, which means matching composition, temperature, and pressure. Of course, the feed stream must be at a slightly higher pressure than the feed tray to allow flow into the column. So, a liquid feed would be brought close to saturation before entering the column. If the vapor effluent from a reactor were to be fed directly to a column, it would be cooled close to saturation. Two-phase feeds are also possible. If the column diameters above and below the feed are different enough to make it impossible to have a column of uniform diameter, the feed conditions could be adjusted to equalize the top and bottom diameters. Therefore, in extreme situations, subcooled liquid or superheated vapor feeds might be desirable.

The McCabe-Thiele method can be used to determine the number of equilibrium stages needed for a separation. If the desired top and bottom compositions are specified (or alternatively, fractional recoveries) and a reflux ratio chosen, the upper operating line can be drawn, and the lower operating

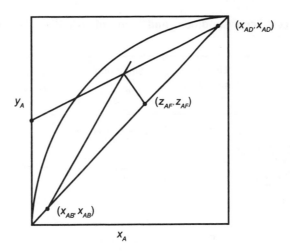

Figure 3.14 Operating lines intersecting feed line

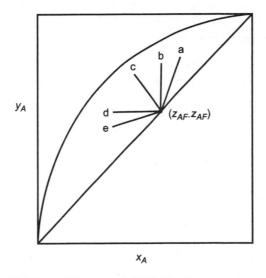

Figure 3.15 Different feed (q) lines for different feeds: (a) subcooled liquid, (b) saturated liquid, (c) vapor-liquid mixture, (d) saturated vapor, (e) superheated vapor

line is drawn from the intersection of the upper operating line and the q-line. Then, as is illustrated in Figure 3.16, the number of stages can be stepped off starting at the top with alternating horizontal lines and vertical lines. A horizontal line to the equilibrium curve is the solution to the equilibrium on a tray, and the vertical line to the operating line is the material balance on that tray (or from the top to any tray above the feed, or from the bottom to any tray below the feed). Each step is an equilibrium stage, so the total number of steps is the number of equilibrium stages required. It is observed that the vertical lines change from the upper operating line to the lower operating line when the step straddles the intersection of the q-line and the operating lines. This represents the optimum feed location that minimizes the number of stages. In Figure 3.16, the feed is on Stage 4 (always count from the top), and there are >7 stages. In the design phase, fractional stages can exist. Before the column is constructed, the extra stage would be added, and it would be planned to operate at a slightly different reflux ratio to make the bottom step intersect the point (x_{AB}, x_{AB}).

Table 3.3 Mole Balances, Feed Condition, and Slope of q-line for Different Distillation Column Feed Conditions

Feed Condition	q	Line in Figure 3.15	Slope of q line $\dfrac{q}{q-1}$	$\dfrac{H-h_F}{H-h}$	Liquid Mole Balance	Vapor Mole Balance
Subcooled liquid	$q>1$	a	$1<\dfrac{q}{q-1}<\infty$	$h_F<h$	$\bar{L}>L+F$	$V<\bar{V}$
Saturated liquid	$q=1$	b	$\dfrac{q}{q-1}=\infty$	$h_F=h$	$\bar{L}=L+F$	$V=\bar{V}$
Vapor-liquid mixture	$0<q<1$	c	$-\infty<\dfrac{q}{q-1}<0$	$h<h_F<H$	$\bar{L}=L+L_F$	$V=\bar{V}+V_F$
Saturated vapor	$q=0$	d	$\dfrac{q}{q-1}=0$	$h_F<H$	$\bar{L}=L$	$V=\bar{V}+F$
Superheated vapor	$q<0$	e	$0<\dfrac{q}{q-1}<1$	$h_F>h$	$\bar{L}<L$	$V>\bar{V}+F$

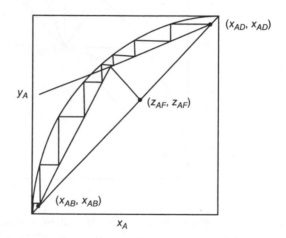

Figure 3.16 Complete McCabe-Thiele construction

A total condenser and a total reboiler are not equilibrium stages, since saturated vapor at the dew point is condensed to saturated liquid at the bubble point (total condenser), or vice versa (total reboiler). A partial condenser and a partial reboiler are equilibrium stages, since the condensation or vaporization is "partial" and there are equilibrium phases in equilibrium, just as in a flash operation. Partial condensers are used if a vapor product is needed or if there are noncondensable components. In appropriate applications, there can be a partial vapor-liquid condenser, in which noncondensables are removed as vapor, but a condensable, desired product is removed as a liquid. All of these configurations are illustrated in Figure 3.17. There is also a reflux drum present with a controlled liquid level to smooth out fluctuations. The reflux pump is necessary, because in an actual process, the condenser, reflux drum, and reflux pump are likely to be located close to the ground, so the pump provides pressure to overcome the head to return the reflux to the top of the column.

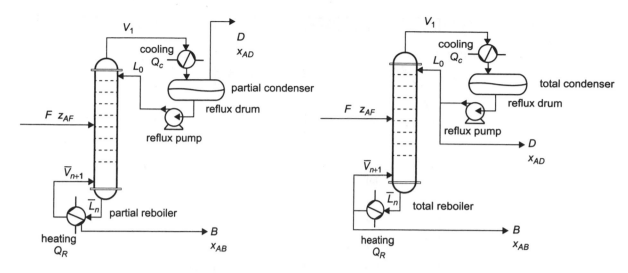

Figure 3.17   Different types of reboilers and condensers

Despite the limitations of the McCabe-Thiele method, a complete understanding of the graphical representations provides an understanding of distillation, including distillation with more than two components and distillation systems for which the constant molar overflow assumption is not valid.

In a chemical plant, about the only variable that can be adjusted is a flowrate. In a distillation column, the flowrate that is adjusted is the reflux stream, which is typically described in terms of the reflux ratio, L/D. Figure 3.18 illustrates four different reflux ratios for a saturated liquid feed; however, the discussion that follows applies to any feed condition. Only the upper section of the column is shown. In Figure 3.18(a), the upper operating line intersects the equilibrium line and the q-line at the same point. If stages are stepped off, the intersection point is never reached. This is the minimum reflux ratio. The operating line in Figure 3.18(a) represents the minimum reflux ratio, which requires an infinite number of stages. However, if the reflux ratio is increased slightly, which lowers the y-intercept, as illustrated in Figure 3.18(b), a finite number of stages can be counted. As the reflux ratio increases, the intercept decreases, the steps get larger, and fewer equilibrium stages are needed. When the intercept equals zero, the reflux ratio is infinite, called *total reflux*, which is illustrated in Figure 3.18(d). At total reflux, there are no product streams and no feed. This is an operational limit, which is used to start up a distillation column and bring it to steady state. Therefore, the McCabe-Thiele diagram illustrates a key concept in distillation, that increasing the reflux ratio reduces the number of equilibrium stages required to obtain the same product specifications. A concept introduced in Chapter 1 is that just about the only way to adjust anything in a chemical process is to open or close a valve. In distillation, there would be valves included to adjust the ratio of L_0/D in Figure 3.17. This is how the reflux ratio is adjusted.

As the boiling points of the two components being distilled become closer, the equilibrium curve moves toward the diagonal line, since the equilibrium liquid and vapor mole fractions become closer together. Figure 3.19 illustrates that the minimum reflux ratio must be higher for a separation involving closer boilers, since, for the same feed, the y-intercept is smaller. Therefore, the actual reflux ratio must also increase for closer boilers.

The feed composition also affects the reflux ratio. The q-line moves to the left if the feed is dilute in the more volatile component. Figure 3.20 illustrates how a more dilute feed in the more volatile component requires a larger reflux ratio. Once again, only the minimum reflux ratio is shown.

The feed condition also affects the required reflux ratio, as is illustrated in Figure 3.21 for the minimum reflux ratio for each feed condition. Subcooled liquid requires the lowest reflux ratio, while superheated vapor requires the highest reflux ratio.

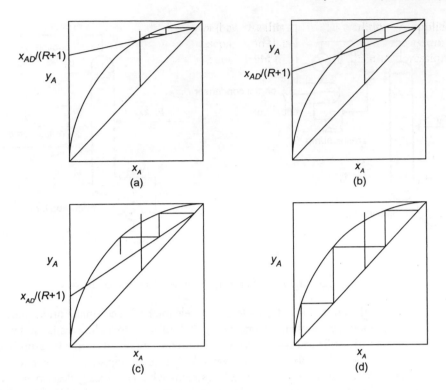

Figure 3.18 Effect of increasing reflux ratio on number of stages. Part (a) reflects minimum reflux, and Part (d) reflects total reflux.

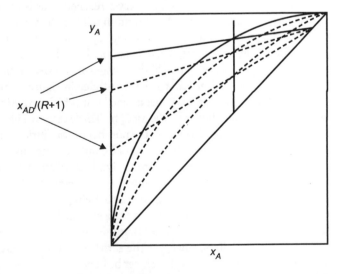

Figure 3.19 Effect of more difficult separation (closer boilers) on reflux ratio

While it is difficult to illustrate, as the desired distillate mole fraction approaches 1.0 for a fixed number of stages, the slope of the upper operating line increases, so the intercept decreases, which means that a larger reflux ratio is needed.

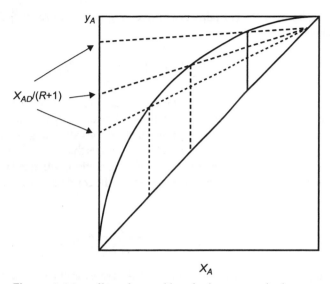

Figure 3.20   Effect of more dilute feed in more volatile component on reflux ratio

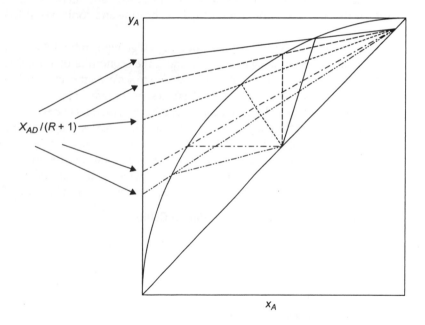

Figure 3.21   Effect of feed condition on reflux ratio

As the reflux ratio increases, all of the internal flowrates increase. As will be seen later, there is an associated increase in column diameter, but, as already stated, fewer equilibrium stages are required. However, as the internal flowrates increase, the amount of liquid that must be condensed and the amount of vapor that must be reboiled both increase, thereby increasing the heat duties of the condenser and reboiler. For a total condenser, the energy balances is

$$VH_V + Q_C = Lh_L + Dh_D \tag{3.48}$$

Since $h_L = h_D$, because the enthalpy does not change when a stream splits, and since $V = L + D$, Equation (3.48) becomes

$$Q_C = -V(H_V - h_D) \approx -V \sum_{i=1}^{N} x_{iD} \Delta H_{vap,i} \tag{3.49}$$

which is how the condenser heat duty can be calculated. The approximation in the last equality in Equation (3.49) includes the assumptions of ideal gas behavior and ideal liquid solutions. If these conditions are not valid, a standard thermodynamic method for calculating the enthalpies of the vapor and liquid mixtures would be used. Equation (3.49) also shows that as the flowrate of the vapor stream increases, the condenser heat duty increases. For a given separation, with fixed feed and outlet flowrates, the value of V increases as reflux ratio increases; therefore, the condenser heat duty also increases.

A similar development for the reboiler yields

$$Q_R = \overline{V}(H_{\overline{V}} - h_{\overline{L}}) \approx \overline{V} \sum_{i=1}^{N} x_{iB} \Delta H_{vap,i} \tag{3.50}$$

where, once again, it is seen that as the internal flowrates increase, the reboiler heat duty also increases.

The energy balances clearly show that as the reflux ratio increases, the heat duties increase, which means that the operating cost (virtually all heating and cooling utilities) of a distillation column increases.

Table 3.4 summarizes the reasons for needing a high reflux ratio. Table 3.5 summarizes the effect of increasing the reflux ratio. Figure 3.22 illustrates the economics of a distillation column, and it is noted that Figure 3.22 is an illustration and not drawn to scale. As the reflux ratio increases, the number of trays decreases, so the cost of the column decreases. As the reflux ratio increases further, the diameter increases, so the cost of the column increases. As the reflux ratio increases, the cost of energy (utilities) increases, and the cost of the condenser and reboiler also increases. If the equipment cost is put on the same basis as the operating cost (utilities), which is done by assuming that the equipment cost is equivalent to a loan with periodic payments, the resulting curve suggests that there is an optimum reflux ratio. EAOC stands for Equivalent Annual Operating Cost, which is the sum of the actual operating costs (utilities) and the periodic payment on the initial equipment cost. The location of this optimum value

Table 3.4   Reasons for Designing a Distillation Column with a Large Reflux Ratio

Separation is between close boilers
Feed dilute in more volatile component
Feed contains more vapor
Fewer stages desired (perhaps column is too tall)
Energy is inexpensive

Table 3.5   Economic Effect of Increasing Reflux Ratio

Design Parameter Effect	Economic Effect
Number of stages (column height) decreases	Equipment cost decreases
Diameter increases	Equipment cost increases
Condenser/reboiler heat duties increase	Operating cost (utilities) increase
	Condenser and reboiler cost increase

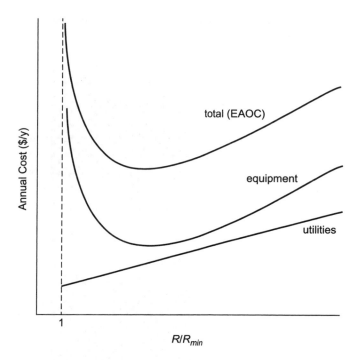

Figure 3.22   Optimization of distillation column

for R/R_{min} is a strong function of the cost of energy. In the mid-20th century, when energy was inexpensive, optimum values in the 1.5 to 1.8 range were recommended. In the latter part of the 20th century, when energy prices increased, optimum values in the 1.2 to 1.5 range were recommended. In the early 21st century, energy prices have fluctuated significantly, so it is important to remember that the optimum value changes with energy prices.

Packed Columns

A schematic of a packed column is shown in Figure 3.23. The concept is that liquid coats the packing, providing vapor-liquid surface area for mass transfer, similar to the vapor-liquid interface created by bubbles in tray columns. The packing can be random or structured. Random packing consists of small objects, for which there are a multitude of shapes, and the details are discussed later. These objects are randomly placed in a large, empty column. In contrast, structured packing consists of sections with a solid/void structure that are layered into the column. The key parameter is the interfacial area, introduced in Section 3.1.4. Separations in packed columns are continuous differential separations rather than staged separations. In continuous differential separations, equilibrium is assumed to exist at the vapor-liquid interface. The height of the column becomes the design parameter, as opposed to the number of stages.

The height can be obtained from the interfacial area/cross-sectional area, S, by defining a mass transfer area, a, that is specific to each type of packing, having units of mass transfer area/ packed volume. The packed volume includes the void fraction, which is specific to the packing dimensions and shape. Therefore,

$$S = aZ \tag{3.51}$$

where Z is the height of the column.

If the mass transfer area, a, of a packing is known, then the height, Z, can be calculated from S in Equation (3.51). Manufacturers provide specifications on packings, and there is an extensive

L_{in}, x_{in} V_{out}, y_{out}

L_{out}, x_{out} V_{in}, y_{in}

Figure 3.23   Model of a packed column

literature available for some of the most common packing materials. The interfacial area/cross-sectional area, S, can be calculated from the height and number of transfer units. For this explanation, vapor transfer units and heights are used, as described in Equation (3.24). The height of a transfer unit $H_V = V/F_y$ in Equation (3.22) can be calculated if the internal flowrates are known, since there are correlations available for the mass transfer coefficient F_y. The question is how to evaluate the integral for the number of transfer units, N_V, in Equation (3.23). For this type of analysis, a McCabe-Thiele diagram is drawn, just as for staged separations; however, stages are not stepped off. The integral must be evaluated separately above the feed and below the feed, so one limit on each integral is the feed composition. As the values of mole fraction are varied between the feed and the distillate (top of column) and the feed and bottom (bottom of column), Equation (3.16) can be used with the material balance line (operating line) to determine pairs of corresponding values of y and y_i. The graphical representation of the simultaneous solution for this problem is illustrated in Figure 3.24. Consistent with the two-film model in Figure 3.3, points on the operating line represent passing stream mole fractions at every point in the column. Similarly, points on the equilibrium curve represent the interface composition, assumed to be in equilibrium, at every point in the column. This set of pairs of (x, y) and (x_i, y_i) allows the integral in Equation (3.23) to be evaluated numerically. The total height of the column is the sum of the heights of the upper and lower sections, and the feed height is clearly defined. An important concept to understand is that Figure 3.24 clearly shows that equilibrium occurs at the interface at every location in the column, but that the equilibrium values are different at different values of the column height.

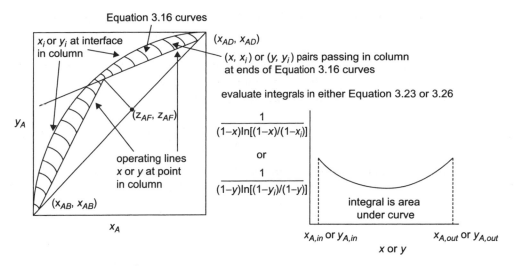

Figure 3.24   Solution of two-film model and material balances in distillation column

3.2.2.2   Mass Separating Agents

The McCabe-Thiele method can also be used for separations with mass separating agents. There are some differences. In binary distillation, both components are volatile. On any stage, the more volatile component is transferred from the liquid phase to the vapor phase, while the less volatile component is transferred from the vapor phase to the liquid phase. For mass separating agents, the idealized situation is that a solute is transferred from one phase to a different phase, where both solvents are immiscible. The phase pairs can be gas-liquid (absorption and stripping), liquid-liquid (extraction), solid-liquid (leaching, washing, adsorption), or solid-gas (adsorption). (The difference between gas [e.g., air] and vapor is that the former does not condense at or near typical operating conditions, whereas the latter can condense at typical operating conditions.)

The rigorous method is to solve Equations (3.3) and (3.4) simultaneously for every stage. For gas-liquid systems (absorption and stripping), energy balances might also be needed, since dissolving a gas into a liquid (*absorption*), HCl into water, for example, produces a heat of solution and a noticeable temperature change. For separations such as extraction and *adsorption* from a liquid, heat effects are usually negligible. If there are negligible heat effects and if the solvents are completely immiscible, the McCabe-Thiele method is applicable. If there is only one solute and two immiscible solvents and if the mole fractions are small enough, the total flowrates are approximately constant. Under these circumstances, Equation (3.4) can be rewritten as

$$\frac{y_{out} - y_{in}}{x_{in} - x_{out}} = \frac{L}{V} \tag{3.52}$$

where the subscript i has been dropped, since there is only one solute. This is the equation of a straight line on a plot of y versus x with slope L/V connecting the points representing the inlet and outlet mole fractions. This is illustrated in Figure 3.25, assuming countercurrent operation, with an arbitrary equilibrium curve. For multistage separations, countercurrent is the norm. If the two phases flow in the same direction, called *cocurrent separations*, once equilibrium is attained in the first stage, all subsequent stages would be useless, since the feed to those subsequent stages would already be at equilibrium.

Another difference between distillation and mass separating agents is that the operating line can be on either side of the equilibrium curve. As illustrated in Figure 3.25, if the solute transfers from the V phase to the L phase, $x_{out} > x_{in}$, and $y_{out} < y_{in}$, since each point on the operating line represents streams passing in the opposite direction. If the solute transfers from the L phase to the V phase, $x_{in} > x_{out}$, and $y_{in} < y_{out}$. If there are multiple stages, Equation (3.52) can be rewritten from one end of the cascade of stages to any intermediate stage

$$\frac{y - y_{in}}{x - x_{out}} = \frac{L}{V} \tag{3.53}$$

which is the equation of the operating line for the separation. Once this operating line is known, the stages can be stepped off, as illustrated in Figure 3.26.

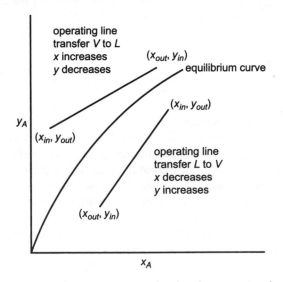

Figure 3.25   McCabe-Thiele construction for absorber (V to L) and stripper (L to V)

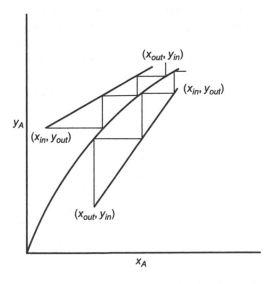

Figure 3.26   Stepping to count stages for absorber and stripper

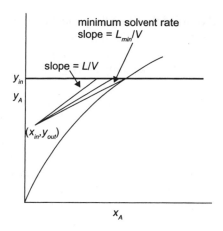

Figure 3.27   Minimum solvent rate for absorber

Consider a separation involving transfer from the V phase to the L phase. The inlet mole fractions, y_{in} and x_{in}, would be known along with the inlet flowrate of the V phase, V. The purpose of the separation is to remove solute from the V phase, so the inlet mole fraction would be known. The inlet mole fraction of the L phase would presumably be known and be very small, if not zero. There would probably be a target amount of solute to be removed or a target outlet mole fraction, so y_{out} would also be known. Figure 3.27 shows that the flowrate L can have multiple values for a fixed value of V, but as L gets smaller, resulting in a smaller slope, the operating line gets closer to the equilibrium curve, resulting in more steps, meaning more stages are needed. This is analogous to the trade-off between reflux ratio and number of stages in distillation. The destination solvent rate replaces the reflux ratio. There is also a minimum solvent rate, which is determined either by the intersection of y_{in} with the equilibrium curve or by a tangent, depending on the curvature of the equilibrium curve.

For continuous differential separations, the operating line and equilibrium curves are similar to those in staged separations, since there is an equilibrium relationship and since the operating line represents the tray-to-tray mass balances. The method described in Section 3.2.2.1 would be used to evaluate the integral for the number of transfer units.

3.2.3   Dilute Solutions—The Kremser and Colburn Methods

Under the following conditions, an analytical solution can be found for countercurrent separations. The assumptions are

- Dilute solutions
- Total stream flowrates remain constant
- Equilibrium relationship is linear
- Isothermal operation
- No energy associated with transfer between phases

The results are presented without derivation. The result for transfer from the V phase to the L phase for staged separations is called the *Kremser method* (Kremser, 1930; Souders and Brown, 1932) and is

$$\frac{y_{out} - mx_{in}}{y_{in} - mx_{in}} = \frac{1 - A}{1 - A^{N+1}} \tag{3.54}$$

where $A = L/mV$ and is called the *absorption factor* (because this method is traditionally applied to gas absorption into a liquid, but it is valid for any separation that is consistent with the assumptions), N is the number of stages, and m is the partition coefficient from the equilibrium expression $y = mx$. Equation (3.54) can be written explicitly for N as

$$N = \frac{\ln\left[\left(1 - \frac{1}{A}\right)\left(\frac{y_{in} - mx_{in}}{y_{out} - mx_{in}}\right) + \frac{1}{A}\right]}{\ln A} \tag{3.55}$$

There is no explicit relationship for A. For transfer from the L phase to the V phase, the equivalent relationship is shown in Table 3.6. When $A = 1$, Equations (3.54) and (3.55) are indeterminate. The results for such a case, obtained from application of L'Hôpital's rule, are also presented in Table 3.6.

For continuous differential separations with transfer from the V phase to the L phase, the relationships are attributed to Colburn (1939) and are

$$\frac{y_{in} - mx_{in}}{y_{out} - mx_{in}} = \frac{\exp\left[N_{oV}\left(1 - \frac{1}{A}\right)\right] - \frac{1}{A}}{1 - \frac{1}{A}} \tag{3.56}$$

and

$$N_{oV} = \frac{\ln\left[\left(1 - \frac{1}{A}\right)\left(\frac{y_{in} - mx_{in}}{y_{out} - mx_{in}}\right) + \frac{1}{A}\right]}{1 - \frac{1}{A}} \tag{3.57}$$

where N_{oV} is defined in Table 3.2.

Table 3.6 Expressions for Kremser and Colburn Methods

Direction of Solute Transfer	Explicit in Mole Fraction	Explicit in N, N_{oV}, or N_{oL}	If $A = 1$	Comments
$V \rightarrow L$ staged	$\dfrac{y_{out} - mx_{in}}{y_{in} - mx_{in}} = \dfrac{1 - A}{1 - A^{N+1}}$	$N = \dfrac{\ln\left[\left(1 - \frac{1}{A}\right)\left(\frac{y_{in} - mx_{in}}{y_{out} - mx_{in}}\right) + \frac{1}{A}\right]}{\ln A}$	$\dfrac{y_{out} - mx_{in}}{y_{in} - mx_{in}} = \dfrac{1}{1 + N}$	$A = L/mV$
$L \rightarrow V$ staged	$\dfrac{x_{out} - y_{in}/m}{x_{in} - y_{in}/m} = \dfrac{1 - S}{1 - S^{N+1}}$	$N = \dfrac{\ln\left[\left(1 - \frac{1}{S}\right)\left(\frac{x_{in} - y_{in}/m}{x_{out} - y_{in}/m}\right) + \frac{1}{S}\right]}{\ln S}$	$\dfrac{x_{out} - y_{in}/m}{x_{in} - y_{in}/m} = \dfrac{1}{1 + N}$	$S = mV/L$
$V \rightarrow L$ continuous differential	$\dfrac{y_{in} - mx_{in}}{y_{out} - mx_{in}}$ $= \dfrac{\exp\left[N_{oG}\left(1 - \frac{1}{A}\right)\right] - \frac{1}{A}}{1 - \frac{1}{A}}$	$N_{oV} = \dfrac{\ln\left[\left(1 - \frac{1}{A}\right)\left(\frac{y_{in} - mx_{in}}{y_{out} - mx_{in}}\right) + \frac{1}{A}\right]}{1 - \frac{1}{A}}$	$\dfrac{y_{out} - mx_{in}}{y_{in} - mx_{in}} = \dfrac{1}{1 + N_{oV}}$	$A = L/mV$
$L \rightarrow V$ continuous differential	$\dfrac{x_{in} - y_{in}/m}{x_{out} - y_{in}/m}$ $= \dfrac{\exp\left[N_{oL}\left(1 - \frac{1}{S}\right)\right] - \frac{1}{S}}{1 - \frac{1}{S}}$	$N_{oL} = \dfrac{\ln\left[\left(1 - \frac{1}{S}\right)\left(\frac{x_{in} - y_{in}/m}{x_{out} - y_{in}/m}\right) + \frac{1}{S}\right]}{1 - \frac{1}{S}}$	$\dfrac{x_{out} - y_{in}/m}{x_{in} - y_{in}/m} = \dfrac{1}{1 + N_{oL}}$	$S = mV/L$

Table 3.6 summarizes the relationships presented for transfer from the V phase to the L phase and also presents the relationships for transfer from the L phase to the V phase. In the latter case, S is the stripping factor, which is the inverse of the absorption factor.

The Kremser and Colburn results are often presented in graphical form, and they are shown in Figures 3.28 and 3.29, respectively. Although it is easy to solve the equations, even for A, similar to the McCabe-Thiele method for distillation, a thorough understanding of these graphs provides a full conceptual understanding of separations involving mass separating agents.

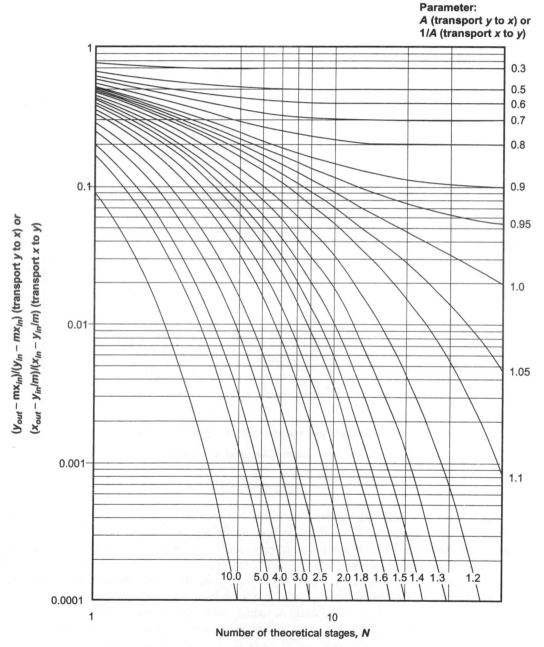

Figure 3.28   Graph for Kremser equation for staged separations

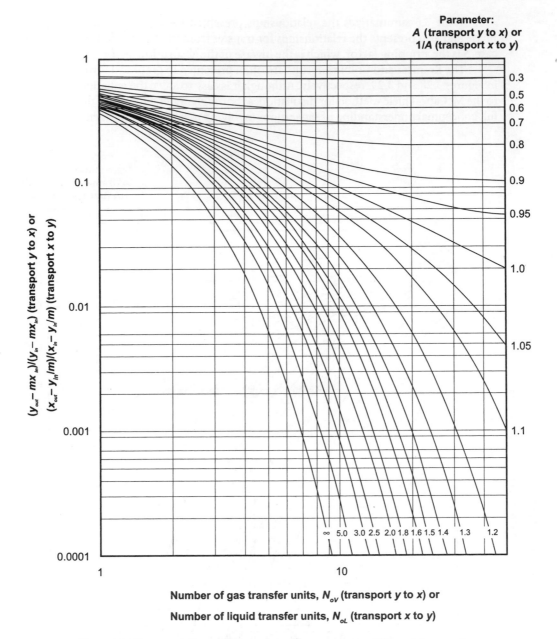

Figure 3.29   Graph for Colburn equation for continuous differential separations

Consider a separation with transfer from V to L. It is desired to remove solute from the V phase such that $y_{out}/y_{in} = 0.1$ using a pure solvent, so $x_{in} = 0$. Point a on Figure 3.30, which is a blown-up portion of Figure 3.28, shows one possible solution to the problem, $N \approx 5.0$ and $A = 1.2$. Moving along the horizontal line in either direction shows that the same separation can be accomplished by increasing N and decreasing A, and vice versa. Since $A \propto L$, this illustrates the trade-off between the number of stages and the solvent flowrate. If it is desired to improve the separation such that $y_{out}/y_{in} = 0.005$, there are two possible methods. One is to increase N at constant A, adding more stages, which makes sense. This is illustrated by following the $A = 1.2$ line down to Point b on Figure 3.30. Another possibility is to increase A

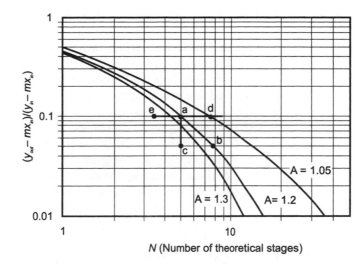

Figure 3.30   Basic concepts of separations involving mass separating agents

at constant N, which is illustrated by the vertical line down to Point c on Figure 3.30. There are two ways to increase A, increase the solvent flowrate L, which should make sense. Another method is to decrease m. Since $y = mx$ at equilibrium, decreasing m decreases the ratio y/x, which means that the L-phase mole fraction at equilibrium increases and/or the V-phase mole fraction decreases at equilibrium, favoring the L-phase. For the specific application of gas absorption, an expression for m might be $m = P^*/P$. If the pressure increases, m decreases, which makes sense since increasing the pressure favors the liquid phase. If the temperature decreases, P^* decreases, so m decreases, which makes sense since decreasing the temperature favors the liquid phase. For a given target separation, if a certain A value is needed, an unfavorable m value, which is a large value for absorption, can be overcome by increasing L as much as needed. While this works in principle, there could be operability issues.

To generalize, even though all separations involving mass separating agents do not follow the assumptions in the Kremser or Colburn methods, a thorough understanding of those graphs provides a thorough understanding of the trends in separations involving mass separating agents. To reiterate what is illustrated in Figure 3.30, Point a shows a target separation, as defined by a value on the y-axis (which is the same y-axis as in Figures 3.28 and 3.29). It is being accomplished with a certain number of stages (or transfer units) at a certain A (or S) value. The same level of separation, defined by the same value of the y-axis, can be achieved with a lower A value and more stages (Point d) and a larger A value and fewer stages (Point e). This illustrates a trade-off between the absorption (stripping) factor and the number of stages (transfer units). Within the absorption (stripping) factor, for the purposes of this example, it is assumed that the equilibrium constant cannot be changed and that the value of L (V) is fixed by upstream demands (meaning that the flowrate of material that must be processed in the separator is fixed). A higher A (S) value corresponds to a higher liquid flowrate (vapor flowrate). This means that the trade-off between absorption (stripping) factor and the number of stages (transfer units) is actually a trade-off between liquid (vapor) flowrate and the number of stages (transfer units). Since A (S) is the key parameter, this also demonstrates how to overcome an unfavorable equilibrium condition. For transfer from V to L, it would be desirable for m to be less than unity, meaning the mole fraction in the liquid phase is always less than that in the vapor phase. This means that the equilibrium favors the liquid phase. However, if m is greater than unity, which alone reduces the value of A (S), the value of A (S), the unfavorable equilibrium can be offset by increasing the value of the destination solvent, L (V). There are equipment limitations to the relative flowrates of the two phases that are dependent of the specific phases involved, and these will be discussed later in the chapter. Finally, at a constant number of stages (transfer units), if A (S) is increased, the value of the y-axis decreases, which means a better separation. This is illustrated by the vertical line through Points a and c in Figure 3.30.

If $A < 1$, where A is the absorption factor, $L/(mV)$, there is a limiting behavior. It can be seen that for $A < 1$, increasing the number of stages does not improve the separation beyond an asymptotic limit. Normally, it would be expected that increasing the number of stages would improve the separation, but this is only true for $A \geq 1$. The reason can be understood by looking at the McCabe-Thiele diagrams shown in Figure 3.31. If the equilibrium line is defined by $y = mx$, and the slope of the operating line is L/V, as shown in Section 3.2.2, then, if $A < 1$, $L/V < m$, so the operating line approaches the equilibrium line at the bottom of the column (see also Figure 3.25), which is the most concentrated part of the column. Therefore, since the solution is approaching saturation at the concentrated portion of the column, where solvent capacity is needed most, the ability to transfer solute is limited, causing the observed asymptote. If $A \geq 1$, $L/V > m$, equilibrium is approached at the less concentrated portion of the column. For transfer in the opposite direction, while the operating line is below the equilibrium line, the argument is exactly the same.

For continuous differential separations, the discussion just presented is identical, except that the number of transfer units replaces the number of stages. Once again, it is important to remember that the number of transfer units is not numerically equal to a number of stages. However, as the number of transfer units increases, the required interfacial area increases (Equation [3.24]). For a gas-liquid separation, this means that the height of the packed column increases. Similarly, as the number of trays increases, the height of a tray column increases.

For other separations, such as liquid-liquid extraction, if the equilibrium is linear and the solutions dilute, the Kremser and Colburn methods can be used, and all of the same trends are true. Quite often, liquid-liquid equilibrium obtained experimentally is expressed in terms of mass fraction, so the flowrates become mass flowrates. In most textbooks, the letters used to represent these flowrates are different to conform with chemical engineering jargon. The output stream from the feed stream (F) is called the *raffinate* (R). The destination solvent input stream is denoted S, and its output stream is called the *extract* (E). To avoid confusion of letters, an alternative method is to retain the symbols L and V and use L for the heavier (more dense) phase (often water) and V for the lighter (less dense) phase (usually organic). In reality, as long as the equilibrium expression is written correctly, the choice of symbols used is irrelevant.

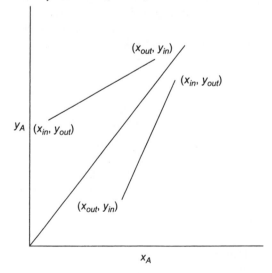

Figure 3.31   Approach to equilibrium at more concentrated end of column

---

Example 3.3

Consider a gas-liquid separation in which acetone in air is absorbed into water to purify the exhaust air. The equilibrium is described by Raoult's law, and the vapor pressure is given by

$$\ln P^*(\text{atm}) = 10.92 - \frac{3598.4}{T(\text{K})} \tag{E3.3a}$$

An absorber is to be designed to treat 100 kmol/h of air containing a mole fraction of 0.05 acetone and reduce it to a mole fraction of 0.0001. Pure water is used as the solvent at a rate of 50 kmol/h.

  a. At a temperature of 35°C and a pressure of 1.25 atm, how many equilibrium stages are needed in a tray tower?

  b. At the same conditions as in Part (a), how many transfer units are needed in a packed tower?

  c. For the tray tower, suggest another set of conditions that would accomplish the desired separation. Any parameter may be changed other than the inlet gas conditions.

  d. Suppose it is suggested that an existing tray tower with 7 stages be used for this separation. At the same temperature and pressure, what liquid flowrate is needed?

  e. Suppose it is suggested that an existing packed tower with 15 transfer units be used with all other operating conditions unchanged. What outlet mole fraction is expected?

Solution

  a. Assuming Raoult's law, $m = P^*/P = 0.375$. Since the transfer is from L to V, the parameter is A, and $A = 50/0.375/100 = 1.333$. From the equation in Table 3.6, explicit in N, for $V \rightarrow L$ transfer, $N = 8.98$. Since there are no fractional stages, the value would be rounded up to 9 stages.

  b. With all of the parameters identical to the solution to Part (a), the equation in Table 3.6, explicit for N_{oV}, for $V \rightarrow L$ transfer, yields $N_{oV} = 10.34$. This value would not be rounded, because, to get the column height, it would be multiplied by H_{oV}, and the resulting value of the height does not have to be an integer.

  c. There are an infinite number of solutions. For example, if the number of stages were increased to 10, A would become about 1.28. There are then an infinite number of temperature, pressure, and L values that could equal this value.

  d. In this case, the desired separation, that is, all mole fractions on the y-axis of the Kremser graph (Figure 3.28), are known, and the number of stages is known. The unknown value is A. The equation for $V \rightarrow L$ transfer in Table 3.6 explicit in the mole fraction function and containing N must be solved for A. The value is 1.504. With m and V known, solving for L yields $L = 56.4$ kmol/h. This makes sense, since, with fewer stages, the destination solvent rate must be increased.

  e. In this case, the value of A and the number of stages are known. The unknown value is the mole fraction function. The equation for $V \rightarrow L$ transfer in Table 3.6 explicit in the mole fraction function and containing N_{oV} must be solved for y_{out}. The result is $y_{out} = 8.41 \times 10^{-6}$. This makes sense, since, with more transfer units with everything else held constant, there should be more solute removal from the original phase.

3.3   EQUIPMENT

3.3.1   Drums

Drums are used in partial vaporization/condensation operations and in distillation columns, among other uses. A brief review of these drums is presented here, and more details of their design are presented in Chapter 5.

As discussed in Section 3.2.1, after a partial vaporization/condensation, the vapor-liquid mixture must be allowed to disengage so that separate vapor and liquid streams can be removed continuously. These drums, often called *knockout drums*, are usually vertically oriented with a typical height-to-diameter ratio between 2.5/1 and 5/1.

Distillation columns have reflux drums, as illustrated in Figure 3.17. The purpose of the reflux drum is to smooth out fluctuations, particularly in the reflux stream, by maintaining and controlling the liquid level in the drum. It is also important to understand that the reflux drum and reflux pump, while drawn at the top of the column, are typically located closer to the bottom of the column. It is usual to locate equipment such as pumps closer to the ground because it makes maintenance easier, and often the reflux drum will be located directly above the reflux pump, which is located at ground level. The height of the drum above ground level is often determined on the basis of the required net positive suction head ($NPSH_R$) of the pump (see Chapter 1, Section 1.5.2). Therefore, the reflux pump must be designed to supply the head necessary (plus a small frictional loss) for the reflux stream to flow from ground level to the top tray of the tower. Reflux drums are usually horizontally oriented, with a typical length-to-diameter ratio between 2.5/1 and 5/1.

While not specifically related to distillation equipment, drums are also used when mixing liquid streams, with the most common example being mixing recycled liquid and feed liquid. The liquid level in these drums is controlled by varying the fresh feed flowrate. These mixing drums serve a similar purpose as reflux drums; the downstream flowrate is maintained by controlling the liquid level in the drum. Mixing drums are usually vertically oriented, with a height-to-diameter ration between 2.5/1 and 5/1.

3.3.2   Tray Towers

3.3.2.1   Types of Trays, Flow Patterns, Downcomers, and Weirs

The simplest type of tray in a distillation column is a sieve tray, which is a plate containing holes, usually less than 1-in diameter. The purpose of the holes is for the vapor to flow up through the holes and form bubbles that pass upward through a level of liquid on the stage. The presence of bubbles creates surface area for mass transfer between the phases. The liquid flows across the tray, and the liquid level on the tray is maintained by a weir. The liquid flows over the weir and down to the tray below through a channel that is called a *downcomer*. The simplest flow pattern is across one tray and then across the tray below in the opposite direction. The liquid flows down due to gravity, while the vapor flows upward due to the pressure drop, since it has been established that both the temperature and pressure are higher at the bottom of the column. Figure 3.32 illustrates the internals of a distillation column.

The holes in the trays can also have caps or valves. The former are bubble cap trays, and the latter are valve trays. Bubble cap trays were the earliest ones used, but sieve and valve trays have replaced them (Kister, 1992). Figure 3.33 illustrates various types of caps and one valve. Bubble caps make smaller bubbles than sieve trays, which can be seen by the design of the holes in the cap. However, they are more expensive than sieve trays. Valve trays have a cap on the hole without the extra holes for bubbles. Bubble caps and valves move up and down with the vapor flowrate, so they are more amenable than sieve trays to scale-down. Since, at low vapor flowrates, the bubble cap or valve covers the top of the hole, bubble cap and valve trays are less prone than sieve trays to a phenomenon called weeping. Weeping is where the liquid falls through the holes in the trays due to low vapor flowrate, and it is an undesirable situation. However, bubble cap trays are more expensive than valve and sieve trays, and they are more prone to fouling, because the small holes can collect solids, and the smaller bubbles mean a higher pressure drop. While sieve trays are more prone to weeping, they are easier to clean during plant shutdown and are recommended for services in which fouling or corrosion is anticipated. There are also fixed valve trays, in which the valves covering the holes do not move. They are not prone to sticking shut and erosion, and they are better suited for process upsets. More details are available elsewhere (Hebert and Sandford, 2016).

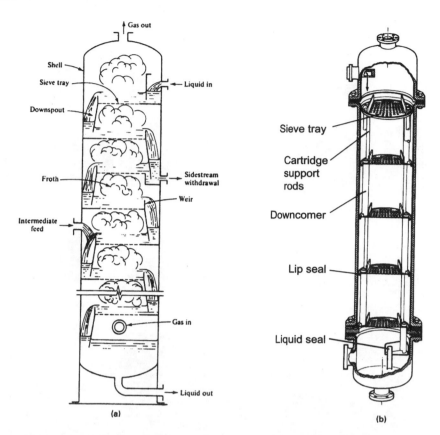

Figure 3.32 Details of internal construction of a distillation column (Couper et al. [2012])

Figure 3.32 suggests a cross-flow pattern of liquid on alternative trays. This is the most common flow pattern, because of its simplicity. However, as the flow path of fluid becomes long (for large diameter trays) it may be desirable to reduce the flow path resulting in other flow patterns (double-pass trays, triple-pass trays, etc.). One reason for this is the difficulty in building a large-diameter column with perfectly horizontal trays, because very large-diameter trays are "field erected" in sections; they are not in one piece. Figure 3.34 shows one alternative, a two-pass tray, in which the flow is edge to center and back. One advantage of this flow pattern is that the flow is split, so that there is less liquid loading on a given tray. Another advantage is that variation in the height of liquid above the tray is more uniform and bubbles do not bypass the part of the column near the weir. The two-pass flow pattern is more common for large-diameter trays and high liquid rates. Its main disadvantages are the added complexity and cost.

As will be discussed later, the weir height, which is the main component of the level of liquid on the tray, affects the column pressure drop, since the vapor flowing upward must overcome the liquid head on each tray. While this suggests a small weir height is desirable, for a large-diameter column, it may be difficult to guarantee completely level trays, which may lead to puddles and dry spots on the tray causing poor liquid-vapor contact and low efficiencies. Therefore, weir heights less than 2 in are rare, although possible, when a low pressure drop is required. Another limitation is entrainment, which is a mist of liquid formed from the froth on the tray being carried up to the tray above. This results in remixing of liquid from the tray below to the tray above, counteracting

(a)

(b)

Figure 3.33   (a) Different bubble cap designs, (b) a valve for a valve tray (Reprinted and electronically reproduced by permission from Wankat [2017])

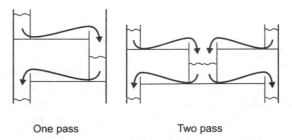

One pass   Two pass

Figure 3.34   Comparison of one-pass and two-pass trays. (Pilling [2006], reproduced by permission of Sulzer Chemtech US)

any change in concentration that occurred in the liquid phase between trays. Clearly, this is counterproductive to the objective of the separation. If the weir height is kept low, entrainment is less likely. Typical weir heights are in the 2-in to 4-in range and are always less than one-half the distance (spacing) between the trays.

3.3.2.2 Flooding: Entrainment, Tray Spacing, Column Height, and Column Diameter

Determining the number of trays is only the first step in designing a tray column. The height is based on the tray spacing. The diameter is based on a concept known as *flooding*, which can be caused by excessive entrainment. Additionally, the tray spacing affects flooding.

Entrainment is the situation where the upward-flowing vapor carries liquid from the tray below to the tray above. Effectively, this results in a mixing of liquids at different compositions, negating or reducing the separation that has occurred. Flooding can be caused by excessive entrainment.

The column diameter is based on flooding. Another way to think about flooding is that, if the upward vapor velocity is too large, the drag force on the liquid exceeds gravity, and the liquid does not fall through the column. A precursor to flooding is called *loading*, and this is manifested by a rapid increase in the pressure drop in the column. Based on the relationship between mass flowrate and velocity developed in Chapter 1, $\dot{m} = \rho A v$, for a given mass flowrate, the velocity can be reduced by increasing the cross sectional area of the column, that is, by increasing the column diameter. Typically, columns operate in the range of 75% to 80% of flooding; therefore, to determine the diameter of the column, the flooding velocity must be calculated.

Tray spacing also affects flooding. The typical tray spacing is between 18 and 24 in. This allows for easy access for maintenance. Tray spacings up to 36 in can be found, and, for good reason, smaller tray spacings are possible. The column height is the number of trays times the tray spacing plus additional height at the top and bottom of the column. In addition, the column is usually raised off of the ground by means of a column skirt. One reason is that the bottom product leaves the column as saturated liquid, and additional head is needed to avoid the NPSH issues discussed in Chapter 1 if a pump is needed to provide additional pressure to move the bottom product to its next location.

Flooding velocities are calculated from empirical correlations based on experimental results. The most commonly available result is for sieve trays. There are so many different kinds of bubble caps and valves that there is no single, universally applicable empirical correlation available for these types of trays. Sieve trays, on the other hand, are simple and generic. The correlation attributed to Fair and Matthews (1958) is one of the most commonly cited, and it is presented graphically in Figure 3.35. In Figure 3.35, the x-axis is a parameter usually called F_{lv}, which includes the ratio of **mass** flowrates through the column. The y-axis is a flooding capacity factor that allows calculation of the flooding velocity. Since it has been established that, in the most general case, these flowrates are different on every tray, process simulators perform the flooding calculation on every tray and provide a profile of diameters. This is called *tray sizing*, and the user input must include the tray spacing. There is also active area versus total area. The active area does not include the cross-sectional area for the downcomers and is the active area for the upward vapor flow. Typically, the active area is 85% to 90% of the total area, the column diameter is calculated on the basis of the total area, and this parameter is also input to a simulator. Typically, once the tray sizing profile is obtained from a simulator, a diameter is chosen, and a tray rating is performed for the given diameter, which results in a percentage-of-flooding profile for the column. As long as there are no trays that are flooding or are close to weeping (<40% of flooding for sieve trays), that diameter is appropriate. If a constant diameter column results in operation of some trays above the 80% flooding limit and some below the weeping limit, then the use of different column diameters above and below the feed may have to be considered. This situation is not uncommon for high vacuum operation.

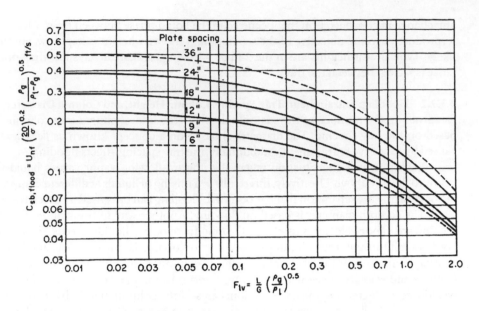

Figure 3.35 Capacity factor for flooding of sieve trays from Fair and Matthews (1958) (Reprinted by permission from Petroleum Refiner, copyright 1958, Gulf Pub. Co.)

To estimate the column diameter without a simulator, McCabe-Thiele approximate results are used. In this method, the flowrates are constant in each section of the column, so there are only two possible calculations, above and below the feed. An equation has been suggested to determine the limiting (larger diameter) section (Wankat, 2017). Intuitively, it may seem that the section with the larger molar flowrates will require a larger diameter. However, while the chemistry of phase equilibrium is based on molar flows and compositions, equipment sizing is based on mass flowrates. Therefore, the molecular weight also affects the mass flowrate. Additionally, based on the relationship between mass flowrate and velocity, $\dot{m} = \rho A v$, the density affects the velocity, so a large temperature difference can be significant. Finally, especially in vacuum columns, the pressure drop can be a large fraction of the column pressure, which means that the vapor density can change significantly through the column. In general, the most desirable feed conditions are saturated liquid, saturated vapor, or a vapor-liquid mixture, because the feed should match column conditions (saturated). This suggests that the molar flowrates below the feed will be larger than those above the feed. Quite often, the higher boiler, which is at the bottom of the column, has the larger molecular weight. Thus far, it seems like designing for the bottom of the column is limiting. However, the pressure and temperature at the bottom of the column are both larger. For an ideal gas, the density $\rho = PM/RT$, where M is the molecular weight, so, depending on the exact numbers, the effects of temperature and pressure may offset. For vacuum columns, the pressure and therefore density will be significantly smaller above the feed. While it is usually most desirable to design a column with a uniform diameter, vacuum columns often have different diameters, larger above the feed and smaller below the feed, because the velocity above the feed is so much larger than the velocity below the feed. In a standard column, if the diameters above and below the feed are not compatible, that is, one section is so far below flooding to create weeping, the feed can be made into a vapor-liquid mixture to equalize the mass flowrates in both sections. In multicomponent distillation columns, flowrates change on every tray, and there are methods such as pump-arounds that can be used to avoid localized flooding (Wankat, 2017). At the level of the estimated diameter calculation, it is probably best to determine the diameter both above and below the feed.

The curves in Figure 3.35 have been fit to the equation (Wankat, 2017):

$$\log_{10} C_{sb,f} = -a - b\log_{10} F_{lv} - c\left[\log_{10} F_{lv}\right]^2 \tag{3.58}$$

where the constants are given in Table 3.7 for the different tray spacings.

The parameter F_{lv} is defined as

$$F_{lv} = \frac{LM_L}{VM_V}\left(\frac{\rho_V}{\rho_L}\right)^{0.5} \tag{3.59}$$

where L and V are molar flowrates, and M_L and M_V are molecular weights used to convert the molar flowrates to mass flowrates.

The nomenclature in Equation (3.59) and Figure 3.35 must be understood. First of all, throughout the history of separations, the variables V and G have been used somewhat interchangeably. V stands for vapor, while G stands for gas. V is being used in this chapter for all gas-vapor phases, but the original correlation of Fair and Matthews (1958) used G. On the x-axis of Figure 3.35, it is assumed that L and G are in mass units; Equation (3.59) expresses this clearly. On the y-axis of Figure 3.35, the parameter $C_{sb,flood}$, which is abbreviated here as $C_{sb,f}$, is

$$C_{sb,f} = u_f\left(\frac{20}{\sigma}\right)^{0.2}\left(\frac{\rho_L - \rho_V}{\rho_V}\right)^{0.5} \tag{3.60}$$

where u_f is the flooding velocity **in ft/sec**, and σ is the surface tension of the liquid phase **in dyne/cm²**. The units of Equation (3.60) are exactly as stated; there is no metric equivalent available. Since the surface tension term is raised to a small power, only a small amount of error is introduced by assuming the entire term is unity.

The algorithm to determine the column diameter required is as follows:

1. Determine all internal flowrates in molar units. For distillation, this involves knowing the reflux ratio and the feed condition. For mass separating agents, the flowrates are likely given. Convert these to mass units. For distillation, it is usually assumed that the molecular weights in the vapor and liquid phases are equal, because the phases are not that different in composition. For mass separating agents, especially in dilute solutions, the molecular weight is approximately that of the solvent. (This may not be true for non-dilute vapor phases.)

2. Determine the liquid and vapor densities. The liquid density must be found from a reference. The vapor density can be estimated using the ideal gas law or another equation of state. The top and bottom temperatures and pressures are used, depending on which section the calculation is based.

Table 3.7   Constants in Equation (3.58) (Wankat, 2017)

Tray Spacing (in)	a	b	c
6	1.1977	0.53143	0.18760
9	1.1622	0.56014	0.18168
12	1.0674	0.55780	0.17919
18	1.0262	0.63513	0.20097
24	0.94506	0.70234	0.22618
36	0.85984	0.73980	0.23735

3. Calculate F_{lv} using Equation (3.59).

4. Use Equation (3.58) or Figure 3.35 to determine $C_{sb,f}$.

5. Calculate u_f (which will be in ft/sec) from Equation (3.60). If metric units are desired, convert it to m/s.

6. Calculate the actual velocity (u_{actual}) as a percentage of the flooding velocity.

7. Calculate the column active area from

$$A = \frac{VM_V}{\rho u_{actual}} \tag{3.61}$$

where VM_V is the mass flowrate $\dot{m}$.

8. Determine the actual column area as actual area = active area/fraction active area.

9. Determine the column diameter from the actual area.

Example 3.4

A distillation column that has a total condenser and a partial reboiler is fed 500 kmol/h of an equimolar mixture of pentane and hexane. The feed is saturated liquid. It is necessary to produce outlet streams containing 95 mol% pentane and 99 mol% hexane. The distillation column contains a partial reboiler and a total condenser and operates at 1.4 times the minimum reflux ratio, which is a reflux ratio of 1.06. It has been determined that the column requires 15 equilibrium stages with the feed on Stage 5. The pressure at the bottom of the column is 2 bar. Assume that the McCabe-Thiele approximation is valid.

a. Determine the exit flowrates, D and B.

b. Determine all of the internal flowrates.

c. Determine the required column diameter for sieve trays with a 24-in tray spacing operating at 75% of flooding with 88% active area.

Data provided:

	Antoine Coefficients			
	A	*B*	*C*	SG
Pentane	6.85296	1064.84	232.012	0.626
Hexane	6.87601	1171.17	224.408	0.655

Solution

Figure E3.4 illustrates this column.

a. The outlet flowrates are obtained from an overall mole balance and a pentane mole balance.

$$500 = D + B \tag{E3.4a}$$

$$500(0.5) = 0.95D + 0.01B \tag{E3.4b}$$

The result is that $D = 261$ kmol/h and $B = 239$ kmol/h.

b. The internal flowrates are calculated from the top to the bottom, since the reflux ratio is the additional piece of information needed to start the calculation. Since $R = L/D$, and R and D are known, $L = (1.06)(261) = 277$ kmol/h. From a mass balance on the condenser, $V = L + D$, $V = 538$ kmol/h. Since the feed is saturated liquid, it is assumed that all of the feed goes to the bottom of the column. Therefore, $\bar{L} = L + F = 777$ kmol/h, and $V = \bar{V} = 538$ kmol/h. If the feed contained a vapor-liquid mixture, then the vapor flowrate above the feed would be

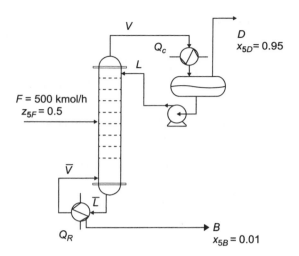

Figure E3.4 Illustration of Example 3.4

larger than that below the feed by the flowrate of vapor in the feed. If the feed was saturated vapor, then the liquid flowrate would be assumed identical between the top and bottom, and the vapor flowrate above the feed would be the vapor flowrate below the feed plus the feed. These flowrates are all needed for the flooding calculation to determine the column diameter.

c. Since all of the liquid feed goes to the bottom of the column, the initial design will be for the bottom of the column, because it is assumed that the flowrates will be higher below the feed, requiring a larger diameter. For completeness, the top diameter will also be determined.

To use Equation (3.59), the vapor density must be calculated. Ideal gas will be assumed, so that

$$\rho_V = \frac{PM}{RT} = \frac{(2\,\text{bar})(96\,\text{kg/kmol})}{(0.08314\,\text{m}^3\text{bar/kmol/K})(273.15+92.62\,\text{K})} = 6.31\,\text{kg/m}^3 \qquad \text{(E3.4c)}$$

In Equation (E3.4c), the molecular weight of hexane is used for the bottom of the column, and the bottom temperature is obtained from Antoine's equation at 2 atm for pure hexane. This is clearly an assumption. On the bottom tray, since there is almost pure hexane, the assumption may be valid. However, the temperature decreases on every tray above the feed, as does the pressure and the molecular weight (since pentane has a lower molecular weight). So, without knowing the pressure drop, the pressure is assumed to be uniform throughout the column. The small change in vapor density within the column is not significant relative to the other assumptions in this approximate calculation. Additionally, the bottom temperature is really the bubble point of the mixture. Overall, this method is clearly an estimation. In a process simulator, for example, the calculation that follows is performed rigorously on every tray.
From Equation (3.59),

$$F_{lv} = \frac{\bar{L}M_L}{\bar{V}M_V}\left(\frac{\rho_V}{\rho_L}\right)^{0.5} = \frac{777}{538}\left[\frac{6.31}{626}\right]^{0.5} = 0.145 \qquad \text{(E3.4d)}$$

The molecular weight is omitted from the calculation in Equation (E3.4d), because it is assumed that the compositions of vapor and liquid on a tray are about the same, so the average molecular weights are about the same. This only holds for distillation. For absorbers and strippers, the molecular weight must be included, and a good starting approximation is the molecular weight of the solvent.

Applying Equation (3.58) for 24-in tray spacing, using the values in Table 3.7, yields $C_{sb,f}$ = 0.3054. The flooding velocity is then calculated from Equation (3.60):

$$u_f = C_{sb,f}\left(\frac{\sigma}{20}\right)^{0.2}\left(\frac{\rho_V}{\rho_L-\rho_v}\right)^{0.5} = 0.3054\left(\frac{6.31}{655-6.31}\right)^{0.5} = 3.025 \text{ ft/sec} \qquad \text{(E3.4e)}$$

The term involving surface tension is assumed to be unity. In the worst case, if water was involved, with a surface tension of 72 dyne/cm^2 at room temperature, the term would be $(72/20)^{0.2}$ = 1.3. For organic molecules, the surface tension at room temperature is on the order of 30 dyne/cm^2, so omitting this term is within the approximations of this calculation. At 75% of flooding, the actual vapor velocity is u = 0.75(3.025 ft/sec)(0.3048 m/ft) = 0.692 m/s. Now, the active area is calculated from

$$A_{active} = \frac{\dot{m}}{\rho u} = \frac{\bar{V}M}{\rho u} = \frac{(538 \text{ kmol/h})(96 \text{ kg/kmol})}{(6.31 \text{ kg/m}^3)(0.692 \text{ m/s})(3600 \text{ s/h})} = 3.28 \text{ m}^2 \qquad \text{(E3.4f)}$$

So, the actual area, A_{actual}, is 3.28/0.88 = 3.73 m^2. Solving for the diameter,

$$d = \left(\frac{4A_{actual}}{\pi}\right)^{0.5} = \left(\frac{4(3.73 \text{ m}^2)}{\pi}\right)^{0.5} = 2.18 \text{ m} \qquad \text{(E3.4g)}$$

If the entire calculation was repeated for the top, using the molecular weight of pentane (72 kg/kmol) and the top section density (5.23 kg/m^3, calculated using the top temperature of 58.05°C) and flowrates, the top diameter would be 1.73 m. To determine whether the top section will perform correctly, the actual velocity above the feed is calculated assuming that the bottom active area is the same as the top active area, that is, the column is at uniform diameter,

$$u_{actual} = \frac{VM}{\rho A_{active}} = \frac{(538 \text{ kmol/h})(72 \text{ kg/kmol})}{(5.23 \text{ kg/m}^3)(3.28 \text{ m}^3)(3600 \text{ s/h})} = 0.63 \text{ m/s} \qquad \text{(E3.4h)}$$

and the percentage of flooding in the top section is (0.63 m/s)/[(0.3048 m/ft)(3.025 ft/sec)] = 0.68. Since this is not small enough for weeping to occur, then a 2.18 m diameter column will work.

3.3.2.3 Tray Efficiency

Real trays are not 100% efficient, which means that they are not equilibrium trays. In a column, the tray efficiencies must be known to determine the number of actual trays. Individual tray efficiencies can be defined on the basis of the fractional approach to equilibrium, and these are called Murphree efficiencies:

$$E_{MV} = \frac{y_{out} - y_{in}}{y_{out}^* - y_{in}} \qquad (3.62)$$

$$E_{ML} = \frac{x_{out} - x_{in}}{x_{out}^* - x_{in}} \qquad (3.63)$$

where the numerators represent the actual mole fraction change on a tray, and the denominators represent the mole fraction change on a tray if equilibrium is reached on the tray. The subscripts V and L represent a vapor-phase and a liquid-phase basis, respectively. While these definitions make intuitive sense, they are not very useful for a variety of reasons. One reason is that they differ for every tray, and experimental results would be needed for every tray before the design of a column.

Another method for estimation is to use empirical results gathered over many years of practice, and these results were presented by O'Connell (1946). The results are presented in Figure 3.36, which are for overall, average column efficiencies. There are different correlations for distillation columns and for absorbers/strippers. The parameter for distillation is $\alpha\mu$, where α is the relative volatility of the key components (only two components in binary distillation), and μ is the viscosity of the feed in cP (centipoise). The efficiency decreases as the relative volatility increases because more mass must be transferred to reach equilibrium, and the efficiency decreases as the viscosity increased because mixing is more difficult, which lowers the mass transfer rate. A curve fit of the points shown (in percent) has been suggested (Couper et al., 2012, page 471):

$$E_o = 53.977 - 22.527\log_{10}(\alpha\mu) + 3.07[\log_{10}(\alpha\mu)]^2 - 11.0[\log_{10}(\alpha\mu)]^3 \tag{3.64}$$

An alternative curve fit has also been suggested (Wankat, 2017):

$$E_o = 52.782 - 27.511\log_{10}(\alpha\mu) + 4.4923[\log_{10}(\alpha\mu)]^2 \tag{3.65}$$

For absorbers, the parameter is HP/μ, where H is Henry's law constant (in lbmol/ft^3/atm), P is the pressure (in atm), and μ is the viscosity (in cP).

Example 3.5

Estimate the efficiency of the distillation column in Example 3.4. The feed viscosity is 0.25 cP. Then, calculate the number of actual trays in the column and the column height.

Solution

The relative volatilities must be calculated first. The relative volatility is $\alpha = K_5/K_6$. Assuming constant pressure in the column, $\alpha = P_5^*/P_6^*$. Using the top temperature of 58.05°C, α_{top} = 2.83, and using the bottom temperature of 92.62°C, α_{bottom} = 2.46. The geometric average of the top and bottom relative volatilities is usually used, which yields α = 2.64; however, since the two numbers are so close, the algebraic average is virtually the same number. Using Equation (3.64) yields 58.2%, and using Equation (3.65) yields 57.9%. So, the actual number of trays is based on 14 equilibrium trays, since one of the equilibrium stages is the partial reboiler. The result is about 24 trays. For a 24-in tray spacing (0.6096 m), the column height is 14.6 m, which is a height-to-diameter ratio of 14.6/1.94 = 7.23, which is well within the limit of typical operation.

3.3.2.4 Pressure Drop

As stated previously, liquid flows down through a column due to gravity, and vapor flows upward due to a pressure difference. While the top and bottom pressures are determined by the bubble points of the top and bottom trays, they are not independent of the hydraulics on the trays. Each tray has a pressure drop that the vapor must overcome, and the total pressure drop is the sum of the pressure drops on the trays. Therefore, it is necessary to understand something about the tray hydraulics to understand the pressure drop on each tray.

Figure 3.37 is a schematic of the liquid flow across a tray. The vapor flowing upward must overcome the downcomer head (h_{dc}), which is the sum of all of the other heads shown in Figure 3.37,

$$h_{dc} = h_{weir} + h_{crest} + h_{grad} + h_{du} + h_{dry} \tag{3.66}$$

where
h_{weir} = weir height, the dominant term
h_{crest} = crest over weir, the level of liquid above the top of the weir
h_{grad} = hydraulic gradient, difference in level from the downcomer side and the weir side
h_{du} = frictional loss of liquid flowing out of downcomer
h_{dry} = the "dry" head, the head to get vapor through holes in tray

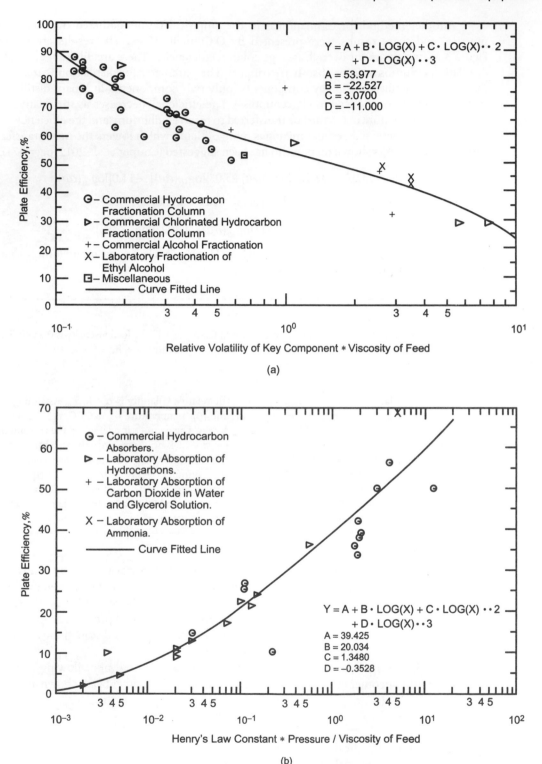

Figure 3.36   Efficiencies of distillation columns and absorber-strippers (Reproduced with permission from Couper et al. [2012])

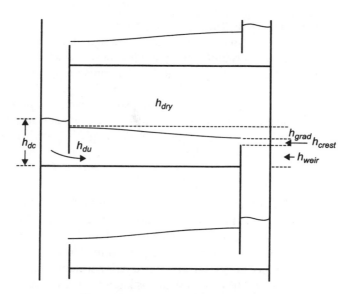

Figure 3.37 Pressure heads on trays

If the weir height is assumed to be the dominant contribution to the pressure drop, and assuming that all weir heights are the same, the column pressure drop becomes

$$\Delta P = \left(N_{above} \rho_{above} + N_{below} \rho_{below} \right) g h_{weir} \tag{3.67}$$

where the subscripts *above* and *below* refer to the location in the tower relative to the feed. The liquid densities are used as a conservative estimate, since the portion of the tray containing a froth (two-phase, vapor-liquid mixture) is unknown in an initial design. Since the estimation methods presented here are based on the properties of the more volatile component above the feed tray and the less volatile component below the feed tray, their liquid densities would be used for the appropriate section of the column.

Example 3.6

Estimate the pressure drop in the column of Examples 3.4 and 3.5 if the weir height is 4 in.

Solution

Equation (3.67) defines the pressure drop. Since there are 14 equilibrium trays with the feed on Tray 5, when applying the efficiency to convert to 24 actual trays, the feed is located proportionally at the same location. So the feed is on Tray 5(24/14) = 8.57 ≈ 9. Given that Equation (3.67) is an estimate, there would be little difference in rounding down to the feed on Tray 8. Using the densities in Example 3.4, Equation (3.67) becomes

$$\begin{aligned} \Delta P &= [9(626 \text{ kg/m}^3) + 15(655 \text{ kg/m}^3)](9.81 \text{ m/s}^2)(4 \text{ in})(0.0254 \text{ m/in}) \\ &= 15{,}408 \text{ Pa} = 15.4 \text{ kPa} = 0.154 \text{ bar} \end{aligned} \tag{E3.6a}$$

In Example 3.4, the pressure was assumed constant at 2 atm throughout the column. Applying this pressure drop would change the calculated diameter slightly.

3.3.2.5   Condensers and Reboilers

The impact of the reboiler and condenser on the design and operation of a distillation column is often overlooked in basic textbooks. They are just heat exchangers, the design of which was discussed in Chapter 2. Figure 3.38 shows the T-Q diagrams for a typical condenser and reboiler. It is assumed that the streams being reboiled and condensed are pure components, hence the horizontal lines representing constant temperature. If those streams were not pure, the lines would be slightly inclined, representing the difference between the bubble-point and dew-point temperatures. It is also assumed that the condensate is not subcooled, which might happen in some applications.

For the reboiler, the energy balances and design equations are

$$Q_R = U_R A_R \Delta T = U_R A_R (T_s - T_p) = \overline{V} \lambda_p = -\dot{m}_s \lambda_s \tag{3.68}$$

where the subscript R denotes the reboiler, the subscript p denotes the process stream, and the subscript s denotes steam, which is a standard source of heat. The overbars indicate the bottom of the column. Built in to Equation (3.68) is the assumption that all phase changes are between saturated liquid and saturated vapor. The log-mean subscript is omitted from the temperature difference, because the temperature difference between the two streams is constant at all points, and thus $\Delta T_{lm} = \Delta T$.

Once the flowrate of the process stream is known, since the latent heat can be found in the literature, the heat duty, Q, is known. If the heat transfer coefficient and the available steam temperature are known, then there is a unique correspondence between T_p and A. So, a reboiler can be designed for any desired value of T_p. There is an exact pressure for this value of T_p, based on Antoine's equation. Care must be taken to keep $(T_s - T_p)$ from being too large to avoid film boiling, as discussed in Chapter 2.

The pressure in the condenser is obtained from the pressure in the reboiler minus the column pressure drop. The column pressure drop is not arbitrary but is related to the weir height, as discussed the previous section. From Antoine's equation, the condensing temperature is then known. In the condenser, the energy balances and design equation are

$$Q_c = U_c A_c \Delta T_{lm} = U_c A_c \frac{(T_c - T_{cw,in}) - (T_c - T_{cw,out})}{\ln\left(\dfrac{T_c - T_{cw,in}}{T_c - T_{cw,out}}\right)} = -V\lambda_p = \dot{m}_{cw} C_{p,cw}(T_{cw,out} - T_{cw,in}) \tag{3.69}$$

where the subscript c denotes the condenser or condensation temperature, and the subscript cw denotes cooling water, which is the most desirable heat sink in a condenser. Therefore, if the heat transfer coefficient is known, there is only one condenser area that will satisfy Equation (3.69) for the calculated T_c.

When simulating a column, the pressure drop is an input variable. However, it is not known until the trays are designed, but the pressure drop is dependent on the number of trays, which is not known until the column is simulated. Therefore, column design is an iterative process, and this is not the only source of iterations. As a first approximation, a pressure drop could be assumed

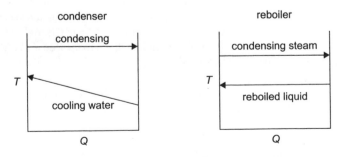

Figure 3.38   Typical T-Q diagrams for condenser and reboiler

and the weir height calculated after determining the number of trays and feed location. If the weir height is reasonable, the result may be sufficient for a first step. If not, then the pressure drop must be changed to correspond to a reasonable weir height.

The condenser and reboiler also affect the pressure in the column. In a chemical plant, the least expensive source of cooling in the condenser is usually process cooling water. Air can be used, but the heat transfer coefficient from a gas is much smaller than from a liquid. While the following values may differ from plant to plant and seasonally, typically, process cooling water is available at 30°C and is returned at 40°C. (The returned cooling water can be cooled back to 30°C by evaporating about 2%, so, other than makeup for the 2% evaporated, it is a closed loop.) If a 10°C approach temperature between the cooling water and the condensation temperature is desired (Figure 3.38), then the top column pressure must be high enough for the boiling point of the top component to be 50°C (bubble point of mixture at top). Quite often, distillation of low-boiling components, such as ethane-ethylene or propane-propylene, are run at significant pressures to make the top temperature around 50°C.

Columns can also be run at vacuum by using either a steam ejector (covered in Chapter 5) or a vacuum pump at the top of the column. There are two possible reasons for running a column under vacuum. One is that a component in the distillation is temperature sensitive. For example, acrylic acid spontaneously polymerizes at 90°C. To keep the temperature below 90°C, the pressure must be 0.16 bar. So, in the purification of acrylic acid from acetic acid, the undesired product of a side reaction, the column is run at vacuum. (In this particular example, acrylic acid is the higher boiler, so it is the bottom pressure that is at 0.16 bar!) A second reason is the available steam temperature for high boilers. Typically, the highest pressure, saturated steam available for a reboiler is around 250°C. For example, phthalic anhydride's normal boiling point is 295°C. Therefore, the column that purifies phthalic anhydride is run at around 0.4 bar, where phthalic anhydride boils at around 240°C.

Example 3.7

Calculate the heat duty for the condenser and the reboiler and their required heat transfer areas. In the condenser, cooling water is used, entering at 30°C and leaving at 40°C. In the reboiler, low-pressure steam is used, which is condensed at 160°C.
Data provided:

	$\lambda(\Delta H_{vap})$ (kJ/kmol)		U_o (W/m²/K)
Pentane	25,720	Condenser	1500
Hexane	29,632	Reboiler	3000

Solution

Since the compositions are known at the top and at the bottom of the column from Example 3.4, if ideal mixture is assumed, which is most likely valid at the relatively low pressure of the column with non-polar hydrocarbon components, the heats of vaporization can be weighted by the mole fractions (Figure E3.4).

$$Q_R = \bar{V}\lambda_p = (538 \text{ kmol/h})[0.01(25{,}720 \text{ kJ/kmol}) + 0.99(29{,}632 \text{ kJ/kmol})]$$
$$(h/3600 \text{ s}) = 4422 \text{ kW}$$
(E3.7a)

$$Q_C = -V\lambda_p = -(538 \text{ kmol/h})[0.95(25{,}720 \text{ kJ/kmol}) + 0.05(29{,}632 \text{ kJ/kmol})]$$
$$(h/3600 \text{ s}) = -3872 \text{ kW}$$
(E3.7b)

It is observed that the absolute values of the heat duties are very close numerically. This is expected. They should be on the same order of magnitude. If they are not, it suggests that the feed is very different from the column conditions, probably either very subcooled or very superheated.

Figure E3.7 shows the T-Q diagrams for the condenser and the reboiler. For the condenser,

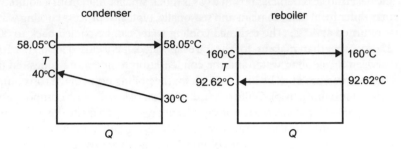

Figure E3.7 T-Q diagrams for condenser and reboiler

$$Q_C = U_o A_o \Delta T_{lm} = 3{,}872{,}000 \ \text{W} =$$
$$(1500 \ \text{W/m}^2/\text{K}) A_o \ \frac{(58.05 - 30) - (58.05 - 40)}{\ln\left[\dfrac{58.05 - 30}{58.05 - 40}\right]} \ \text{K} \tag{E3.7c}$$

so A_o (condenser) = 113.8 m^2,
and for the reboiler,

$$Q_R = U_o A_o \Delta T = 4{,}422{,}000 \ \text{W} = (3000 \ \text{W/m}^2/\text{K}) A_o (160 - 92.62) \text{K} \tag{E3.7d}$$

so A_o (reboiler) = 21.9 m^2.

The utility flowrates are obtained from the respective energy balances. For the condenser,

$$Q_C = \dot{m}_{cw} C_{p,cw} (T_{cw,out} - T_{cw,in}) = 3{,}872{,}000 \ \text{W} =$$
$$\dot{m}_{cw} (4184 \ \text{J/kg/°C})(40 - 30)\text{°C} \tag{E3.7e}$$

so $\dot{m}_{cw}$ = 92.5 kg/s, and for the reboiler,

$$Q_R = \dot{m}_s \lambda_s = 4{,}422{,}000 \ \text{W} = \dot{m}_s (2081.3 \ \text{kJ/kg})(1000 \ \text{J/kJ}) \tag{E3.7f}$$

so $\dot{m}_s$ = 2.12 kg/s.

3.3.3   Packed Towers

Another method for contacting vapors and liquids to accomplish separations is a packed tower. A packed tower is a large, cylindrical vessel packed with inert material. While in tray towers, mass transfer area is created by the gas bubbles within the liquid on the tray, the concept in a packed tower is that liquid coats the solid, packing surface creating mass transfer area for the vapor passing countercurrently. Just as in a tray tower, liquid flows down by gravity and the vapor flows upward due to a pressure gradient.

There are several other differences between packed towers and tray towers. Packed towers flood, but entrainment and weeping do not occur. In general, packed towers have lower pressure drops than tray towers. One problem with packed towers is the potential for flow maldistribution, which is the uneven distribution of liquid across the cross section, which can create regions of uncoated, dry solid, reducing the expected mass transfer area. Since flow maldistribution manifests more often in larger-diameter towers, historically, tray towers were used for larger-diameter (i.e., larger-capacity) columns, such as distillation in oil refineries, and packed towers were used for smaller-diameter (i.e., smaller-capacity) columns, such as removing pollutants from exhaust streams. Flow maldistribution is also related to the column diameter and the packing material diameter. If column diameter/packing diameter >40, maldistribution is more likely. One method that can be used to mitigate flow maldistribution in larger diameter columns is to have the packing in sections with the liquid collected and redistributed across the cross section throughout the column. This is illustrated in Figure 3.39.

3.3.3.1   Packing Types and Shapes

There are two types of packing types, random packing and structured packing. Random packings are randomly dumped into the cylindrical vessel. As illustrated in Figure 3.40, there are more shapes for random packings than can be imagined. The idea is to get the maximum mass transfer area, the largest mass transfer coefficient, and the lowest pressure drop. The random packing material can be plastic, metal, or ceramic, and the choice is based on cost and chemical compatibility.

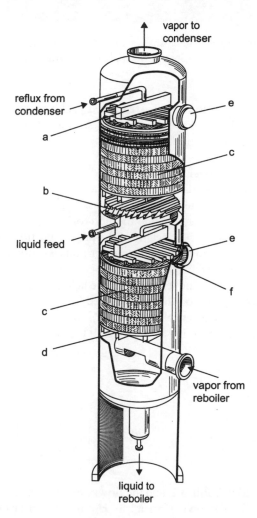

Figure 3.39 Internals of a packed column with structured packing showing redistributor (Stichlmair and Fair [1998])

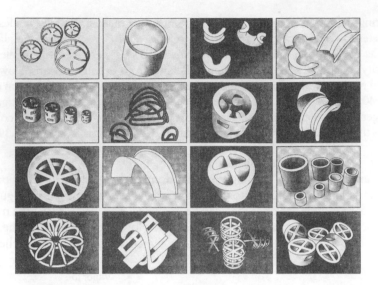

Figure 3.40 Some shapes of random packings (Reprinted by permission from Jiangxi Jintai Special Material LLC, copyright 2005, all rights reserved)

A more recent development is structured packings, which come in sections that slide into the cylindrical vessel. One key advantage of structured packings is a very-low pressure drop. Figure 3.41 illustrates a structured packing.

3.3.3.2   Height and Diameter

Table 3.2 shows four methods for obtaining the height of a packed column. The number of transfer units depends on the equilibrium, as discussed in Section 3.2.2.1. The height of a transfer unit depends on the mass transfer, and one method for calculating the height of a transfer unit is also presented in Section 3.2.2.1. The height of the packed portion of the column is the product of the height of a transfer unit and number of transfer units, as shown in Table 3.2. For a distillation column, the heights above and below the feed are calculated separately and added. For an absorber or stripper, since there is only one section, just one height is calculated.

There is another method that can be used for calculating the height of a packed tower, the height equivalent to a theoretical plate (*HETP*). In this method, the number of trays (also called *plates*) is calculated as if designing a tray tower. If the HETP can be calculated, the tower height is

$$z = N(HETP) \tag{3.70}$$

The expression for *HETP* is

$$HETP = H_{oV} \left[\frac{\ln\left(\dfrac{mV}{L}\right)}{\left(\dfrac{mV}{L} - 1\right)} \right] \tag{3.71}$$

where H_{oV} is defined in Table 3.2, and the flowrates L and V are molar flowrates. The mass transfer coefficient must still be known, since it is in H_{oV}, but the HETP expression replaces the tedious N_{oV} calculation described in Section 3.2.2.1. A possible heuristic is that $HETP \approx 0.5$ m. Another heuristic might be $HETP$ (ft) ≈ 1.5 (packing size, inches).

Figure 3.41 Example of a structured packing (https://www.sulzer.com/en/Products-and-Services/Separation-Technology/Structured-Packings/Gauze-Packings-BX-CY-BXPlus-AYPlus-DC-Hyperfil-and-Multifil)

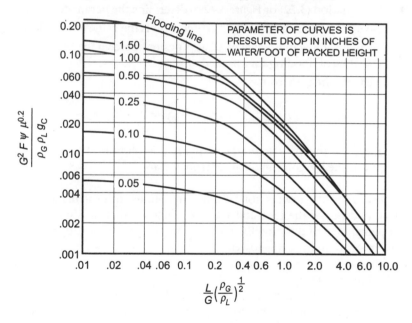

Figure 3.42 Generalized flooding and pressure drop correlation for packed columns (Reprinted with permission from Eckert [1970]. Reproduced with permission by the American Institute of Chemical Engineers)

The diameter of a packed tower is based on a flooding calculation. The concept is identical to that for tray towers, but the empirical correlation is different. A graphical representation of the flooding limit is shown in Figure 3.42. The important line is the flooding line and a curve fit has been presented (Wankat, 2017) that corresponds to the equation

$$\log\left[\frac{G_f'^2 F \psi \mu^{0.2}}{\rho_L \rho_V g_c}\right] = -1.6678 - 1.085\log_{10} F_{lv} - 0.29655[\log_{10} F_{lv}]^2 \tag{3.72}$$

where F_{lv} is defined in Equation (3.59), G_f' is the superficial mass flowrate of vapor at flooding (could be called V_f) in lb/ft²/sec (area unit is cross-sectional area of tower), the densities are in lb/ft³, the viscosity (μ) is in cP, ψ is the inverse of the specific gravity of the liquid (ρ_{water}/ρ_L), F is a packing factor found in Table 3.8, and g_c is the unit conversion 32.2 ft lb$_m$/(lb$_f$sec²). The units must be exactly as stated in this empirical correlation, and they are not consistent with each other. The algorithm is as follows:

1. Determine all internal flowrates in molar units. For distillation, this involves knowing the reflux ratio and the feed condition. For mass separating agents, the flowrates are likely given. Convert these to mass units. For distillation, it is usually assumed that the molecular weights in the vapor and liquid phases are equal, because the phases are not that different in composition. For mass separating agents, especially in dilute solutions, the molecular weight is approximately that of the carrier solvent (non-absorbing component).
2. Determine the liquid and vapor densities. The liquid density must be found from a reference. The vapor density can be estimated using the ideal gas law or another equation of state. The top and bottom temperatures and pressures are used, depending on which section the calculation is based.
3. Calculate F_{lv} from Equation (3.59).
4. Use Equation (3.72) or Figure 3.42 to determine the parameter in the brackets.
5. Solve for G_f', which has units lb/ft²/sec. This is the value of G' at flooding.
6. Calculate the actual value of G' as a desired percentage of the flooding velocity (typically 70%–80% of flooding for a packed tower).
7. Calculate the area from

$$A = \frac{G}{G'} \tag{3.73}$$

where G is the mass flowrate of vapor in the column.
8. Determine the column diameter from the area.

Just as for tray towers, for a distillation column, this calculation should be done both above the feed and below the feed and a diameter chosen that satisfies both results. Since there is no lower percentage of flooding limit (no weeping), choosing an appropriate diameter is easier for a packed tower than for a tray tower.

3.3.3.3   Pressure Drop
The pressure drop can be estimated from Figure 3.42. The units are inches H$_2$O/ft packed height. The curves below the flooding line provide the values for the pressure drop. Once the flooding G' is obtained, the actual value of G' is obtained from the desired flooding percentage. Then, a new y-axis is obtained, and the value of F_{lv} is already known, so the pressure drop is obtained. Then the overall pressure drop is obtained by multiplying that value by the column height.

An empirical correlation for the pressure drop has been presented (Wankat, 2017).

$$\Delta P = \alpha (10^{\beta L'}) \left(\frac{G'^2}{\rho_V} \right) \tag{3.74}$$

where α and β are listed in Table 3.8. L' and G' are superficial mass flowrates in lb/ft²/sec, the density is lb/ft³, and the resulting pressure drop is in inches H$_2$O/ft packed height.

Table 3.8   Parameters for Random Packings

		Nominal Packing Size (in)							
		0.375	0.5	0.625	0.75	1.0	1.25	1.5	2.0
Raschig Rings, Metal, 1/32-in Wall	F	390	300	170	155	115			
	α			1.20					
	β			0.28					
Raschig Rings, Metal, 1/16-in Wall	F		410	290	220	137	110	83	57
	α				0.80	0.42		0.29	0.23
	β				0.30	0.21		0.20	0.14
Raschig Rings, Ceramic	F	1000	580	380	255	155	125	95	65
	α	4.70	3.10	2.35	1.34	0.97	0.57	0.39	0.24
	β	0.41	0.41	0.26	0.26	0.25	0.23	0.23	0.17
Pall Rings, Plastic	F			97		52		32	25
	α					0.22			0.10
	β					0.14			0.12
Pall Rings, Metal	F			70		48		28	20
	α			0.43		0.15		0.08	0.06
	β			0.17		0.16		0.15	0.12
Berl Saddles, Ceramic	F		240		170	110		65	45
	α		1.2		0.62	0.39		0.21	0.16
	β		0.21		0.17	0.17		0.13	0.12
Intalox Saddles, Ceramic	F		200		145	98		52	40
	α		1.04		0.52	0.52		0.13	0.14
	β		0.37		0.25	0.16		0.15	0.10

Source: Reprinted and electronically reproduced by permission from Wankat (2017).

Example 3.8

An air stream flowing at 1500 kmol/h containing benzene vapor is to be scrubbed in a packed bed using 1-in, ceramic Berl saddles using 250 kmol/h of hexadecane as the solvent. The mole fraction of benzene in the inlet air is 0.01, and the required outlet mole fraction is 0.0005. The hexadecane is being returned to the scrubber from a stripper, so the inlet mole faction of benzene in hexadecane is 0.0001. The column may be assumed to operate at 150 kPa and 35°C. Raoult's law may be assumed to define the benzene equilibrium.

   a. Determine the diameter of the packed column for 75% of flooding.
   b. Determine the pressure drop per unit height in the column.

Data provided:

	Antoine Coefficients				
	A	B	C	SG	μ (cP)
Benzene	6.90565	1211.033	220.79		
Hexadecane				0.770	3.4

Solution

a. The method discussed in Sections 3.3.3.2 and 3.3.3.3 is used. Just as in a tray tower, the term F_{lv} must be calculated first. So, the density of the gas phase must be estimated. Since the system is dilute, the density of air can be used. An average molecular weight should be used; however, given the approximate nature of this calculation, using the molecular weight of air is reasonable. So,

$$\rho_V = \frac{PM}{RT} = \frac{(150 \text{ kPa})(29 \text{ kg/kmol})}{(8.314 \text{ m}^3\text{kPa/kmol/K})(273.15 + 35 \text{ K})} = 1.70 \text{ kg/m}^3 \qquad \text{(E3.8a)}$$

From Equation (3.59),

$$F_{lv} = \frac{LM_L}{VM_V}\left(\frac{\rho_V}{\rho_L}\right)^{0.5} = \frac{250 \text{ kmol/h}(226 \text{ kg/kmol})}{1500 \text{ kmol/h}(29 \text{ kg/kmol})}\left[\frac{1.70}{770}\right]^{0.5} = 0.061 \qquad \text{(E3.8b)}$$

From Equation (3.72),

$$\log_{10}\left[\frac{G_f'^2 F\psi\mu^{0.2}}{\rho_L\rho_V g_c}\right] = -1.6678 - 1.085\log_{10}0.061 - 0.29655\left[\log_{10}0.061\right]^2 \qquad \text{(E3.8c)}$$

so that

$$\frac{G_f'^2 F\psi\mu^{0.2}}{\rho_L\rho_V g_c} = 0.1631 \qquad \text{(E3.8d)}$$

and solving for the flooding superficial mass flowrate,

$$G_f' = \left[\frac{(0.1631 \text{ lb/ft}^2/\text{sec})(48.048 \text{ lb/ft}^3)(0.10595 \text{ lb/ft}^3)(32.2)}{45(1000/770)(3.4 \text{ cP})}\right]^{0.5} \qquad \text{(E3.8e)}$$

$$= 1.482 \text{ lb/ft}^2 \text{ sec}$$

where the densities have been converted into the required units, and the value of F, the packing factor, 45, is found in Table 3.8. For 75% of flooding, then $G' = 1.111 \text{ lb/ft}^2/\text{sec}$. The area is found by using Equation (3.73):

$$A = \frac{1500 \text{ kmol/h}(29 \text{ kg/kmol})(2.2 \text{ lb/kg})}{1.111 \text{ lb/ft}^2/\text{sec}(3600 \text{ sec/h})} = 23.92 \text{ ft}^2 \qquad \text{(E3.8f)}$$

so, the diameter is found to be 5.52 ft.

b. The pressure drop is found from Equation (3.74), using the parameters α and β from Table 3.8.

$$\Delta P = 0.39\left(10^{\displaystyle\frac{0.17(250 \text{ kmol/h})(226 \text{ kg/kmol})(2.2 \text{ lb/kg})}{23.92 \text{ ft}^2(3600 \text{ sec/h})}}\right)\left(\frac{1.11^2 \text{ lb/ft}^2/\text{sec}}{0.10595 \text{ lb/ft}^3}\right) \qquad \text{(E3.8g)}$$

$$= 8.00 \text{ in H}_2\text{O/ft packed height} = 0.01965 \text{ atm/ft packed height}$$

This is not a small pressure drop, since the pressure drop in a 50 ft column would be just under 1 atm. It might be better to use a larger packing size to reduce the pressure drop. However, this would reduce the packing factor, thereby increasing the flooding velocity, which increases the column diameter. Ultimately, the decision would be based on economics.

3.3.4 Tray Tower or Packed Tower?

The choice between tray towers and packed towers depends on several variables, and some of the guidelines may be conflicting. Traditionally, packed towers were used for smaller-diameter columns because of the flow maldistribution discussed earlier. Therefore, they were used for lower-capacity operations. However, structured packing combined with liquid flow distributors have made it possible to use packed towers for larger-capacity operations. The lower pressure drop in packed towers makes them uniquely applicable to vacuum operations. Packed towers do flood; however, they do not weep, so it may seem that they are more flexible for operations that might have to be scaled down. However, there is a minimum liquid load (flowrate) for packed towers, and, if distributors are used to avoid flow maldistribution, there is a window of flowrate operation, just as for tray towers, but for different reasons (Stichlmair and Fair, 1998). This is one reason why tray towers are preferred for larger capacity operations.

Tray towers were traditionally used for larger-capacity (larger-diameter) operations. Trays are easier to clean than packing, so trays are more applicable to fouling operations. In a tray tower, access points are built in for maintenance (manways) and cleaning during shutdown. However, in a packed tower, the packing must be removed for the same access. In more advanced distillation operations, such as reactive distillation, adding heat transfer capability to remove heat produced by an exothermic reaction is easier in a tray tower by inserting a liquid take-off at a given tray and pumping the liquid through an external heat exchanger and returning the liquid to another tray.

Ultimately, the choice between a tray tower and a packed tower comes down to the economics. However, constraints for a specific application must also be considered.

3.3.5 Performance of Packed and Tray Towers

In the day-to-day operations of a chemical plant, process conditions change. For example, the rate of production may have to increase or decrease. Reasons for an increase in production include an increased demand for the product or a need to meet contracted customer demand when a similar plant within the company goes off line. These *performance* problems were discussed in the context of fluid flow equipment and heat exchangers in Chapters 1 and 2, respectively.

If a distillation column must be scaled-up (scaled-down) at constant product purities, the material balance requires that inlet and outlet flowrates increase (decrease) by the same amount. The question is how the output changes based on equipment performance. Clearly, there is a limit to scale-up of a distillation column based on the flooding limit. If the inlet flowrate to a distillation column increases at constant composition, and if the outlet compositions are maintained, the reflux rate increases, since the reflux ratio (L/D) stays constant. Therefore, all internal flows increase, making flooding a real possibility. If, for example, a column was designed for 75% of flooding, a 10% to 15% scale-up might be possible, assuming that it is desired to keep the flooding percentage less than 85% to 90% to avoid loading. For scale-down, there might be more flexibility, since weeping does not typically occur until <40% of flooding, with the exact value dependent on the tray type.

However, the situation is not that simple. In Section 2.8, the performance of heat exchangers was discussed, and it was seen that changing the inlet flowrate to a heat exchanger affects the temperatures of both streams and the total heat load. Since it has been shown that the condenser and reboiler affect the pressure of the column, then scale-up or scale-down of a distillation column involves much more than flooding. Therefore, the performance of a distillation column cannot be analyzed without consideration of the performance of both the condenser and reboiler. This is illustrated with a case study.

CASE STUDY

The distillation column illustrated in Figure 3.43 is used to separate benzene and toluene. It contains 35 sieve trays, with the feed on Tray 18. The relevant flows are given in Table 3.9. Your assignment is to recommend changes in the tower operation to handle a 50% reduction in feed. Overhead composition must be maintained or exceed the current specification of 0.996 mole fraction benzene. Cooling water is used in the condenser, entering at 30°C and exiting at 45°C. Medium-pressure steam (185°C, 1135 kPa) is used in the reboiler. The operating conditions of the tower before reduction in feed are in Table 3.9.

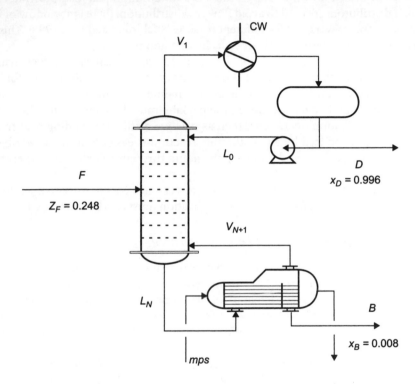

Figure 3.43   Distillation of benzene from toluene

Table 3.9   Operating Conditions for Process in Figure 3.43

Input/Output	Flow (kmol/h)	Mole Fraction Benzene	Temperature (°C)
Inputs			
Feed, F	141.3	0.248	90
Reflux, L_0	130.7	0.996	112.7
Boil-up, V_{N+1}	189.5	0.008	145.3
Outputs			
Distillate, D	34.3	0.996	112.7
Still bottoms, L_N	296.5	0.008	145.3
Bottoms product, B	107.0	0.008	145.3

Among the several possible operating strategies for accomplishing the necessary scale-down are the following:

1. Scale down all flows by 50%. This is possible only if the original operation is not near the lower velocity limit that initiates weeping or reduced tray efficiency. If 50% reduction is possible without weeping, reduced tray efficiency, or poor heat-exchanger performance, this is an attractive option.
2. Operate at the same boil-up rate. This is necessary if weeping or reduced tray efficiency is a problem. The reflux ratio must be increased in order to maintain the reflux necessary to maintain the same liquid and vapor flows in the column. In this case, weeping, lower tray efficiency, or poor heat-exchanger performance caused by reduced internal flows is not a problem. The downside of this alternative is that a purer product will be produced and unnecessary utilities will be used. Increasing the reflux ratio during scale-down increases the utility cost or the column per unit product.

In Example 3.9, the analysis will be done by assuming that it is possible to scale down all flows by 50% without weeping or reduced tray efficiency.

Example 3.9

Estimate the pressure drop through the column. To what weir height does this correspond?

Solution

Interpolation of tabulated data (Perry and Green, 1997) yields the following relationships for the vapor pressures in the temperature range of interest:

$$\ln P^*(\text{kPa}) = 15.1492 - \frac{3706.84}{T(\text{K})} \quad \text{benzene} \tag{E3.9a}$$

$$\ln P^*(\text{kPa}) = 15.3877 - \frac{4131.14}{T(\text{K})} \quad \text{toluene} \tag{E3.9b}$$

It is assumed that the bottom is pure toluene and the distillate is pure benzene. This is a good assumption for estimating top and bottom pressures given the mole fractions specified for the distillate and bottoms. Therefore, at the bottom temperature of 141.7°C, the vapor pressure of toluene, and hence the pressure at the bottom of the column, is 227.2 kPa. At the top temperature of 104.2°C, the vapor pressure of benzene, and hence the pressure at the top of the column, is 204.8 kPa.

Because there are 35 trays, the pressure drop per tray is

$$\Delta P = (227.2 - 204.8)/35 = 0.64 \ \text{kPa/tray} \tag{E3.9c}$$

If it assumed that the weir height is the major contribution to the pressure drop on a tray, then

$$\Delta P = \rho g h \tag{E3.9d}$$

where h is the weir height. Assuming an average density of 800 kg/m³, then

$$h = (640 \ \text{Pa})/[(800 \ \text{kg/m}^3)(9.8 \ \text{m/s}^2)] = 0.08 \ \text{m} \approx 3 \ \text{in} \tag{E3.9e}$$

This is a typical weir height and is consistent with the assumption that the weir height is dominant.

The pressure drop is assumed to remain constant after the scale-down because the weir height is not changed. In practice, there is an additional contribution to the pressure drop due to gas flow through the tray orifices. This would change if column flows changed, but the pressure drop through the orifices is small and should be a minor effect. Examples 3.10 and 3.11 illustrate how the performance of the reboiler and condenser affect the performance of the distillation column.

Example 3.10

Analyze the reboiler to determine how its performance is altered at 50% scale-down.

Solution

Figure E3.10 shows the *T-Q* diagram for this situation. Because the amounts of heat transferred for the base and new cases are different, the *Q* values must be normalized by the total heat transferred in order for these profiles to be plotted on the same scale. The solid lines are the original case (subscript 1). For the new, scaled-down case (subscript 2), a ratio of the energy balance on the reboiled stream for the two cases yields

$$\frac{Q_2}{Q_1} = \frac{\dot{m}_2 \lambda}{\dot{m}_1 \lambda} \tag{E3.10a}$$

If it is assumed that the latent heat is unchanged for small temperature changes, then $Q_2/Q_1 = 0.5$, because at 50% scale-down, the ratio of the mass flowrates in the reboiler is 0.5. The ratio of the heat transfer equations yields

$$\frac{Q_2}{Q_1} = 0.5 = \frac{U_2 A \Delta T_2}{U_1 A \Delta T_1} \approx \frac{\Delta T_2}{43.3} \tag{E3.10b}$$

In Equation (E3.10b), it is assumed that the overall heat transfer coefficient is constant. It is assumed here that $U \neq f(T)$. This assumption should be checked, because for boiling heat transfer coefficients with large temperature differences, the boiling heat transfer coefficient may be a strong function of temperature difference.

From Equation (E3.10b), it is seen that $\Delta T_2 = 21.7°C$. Therefore, for the reboiler to operate at 50% scale-down, with the steam side maintained constant, one possible option is that the boil-up temperature must be 163.4°C. This is shown as the dotted line in Figure E3.10.

From the vapor pressure expression for toluene, the pressure at the bottom of the column is now 372.5 kPa. It would have to be determined whether the existing column could withstand this greatly increased pressure.

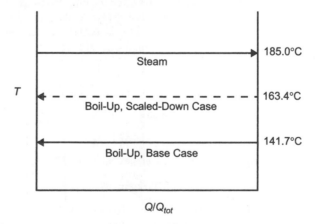

Figure E3.10   Temperature profiles in reboiler for Example 3.10

Example 3.10 shows that the reboiler operation requires that the pressure of the boil-up stream be increased. This is not the only possible alternative. All that is necessary is that the temperature difference be 21.7°C. This could be accomplished by reducing the temperature of the steam to 163.4°C. In most chemical plants, steam is available at discrete pressures, and the steam used here is typical of medium-pressure steam. Low-pressure steam might be at too low a temperature to work in the scaled-down column. However, the pressure and temperature of medium-pressure steam could be reduced if there were a throttling valve in the steam feed line to reduce the steam pressure. The resulting steam would be superheated, so desuperheating would also be necessary. This is usually accomplished by spraying water into the superheated steam. This system is called a *desuperheater*, and the benefit of a desuperheater is to permit flexibility in saturated steam conditions to facilitate process changes and process control. A change in a utility stream is almost always preferred to a change in a process stream, because it does not disrupt process operation at the design conditions.

The condenser is now analyzed in Example 3.11.

Example 3.11

Using the results of Example 3.10, analyze the condenser for the scaled-down case.

Solution

Under the assumption that the pressure drop for the column remains at 22.4 kPa, the pressure in the condenser is now 350.0 kPa. From the vapor pressure expression for benzene, the new temperature in the condenser is 126.0°C. The remainder of the analysis is similar to Example 2.20. The T-Q diagram is illustrated in Figure E3.11. Because an organic is condensing, it is assumed that the resistance on the cooling water side is approximately equal to the resistance on the condensing side. The method presented can be used for any known heat transfer coefficients or any known ratio of heat transfer coefficients. Therefore,

$$\frac{U_2}{U_1} = \frac{\dfrac{1}{h_{o1}} + \dfrac{1}{h_{i1}}}{\dfrac{1}{h_{o2}} + \dfrac{1}{h_{i2}}} = \frac{\dfrac{2}{h_{o1}}}{\dfrac{1}{h_{o1}} + \dfrac{1}{h_{o1}M^{0.8}}} = \frac{2}{1 + \dfrac{1}{M^{0.8}}} \tag{E3.11a}$$

where $M = \dot{m}_2/\dot{m}_1$.

The ratio of the base cases for the energy balance on the condensing stream is

$$\frac{Q_2}{Q_1} = \frac{\dot{m}_2 \lambda}{\dot{m}_1 \lambda} \tag{E3.11b}$$

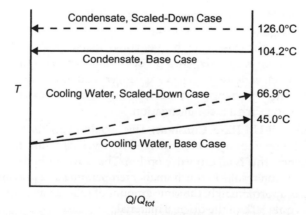

Figure E3.11   Temperature profiles in condenser for Example 3.11

Because the ratio of the mass flowrates is 0.5, and assuming that the latent heat is unchanged with temperature in the range of interest, $Q_2/Q_1 = 0.5$.

The ratio of the base cases for the energy balance and the heat-exchanger performance equation ($Q = UA\Delta T_{lm}$) are, respectively,

$$0.5 = M\frac{T-30}{15} \tag{E3.11c}$$

$$0.5 = \frac{2(T-30)}{(66.42)\left(1+\dfrac{1}{M^{0.8}}\right)\ln\left(\dfrac{96}{126-T}\right)} \tag{E3.11d}$$

Solution of Equations (E3.11c) and (3.11d) yields

$$M = 0.203$$

$$T = 66.9°C$$

Therefore, the cooling water rate must be reduced to 20% of the original rate, much lower than the actual scale-down. The outlet water temperature is increased by about 22°C. This would cause an unacceptable increase in fouling on the cooling water side, since 45°C is typically the maximum allowable return temperature due to fouling. Therefore, in this case, it may be better to use a higher reflux ratio rather than reduce the flows within the column.

Observation indicates that in order to operate the distillation column at 50% scale-down with process flows reduced by 50%, a higher pressure is required and the cooling water will be returned at a significantly higher temperature than before scale-down. The operating pressure of the column is determined by the performance of the reboiler and condenser.

Examples 3.10 and 3.11 reveal a process bottleneck at the reboiler that must be resolved in order to accomplish the scale-down. In the process of changing operating conditions (capacity) in a plant, a point will be reached where the changes cannot be increased or reduced any further. This is called a *bottleneck*. A bottleneck usually results when a piece of equipment (usually a single piece of equipment) cannot handle additional change. In this problem, one bottleneck is the high cooling water return temperature, which would almost certainly cause excessive fouling in a short period of time. Another potential bottleneck is tray weeping due to the greatly reduced vapor velocity. In addition to the solution presented here, there are a variety of other possible adjustments, or debottlenecking strategies, which follow with a short explanation for each.

1. **Replace the Heat Exchangers:** Equation (E3.10b) shows that a new heat exchanger with half of the original area allows operation at the original temperature and pressure. The heat transfer area of the existing exchanger could be reduced by plugging some of the tubes, but this modification would require a process shutdown. This involves both equipment down-time and capital expense for a new exchanger. Your intuitive sense should question the need to get a new heat exchanger to process less material. You can be assured that your supervisor would question such a recommendation.

2. **Keep the Boil-Up Rate Constant:** This would maintain the same vapor velocity in the tower. If the operation occurs near the lower velocity limit that initiates weeping or lower tray efficiency, this is an attractive option. The constant boil-up increases the tower separation. This option results in much smaller temperature and pressure changes but is somewhat wasteful, as approximately the same amount of reboiler and condenser energy is being used to process only 50% of the original material.

3. **Introduce Feed on a Different Plate:** This strategy must be combined with Option 4. The plate should be selected to decrease the separation and increase the bottoms concentration of the lower boiling fraction. This lowers the process temperature and increases the ΔT for heat transfer and the reboiler duty. This may or may not be simple to accomplish depending on whether the tower is piped to have alternate feed plates. As in Option 2, the reflux and coolant inputs will change.

4. **Recycle Bottoms Stream and Mix with Feed:** This strategy must be combined with Option 3. By lowering the concentration of the feed, the concentration of the low boiler in the bottoms can be increased, the temperature lowered, and the reboiler duty increased. This represents a modification of process configuration and introduces a new (recycle) input stream into the process.

The next series of examples involves the performance of an absorber. The same methodology would be used for a stripper.

Example 3.12

An absorber is designed to be 20 m tall, and it is known that H_{oV} is 4 m. The absorber is designed to treat 40 kmol/h of air containing acetone with a mole fraction of 0.002 and reduce the mole fraction to 0.0001. Pure water is the solvent, at a rate of 20 kmol/h. The column temperature is 26.7°C, and the pressure is 1.2 atm. Raoult's law may be assumed to apply, and

$$\ln P^*(\text{atm}) = -\frac{3598.4}{T(\text{K})} + 10.92 \qquad \text{(E3.12a)}$$

It is now observed that the outlet mole fraction of acetone in the air is 0.0002. List as many possible causes of this situation as you can, assuming that only one fault exists at a time. Quantify what is happening in the column for each case, that is, what parameter is off and what is the off-specification value?

Solution

There are six possible single-parameter upsets (off-design conditions) that do not involve malfunctions within the packed bed equipment. All can be understood from the dilute, packed-bed absorber equation in Table 3.6:

$$\frac{y_{in} - mx_{in}}{y_{out} - mx_{in}} = \frac{\exp\left[N_{oV}\left(1 - \frac{1}{A}\right)\right] - \frac{1}{A}}{1 - \frac{1}{A}} \qquad \text{(E3.12b)}$$

which is valid for these dilute solutions. Since the bed height $z = H_{oV}N_{oV}$, $N_{oV} = 5$. For the base case, at $26.7 + 273.15 = 299.85$ K, $P^* = 0.339$ atm. Therefore, $m = P^*/P = 0.283$, and $A = L/(mV) = 20/0.283/40 = 1.77$. Since the water is pure, $x_{in} = 0$ for the base case.

There are only three parameters in Equation (E3.12b), so the problem must be with one of these parameters. It is assumed that N_{oV} is correct, since the most likely way that it could be incorrect is if H_{oV} is incorrect, which would probably be due to an impediment to mass transfer, such as channeling or fouling. However, it was stated that equipment malfunctions were to be ignored.

One possibility is that every parameter on the right-hand side of Equation (E3.12b) is at design conditions but that $y_{out} = 0.004$. This means that the absorber is removing the same fraction of acetone from the air, but that the mole fraction of acetone in the inlet air has doubled.

A second possibility is that every parameter on the right-hand side of Equation (E3.12b) is at design conditions, that y_{in} is at design conditions, but that $x_{in} \neq 0$. Solution of Equation (E3.12b) yields $x_{in} = 0.00352$. How could this happen? It is possible that the water is regenerated in a stripper and that the stripper is exhibiting a malfunction. If there was a stripper, the inlet water would certainly have small amounts of acetone, so a more likely scenario is that the inlet water has more acetone than expected. However, the concept is identical.

The third possibility is that A is not at design conditions. At $y_{in} = 0.002$, $y_{out} = 0.0002$, and $N_{oV} = 5$, Equation (E3.12b) can be solved for $A = 1.275$. Since $A = L/(mV)$, if A has decreased, then either L has decreased, V has increased, or m has increased. Holding two of these parameters at the original values while solving for the third parameter yields either $L = 14.43$ kmol/h, $V = 55.45$ kmol/h, or $m = 0.392$. The first two values make intuitive sense, since a decrease in solvent rate or an increase in gas to be treated will both make the absorber remove less solute, assuming all other parameters remain unchanged. However, the increase in vapor flowrate is 38%. If this really occurred, flooding could be occurring. Since flooding would probably cause the column to malfunction to the extent that there would be no liquid flow downward, it is likely that such a large increase in V is not the cause of the problem.

The value of m can change only if the temperature or pressure changes. Since the new value of $m = P^*/P = 0.392$, the new pressure can be found as $P = 0.865$ atm at the original temperature, and the new temperature can be found as $T = 308.2$ K (35.1°C) at the original pressure. This also makes intuitive sense, since a decrease in pressure or an increase in temperature opposes absorption into the liquid phase. Lower pressure always favors the vapor phase and higher temperature always favors the vapor phase, whether it be pure-phase equilibrium or absorbers and stripper. So, if conditions have changed to be less favorable for absorption into the liquid phase, the removal of acetone from the air will decrease.

If there is a disturbance as in Example 3.12, the next question is how to compensate for the disturbance. In the next set of examples, some methods are suggested. Those that involve manipulation of temperature and pressure are based on the assumption that these parameters can be manipulated. The pressure can be manipulated by designing the upstream pressures to be higher than the column pressure so that valves can be used to set the inputs, or by adding pumps to lower pressure input streams. Heat exchangers can be designed upstream to adjust the inlet temperatures. It must be remembered that, unlike distillation columns, absorbers and stripper conditions are directly related to the feed conditions.

Example 3.13

For Example 3.12, suggest realistic ways to compensate for the disturbance. An example of a non-realistic compensation method is to reduce the air feed (process side), which is probably fixed upstream. It is also assumed that, if the feed composition is not the fault, it cannot be changed from design conditions. Usually, manipulations of the process or process variables are less desirable/possible than changes to utility flows or conditions. Again, assume that only one parameter at a time is manipulated.

Solution

There are numerous answers for each fault. Not all possibilities are discussed here.

If the fault is in flowrate, the other flowrate can be adjusted to maintain a constant A value. For example, if the value of V is 55.54 kmol/h instead of 40 kmol/h, then the value of L could be increased to 20(55.45/40) = 27.73 kmol/h. As discussed earlier, this result is based on the assumption that the column is not flooding.

If the fault is in L, it is assumed that V cannot be adjusted, since that is the process stream to be treated. To keep A constant, the value of m can be adjusted to 0.282(14.43/20) = 0.204. Since $m = P/P^*$, if the pressure can be changed to 1.20(0.282/0.204) = 1.66 atm, the column would perform as designed. The temperature can be changed, which affects P^*, so that the new value would be 0.339(0.204/0.282) =

0.245, which yields a temperature of 291.9 K = 18.75°C. The best solution is probably some combination of temperature and pressure changes that result in smaller changes for each parameter. It is also observed that these values make sense. The separation problem is alleviated by reducing the temperature and/or increasing the pressure, both of which favor the liquid phase.

If the problem is that the inlet liquid contains acetone, then it makes sense that lowering the temperature and/or raising the pressure could compensate for this problem. The solution is mathematically complex. The desired value of the left side of Equation (E3.12b) is 20, since the right side of Equation (E3.12b) should behave as if $x_{in} = 0$. So,

$$\frac{y_{in} - mx_{in}}{y_{out} - mx_{in}} = \frac{0.002}{0.0001} = 20 = \frac{\exp\left[5\left(1-\frac{1}{A}\right)\right] - \frac{1}{A}}{1 - \frac{1}{A}} \tag{E3.13a}$$

must be solved for a new A value. This value is $A = 1.825$. This can be accomplished with either a temperature of 25.9°C, a pressure of 1.24 atm, or $L = 20.65$ kmol/h, once again assuming only one parameter at a time is manipulated. It is observed that these are very small adjustments in operating conditions.

Example 3.14

The column in Example 3.12 is operating normally, but a 10% increase in air to be treated, to 44 kmol/h, is expected. How can column operation be changed to ensure the desired outlet mole fraction of acetone in air is attained? Be quantitative, and assume that only one parameter at a time is manipulated.

Solution

In this case, the desired recovery is to remain constant, so the left side of Equation (E3.12b) is constant. Therefore, the right side of Equation (3.12b) must remain constant, and the only way this can be accomplished at constant N_{oV} is for A to remain constant. Therefore, there are three possibilities. The simplest is to increase L by 10%, to 22 kmol/h. The pressure and temperature can also be adjusted. The new value of $m = 0.257$, which results in either a pressure of 1.32 atm or a temperature of 24.3°C.

It is important to remember the qualitative trends illustrated in Examples 3.13 and 3.14. Even if the Colburn method is not applicable, the trends are identical, but the numbers would change.

3.4   EXTRACTION EQUIPMENT

There is a wider variety of liquid-liquid contacting equipment than vapor-liquid contacting equipment. Both tray and packed columns can be used but are not as common as more modern, proprietary contacting equipment. Given that these equipment details are proprietary, performance details are limited.

Extraction equipment can be classified into four major categories: mixer-settler, static, agitated, and centrifugal.

3.4.1   Mixer–Settlers

A mixer-settler includes two pieces of equipment. The two liquid phases are mixed together, and the mixer can be designed to have a residence time sufficient to ensure equilibrium is attained. Then, the settler, a separate tank, is used to disengage the two immiscible phases. Figure 3.44 shows

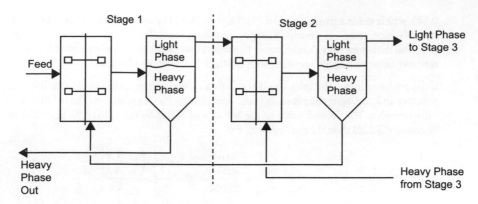

Figure 3.44   Countercurrent mixer-settler schematic

a countercurrent, mixer-settler schematic. Mixer-settlers have high stage efficiency, approaching unity, and are good for viscous fluids, but they are expensive if many stages are needed, since two pieces of equipment are required per stage. Any student who used a separatory funnel in organic chemistry laboratory has actually simulated a mixer-settler.

3.4.2   Static and Pulsed Columns

Static columns include spray columns, tray columns, and packed (random and structured) columns. These are illustrated in Figure 3.45. They have higher capacities and are less expensive than mixer-settlers, but they have lower efficiencies than mixer-settlers and agitated extractors. Other advantages of static columns are low cost and familiar equipment design. Spray columns typically are one theoretical stage. It has been reported that typical packed columns have up to five theoretical stages, while tray columns can have more theoretical stages, but not usually more than eight (Koch and Shiveler, 2015). Packed extraction columns typically have alternating packing and liquid redistribution zones to avoid flow maldistribution, such as channeling, as shown in Figure 3.45(b).

A variation on a static column is a pulsed column, in which short pulses, up and down, are imposed on the flowing fluids. On the up-stroke, the light liquid is dispersed into the heavy phase, and on the down-stroke, the heavy phase in injected into the light phase. This is possible in both tray and packed columns. A subsequent variation on the pulsed column is the Karr extractor, where the trays themselves move up and down, a schematic of which is illustrated in Figure 3.45(d). The trays do not have downcomers and can have as little as 1-in tray spacing, which makes up for the very low efficiencies, which can be as low as the 5% to 10% range.

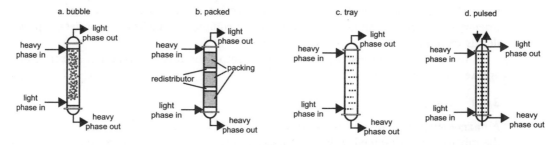

Figure 3.45   Static extraction columns

3.4.3   Agitated Columns

Agitated extraction columns involve adding energy via stirrers or similar agitation devices. The York-Scheibel column (Figure 3.46[a]) is designed to simulate a series of mixer-settlers. This type of column has turbine impellers and baffles to simulate a mixer, and the coalescer simulates a settler. The rotating-disk contactor (Figure 3.46[b]) is similar to the York-Scheibel column, except that rotating disks are used to add energy and provide agitation. These columns can have high efficiencies, but that is dependent on the energy input via the shaft. They can be expensive to purchase and to operate, depending on the trade-off between efficiency (column size) and energy input. Other designs include the Oldshue-Rushton (Regents of the University of Michigan, 2014) column and the Kühni column (Koch Modular Process Systems, 2000–2016). The Kühni column can be adapted for changing physical properties within the column and is amenable to scale-up. An advantage of agitated columns is that a large number of theoretical stages is possible, perhaps more than 20 for the Kühni column. The major disadvantage is high cost.

3.4.4   Centrifugal Extractors

Centrifugal extractors contact the two phases continuously by taking advantage of the density difference between the fluids in a centrifugal field. One of the most common centrifugal extractors is a Podbielniak extractor, shown in Figure 3.47. The heavy liquid flows to the outside, while the light liquid flows to the inside. Then both are collected and removed from the device. A single centrifugal extractor can accomplish the separation of five (perhaps a few more) theoretical stages in a single unit. Due to the high centrifugal forces possible, they can handle liquid pairs with small density differences. Centrifugal extractors are expensive to purchase and to operate. Historically, Podbielniak extractors have been used extensively in the pharmaceutical industry where the volume of products are small and highly pure products are required, for example, in the purification of penicillin. A comparison of extraction equipment is in Table 3.10.

Additional information on extraction equipment is available (Koch and Shiveler, 2015; Koch Modular Processes, 2000–2016).

3.5   GAS PERMEATION MEMBRANE SEPARATIONS

Membrane separations include gas permeation, ultrafiltration, reverse osmosis, nanofiltration (such as reverse osmosis, but for organic molecules), and dialysis. Filtration could be considered a membrane separation, although it is often considered to be a fluid mechanics problem, because the key element is flow through the filter cake, which is like a packed bed. While the focus here is on gas permeation, the equipment involved is similar for all membrane separations.

3.5.1   Equipment

Membrane separation equipment generally falls into three general categories. Flat membranes are possible; however, they do not provide much surface area per unit equipment size, so the applications are limited. Spiral-wound membranes create more membrane surface area per volume of equipment. Figure 3.48 illustrates a spiral-wound membrane, and Figure 3.49 illustrates the underlying flow patterns.

With the advent of technology that allows polymers to be spun into hollow fibers, membrane modules that have the appearance of a shell-and-tube heat exchanger were developed. A hollow-fiber module is illustrated in Figure 3.50. The hollow fibers are illustrated in Figure 3.51. Outside diameters are in the 200 to 1000 μm range, and wall thicknesses are in the 50 to 250 μm

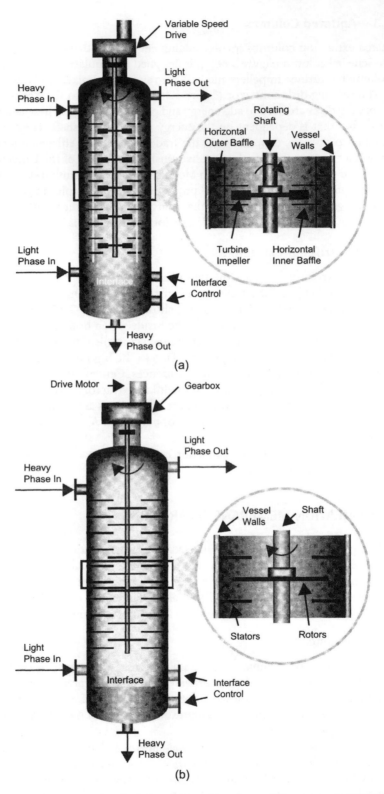

Figure 3.46 (a) York-Scheibel extractor, (b) rotating disk contactor (SCHEIBEL®
column courtesy of Koch-Glitsch LP, Wichita, Kansas)

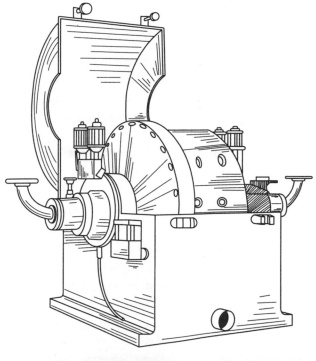

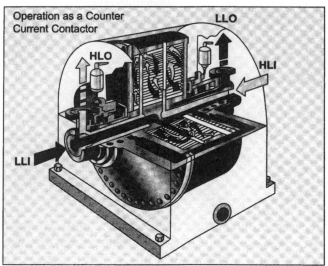

Figure 3.47   Podbielniak extractor photograph and schematic (Courtesy of VertMarkets, Inc.)

range. Most of these membranes are asymmetric, meaning that they have multiple layers with different properties. For example, many of these membranes have a support layer with a thin diffusing layer, with the thin diffusing layer on the order of 0.1 μm. For comparison, human hair is in the range of 30 to 100 μm. So, hollow fibers are about an order of magnitude larger than a human hair with a "hole" in the middle that allows the fiber to approximate a tube. Thousands of hollow fibers can be in a single module.

256 Chapter 3   Separation Equipment

Table 3.10   Summary of Extraction Equipment

Type	Examples	Advantages	Disadvantages
Mixer-Settler		High residence time possible Efficiency can approach one Good for viscous fluids	Expensive for many stages
Static	Spray	Simple, inexpensive	Usually only one theoretical stage
	Tray	2–8 theoretical stages High capacity Low cost Familiar sizing method	Stages have low efficiency
	Packed	2–5 theoretical stages Low cost High capacity Familiar sizing method	
	Pulsating/reciprocating (Karr is similar)	Average cost Low tray spacing, so many actual stages	Average capacity Low stage efficiency: 5%–10%
Agitated	York-Scheibel	High efficiency Average cost	
	Rotating disk contactor	High efficiency Average cost	
	Oldshue-Rushton	High efficiency Average cost	
	Kühni	30 theoretical stages possible Adaptable for scale-up and changing conditions within column	Expensive
Centrifugal	Podbielniak	High efficiency	Expensive Low capacity

The advantage of hollow-fiber membranes is identical to shell-and-tube heat exchangers, which is that a large amount of surface area/unit volume of equipment is possible. While the module appears to mimic a shell-and-tube heat exchanger, the flow patterns do not. In a shell-and-tube heat exchanger, there are two inlet streams and two outlet streams. In a hollow-fiber module, there is one input and two outputs, with the two outputs being the permeate stream and the less permeable (called *retentate*) stream. In fact, the most analogous separation to gas permeation from an input-output perspective is a partial vaporization/partial condensation/flash separation (Section 3.2.1).

3.5.2   Models for Gas Permeation Membranes

The simplest possible model for gas permeation membrane equipment is the "well-mixed" model. The validity of this model is discussed later; however, it is useful to understand the concept behind gas permeation. This model is illustrated in Figure 3.52.

The subscripts p and r indicate permeate and retentate, respectively. The variables p and y indicate pressure and mole fraction, respectively, and F indicates molar flowrate. The upstream side is at a higher pressure than the permeate side, which increases the driving force across the membrane. For the simple model, a two-component system is assumed.

A total balance and a balance on the more permeable component are

$$F_{in} = F_p + F_{out} \tag{3.75}$$

$$F_{in} y_{in} = F_p y_p + F_{out} y_r \tag{3.76}$$

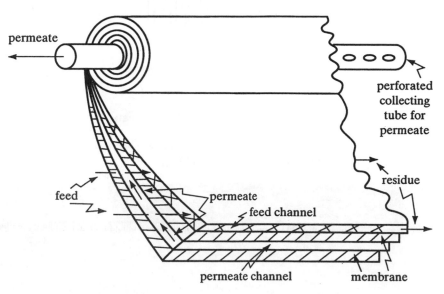

Figure 3.48   Spiral-wound membrane (Reprinted by permission from Geankoplis [2003])

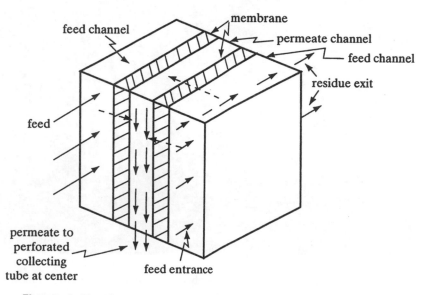

Figure 3.49   Flow patterns in spiral-wound membrane (Reprinted by permission from Geankoplis [2003])

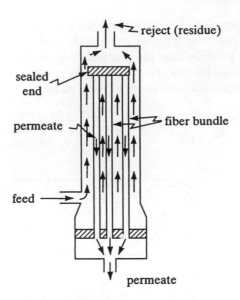

Figure 3.50 Illustration of hollow fiber module (Reprinted by permission from Geankoplis [2003])

Figure 3.51 Hollow fiber membranes (Courtesy of Spintek Corp.)

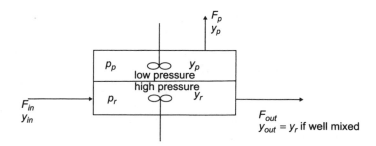

Figure 3.52   Well-mixed model for gas permeation membrane

Rearrangement of Equations (3.75) and (3.76) give

$$y_p = -\frac{(1-\theta)\,y_r}{\theta} + \frac{y_{in}}{\theta}$$

$$(3.77)$$

where $\theta = F_p/F_{in}$, which is often called the *stage cut*. It is a similar term to V/F in the flash-type separations, one outlet flowrate divided by the feed flowrate.

The balances must be combined with rate expressions for transport across the membrane. The rate expressions, assuming ideal gas, are

$$F_p y_p = \frac{P_A A}{t_m}\left[\,p_r y_r - p_p y_p\,\right]$$

$$(3.78)$$

$$F_p\left(1-y_p\right) = \frac{P_B A}{t_m}\left[\,p_r\left(1-y_r\right) - p_p\left(1-y_p\right)\right]$$

$$(3.79)$$

where P_A and P_B are the membrane permeabilities of the more permeable component (A) and the less permeable component (B), respectively, and t_m is the membrane thickness. Tables of permeabilities are available (Geankoplis, 2003; Wankat, 2017). Taking the ratio of Equations (3.78) and (3.79) yields

$$\frac{y_p}{1-y_p} = \alpha_{AB}\,\frac{p_r y_r - p_p y_p}{p_r\left(1-y_r\right) - p_p\left(1-y_p\right)}$$

$$(3.80)$$

where α_{AB} is the ratio of the permeabilities, P_A/P_B. Sometimes, the following rearranged form of Equation (3.80) makes calculations easier.

$$y_r = \frac{y_p\left[\left(\alpha_{AB}-1\right)\left(\dfrac{p_p}{p_r}\right)\left(1-y_p\right)+1\right]}{\alpha_{AB}-\left(\alpha_{AB}-1\right)y_p}$$

$$(3.81)$$

The area, A, can be obtained from either rate expression, Equation (3.78) or (3.79), for a given membrane and thickness, once all flows and mole fractions are known.

Clearly, given the geometry of hollow-fiber membrane modules, the well-mixed assumption is a crude model. Crossflow arrangements are possible, where the feed and retentate are in the middle of the hollow-fiber bundle. Depending on the exact geometry of the equipment, both

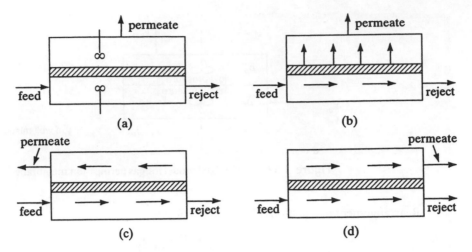

Figure 3.53   Some flow models for gas permeation membranes (Reprinted by permission from Geankoplis [2003])

countercurrent and cocurrent flows are possible, and it is possible to have both in the same piece of equipment. Figure 3.53 illustrates some of these possible flow configurations.

The mathematical analysis of these models involves differential equations, and they parallel both the development for shell-and-tube heat exchangers and continuous differential separators. Turbulent flow is usually assumed in modeling heat exchangers and packed bed, vapor-liquid separators. However, given the very small diameters involved in hollow-fiber membranes, laminar flow is possible, so the parabolic velocity profile may have to be included. Clearly, these models are far more complex than the well-mixed model, and numerical solutions of the differential equations are often required. A summary of some of these models is available (Geankoplis, 2003).

3.5.3   Practical Issues

Gas permeation is typically used for separations involving noncondensable gases. It is more useful for some separations than for others. As the permeate concentration increases, the permeate flowrate decreases. Gas permeation equipment can be staged; however, recompression of the permeate is expensive. The upstream side is typically at a higher pressure than the permeate to increase the driving force, as suggested by Equations (3.78) and (3.79). As stated, as the permeate from subsequent stages increases in concentration, the flowrate decreases. Therefore, gas permeation is more useful as a polishing step, that is, removal of dilute contaminants, than as a coarse separation, such as distillation. This suggests that gas permeation may be better suited for purifying the retentate rather than the permeate.

WHAT YOU SHOULD HAVE LEARNED:

- The separation basis and separating agent for the most common chemical engineering separations

- How to determine the size of tray columns and packed columns based on a percentage of flooding

- How reflux ratio impacts the design of tray and packed distillation columns
- How internal flows impact the design of tray and packed distillation columns
- How the feed condition impacts the design of tray and packed distillation columns
- How the feed composition impacts the design of tray and packed distillation columns
- How the difficulty of separation (difference in relative volatilities) impacts the design of tray and packed distillation columns
- The internals of tray and packed columns
- The impact of the reboiler and condenser on the design and performance of distillation columns
- The economic trade-offs for tray and packed columns
- The key parameters affecting absorption and stripping, and their impact on absorber and stripper design and performance
- The types of equipment used in extraction, their advantages and their disadvantages
- The type of equipment used for gas-permeation membrane separations

NOTATION

Symbol	Definition	SI Units
A	heat transfer area	m^2
A	constant in Antoine's equation	
A	absorption factor	
a	mass transfer area	m^2
B	constant in Antoine's equation	°C
C	constant in Antoine's equation	°C
C	concentration	$kmol/m^3$
$C_{sb,f}$	parameter in flooding calculation	m/s
D_{AB}	diffusion coefficient of solute A in solvent B	m^2/s
d	diameter	m
E_o	overall column efficiency	
F	feed flowrate	kmol/h
F	packing factor in packed bed	

Symbol	Definition	SI Units
F_x, F_y	mass transfer coefficients for liquid (x) or vapor (y)	phase
F_{lv}	parameter in flooding calculation	
G'	superficial gas flowrate in packed bed	kg/m²/s
g_c	unit conversion of 32.2 ft lb/lb$_f$/sec²	
H, h	enthalpy, upper case for vapor, lower case for liquid	kJ/kmol
H, HTU	height of transfer unit	m
$HETP$	height equivalent to theoretical plate	m
H_A	Henry's law coefficient for component A	Pa
h	pressure head	m
K_x, K_y	mass transfer coefficient (x is liquid phase, y is vapor phase)	kmol/m²/s
$L, \overline{L}$	liquid flowrate (over bar is below feed in distillation column)	kmol/h
M	molecular weight	kg/kmol
m	equilibrium/partition coefficient	
N	number of trays	
N	number of transfer units	
N	molar flux	kmol/m²/s
P	pressure	Pa
P_i	membrane permeability of component i	m³/m²/s/kPa
P^*	vapor pressure	Pa
p_i	partial pressure	Pa
Q	heat duty	kW
q	fraction of liquid in distillation column feed	
R	gas constant	m³kPa/kmol/K
R	reflux ratio	
S	interfacial surface area	m²
S	stripping factor	
t_m	membrane thickness	m
T	temperature	K
u	velocity	m/s
$V, \overline{V}$	vapor flowrate (over bar is below feed in distillation column)	kmol/h

Symbol	Definition	SI Units
x	mole fraction in liquid phase	
y	mole fraction in vapor phase	
z	mole fraction in feed stream	
Z	coordinate in direction of column height	

GREEK SYMBOLS

α_{AB}	relative volatility or relative permeability	
α	parameter in calculating pressure drop in packed bed	
β	parameter in calculating pressure drop in packed bed	
γ	activity coefficient	
δ	film thickness	m
θ	stage cut in gas permeation membrane	
λ	latent heat of vaporization	kJ/kmol
μ	viscosity	kg/m/s
ρ	density	kg/m^3
σ	surface tension (must be dyne/cm^2)	
ϕ	fugacity coefficient	
ψ	density of water/density of liquid in packed bed	

SUBSCRIPTS

A, B	components
actual	refers to actual column area
active	refers to active column area
B	bottoms of distillation column
C	refers to condenser
cw	cooling water
D	distillate
F, f	feed
f	refers to flooding conditions
i	refers to component i
i	refers to interface

Symbol	Definition
in	refers to inlet stream
L	refers to liquid stream, liquid content of stream, or liquid-phase transfer units
lm	log mean
m	refers to membrane
min	minimum
OL, OV	refers to overall liquid and overall vapor transfer units or height of transfer unit, respectively
out	refers to outlet stream
p	refers either to process stream or permeate stream
R	refers to reboiler
r	refers to retentate stream
s	steam
V	refers to vapor stream, vapor content of stream, or vapor-phase transfer units
vap	vaporization
1, 2	refers to stage number of different phases

SUPERSCRIPTS

L	liquid phase
I, II	refers to different phases
*	equilibrium value

REFERENCES

Berry, R. I. 1981. *Chem Eng* 88(July 13): 63.

Colburn, A. P. 1939. "A Method of Correlating Forced Convection Heat Transfer Data and a Comparison with Fluid Friction." *Trans AIChE* 29: 174.

Couper, J. R., W. R. Penney, J. R. Fair, and S. M. Walas. 2012. *Chemical Process Equipment, Selection and Design*, 3rd ed. Boston: Elsevier.

Eckert, J. S. 1970. "Selecting the Proper Distillation Column Packing." *Chem Eng Prog* 66(3): 39–44.

Fair, J. R., and R. L. Matthews. 1958. "Better Estimate of Entrainment from Bubble-Cap Trays." *Petrol Refin* 37(4): 153.

Geankoplis, C. 2003. *Transport Processes and Separation Principles*, 4th ed. Upper Saddle River, NJ: Prentice Hall.

Hebert, S., and N. Sandford. 2016. "Consider Moving to Fixed Valves." *Chem Engr Prog* 112(5): 34–41.

Kister, H. 1992. *Distillation Design*. New York: McGraw-Hill.

Koch, J., and G. Shiveler. 2015. "Design Principles for Liquid-Liquid Extraction." *Chem Engr Prog* 111(11): 22–30.

Koch Modular Processes. 2000–2016. Extraction Column Types. http://modularprocess.com/liquid-liquid-extraction/extraction-column-types.

Kremser, A. 1930. "Theoretical Analysis of Absorption Process." *Natl Petrol News* 22(21): 42.

O'Connell, H. E. 1946. "Plate Efficiencies of Fractionating Columns and Absorbers." *Trans Amer Inst Chem Eng* 42: 741.

Perry, R. H., and D. W. Green (Eds.). 1997. *Perry's Chemical Engineers' Handbook*, 7th ed. New York: McGraw-Hill, 2-61–2-75.

Pilling, M. 2006. "Design Considerations for High Liquid Rate Tray Applications." *Proceedings of 2006 AIChE Annual Meeting*, Paper 264e.

Regents of the University of Michigan. 2014. *Encyclopedia of Chemical Engineering Equipment*. http://encyclopedia.che.engin.umich.edu/Pages/SeparationsChemical/Extractors/Extractors.html.

Seader, J. D., E. J. Henley, and D. K. Roper. 2011. *Separation Process Principles. Chemical and Biochemical Applications*, 3rd ed. New York: Wiley.

Souders, M., and G. G. Brown. 1932. "Fundamental Design of High Pressure Equipment Involving Paraffin Hydrocarbons. IV. Fundamental Design of Absorbing and Stripping Columns for Complex Vapors." *Ind Eng Chem* 24: 519–522.

Stichlmair, J. G., and J. R. Fair. 1998. *Distillation. Principles and Practice*. New York: Wiley.

Treybal, R. E. 1980. *Mass Transfer Operations*, 3rd ed. New York: McGraw-Hill.

Turton, R., R. C. Bailie, W. B. Whiting, J. A. Shaeiwitz, and D. Bhattacharyya. 2012. *Analysis, Synthesis and Design of Chemical Processes*, 4th ed. New York: Pearson.

Wankat, P. 2017. *Separation Processes. Includes Mass Transfer Analysis*, 4th ed. Upper Saddle River, NJ: Prentice Hall.

PROBLEMS

Short Answer Problems

1. Sketch a *T-xy* diagram for an ideal, binary mixture.

   a. Illustrate a partial vaporization for a feed of 25 mol% of the more volatile component. Indicate the two phases in equilibrium, the bubble point, and the dew point.

   b. Illustrate a partial condensation for a feed of 50 mol% of the more volatile component. Indicate the two phases in equilibrium, the bubble point, and the dew point.

2. Sketch a *P-xy* diagram for an ideal, binary mixture.

   a. Illustrate a flash for a feed of 75 mol% of the more volatile component. Indicate the two phases in equilibrium, the bubble point, and the dew point.

   b. Illustrate a partial condensation for a feed of 50 mol% of the more volatile component. Indicate the two phases in equilibrium, the bubble point, and the dew point.

3. What are the physical assumptions of constant molar overflow? What are the consequences? To what types of systems does constant molar overflow typically apply?

4. State three reasons for requiring a high reflux ratio. Illustrate these on a McCabe-Thiele diagram.

5. How does the feed condition (saturated, superheated, subcooled) affect the reflux ratio for a given feed mole fraction? Illustrate the answer on McCabe-Thiele diagrams.

6. What are the limiting cases for reflux ratio? Why are neither of these cases practical?

7. What is the best location for the distillation column feed?

8. What happens if the feed to a distillation column is not at the optimum location? Illustrate this on a McCabe-Thiele diagram.

9. A distillation column requires 18 equilibrium stages that are 40% efficient. It has a partial reboiler.

   a. How many trays are in the column if there is a total condenser?

   b. How many trays are in the column if there is a partial condenser?

10. Why are some distillation columns run at elevated pressures? State and explain as many reasons as you can.

11. Why are some distillation columns run at vacuum? State and explain as many reasons as you can.

12. The feed to a distillation column is significantly subcooled. What happens on the feed tray? What is the potential effect on column diameter? What is the effect on the reboiler and condenser duties?

13. The feed to a distillation column is significantly superheated. What happens on the feed tray? What is the potential effect on column diameter? What is the effect on the reboiler and condenser duties?

14. Explain the relationship between reflux ratio and the number of equilibrium stages in a distillation column.

15. What is flooding? What is loading? What is weeping?

16. Discuss the advantages and disadvantages of sieve trays versus bubble cap trays.

17. Why might vacuum columns be designed with a larger diameter above the feed compared to below the feed?

18. Does tray spacing have an effect on column diameter? Explain why or why not.

19. The feed to a column is to be saturated liquid. An initial design based on the column conditions below the feed shows that trays above the feed might be weeping. Suggest possible remedies.

20. Discuss the advantages and disadvantages of packed versus tray columns.

21. An existing distillation column is to be scaled up (increase throughput) without changing the distillate and bottom compositions. What happens to the reflux ratio? What problems might arise in the performance of this column?

22. For the distillation column in Problem 3.21, it is suggested to change the pressure of the column. Explain why this might work.

23. For the distillation column in Problems 3.21 and 3.22, how can the pressure in the column be changed?

24. For the distillation column in Problems 3.21, 3.22, and 3.23, someone suggests that the column pressure can be increased by increasing the feed pressure. What is your response? Explain the rationale for your response.

25. The preliminary design of a distillation column shows that it is flooding. If, for some reason, the diameter cannot be changed, suggest two other design modifications that could alleviate this problem, and explain why each works.

26. A colleague gives you a preliminary tray column design. It shows 12-in tray spacing with 8-in weirs. There are 30 actual trays, and the diameter is 5 ft. Comment on all aspects of this design and suggest changes, if necessary.

27. A colleague gives you a preliminary tray column design. It shows 24-in tray spacing with 6-in weirs. There are 60 actual trays and the diameter is 5 ft. Comment on all aspects of this design and suggest changes, if necessary.

28. What is a packing factor, and how does it affect packed column design?

29. What is the absorption (stripping) factor? How does it affect absorber (stripper) design?

30. State three operating conditions that favor absorption (i.e., make absorption easier). Provide a physical explanation for each condition.

31. State three operating conditions that favor stripping (i.e., make stripping easier). Provide a physical explanation for each condition.

32. An absorber is not performing as specified, meaning that the target outlet vapor composition is higher than desired. State as many possible reasons for this that you can identify, and explain why each reason makes physical sense.

33. A stripper is not performing as specified, meaning that the target outlet liquid composition is higher than desired. State as many possible reasons for this that you can identify, and explain why each reason makes physical sense.

34. Comment on the following statement: *A transfer unit in a packed column is numerically equivalent to an equilibrium stage in a tray column.*

35. What do height of a transfer unit and the number of transfer units indicate?

36. Explain the physical reason why, when the absorption or stripping factor is less than one, increasing the number of stages or increasing packed column height does not increase the recovery of solute.

37. What is the separation basis for distillation? What is the separating agent?

38. What is the separation basis for absorption? What is the separating agent?

39. What is the separation basis for extraction? What is the separating agent?

40. What is the separation basis for leaching? What is the separating agent?

41. What is the separation basis for gas permeation membranes? What is the separating agent?

Problems to Solve

42. For this problem, the feed is 60 mol% dimethyl ether (DME) and 40 mol% methanol. Assume that the equilibrium data provided are correct.

    a. A feed of 200 kmol/h is flashed at 5 bar and 50°C. What are the liquid and vapor flowrates and the mole fractions in the vapor and liquid exit streams?

    A feed of 200 kmol/h is to be fractionated to give a distillate containing 98 mol% dimethyl ether and a bottoms product containing 98 mol% methanol. The feed is 20% vapor and 80% liquid. There is a total condenser and a partial reboiler.

    b. Determine the molar flowrates of the distillate and bottoms.

    c. Calculate all internal flows for operation at a reflux ratio of 0.1.

d. In order to use cooling water in the condenser, assume that the DME must condense no lower than 55°C. Estimate the top pressure for this column.

e. In order to use low-pressure steam in the reboiler, the bottom temperature must be no higher than 148°C. Estimate the bottom pressure for this column.

f. Determine the diameter of a sieve tray tower with 18-in tray spacing to operate at 75% of flooding. The active tray area is 85% of the column area. Determine the diameter both above and below the feed. There are six equilibrium stages with the feed on the third stage.

g. Assume that the tray efficiency is 42%. Estimate the maximum weir height for which trays can be designed to meet the pressure specifications.

h. Suppose you determine that the height to diameter ratio for this column exceeds 20, which is usually the maximum recommended value. Suggest at least one design change that would result in a height to diameter ratio smaller than 20. Explain the rationale for your suggestion and its consequences.

i. Determine the diameter of a packed tower to operate at 70% of flooding using 1.5-in, ceramic Berl saddles. Determine the diameter both above and below the feed.

j. Look up the behavior of the DME-methanol system. How would it impact the answers to this problem?

Vapor Pressure Data

$$\ln P^{*}(\text{bar}) = A - \frac{B}{T(\text{K})}$$

	A	B
DME	10.543	2623.14
Methanol	12.8591	4336.7

Other Data

	Liquid density (kg/m³)	Viscosity (cP)
DME	655	0.5
Methanol	788	0.4

43. Cumene (isopropyl benzene) is manufactured from propylene and benzene. Most of the world's supply of cumene is directly converted into phenol, which is a raw material for a variety of products such as plasticizers. Acetone is also manufactured as a by-product of phenol manufacture. The separation section of a cumene process consists of a flash, to remove the unreacted propylene, followed by a distillation column to separate the benzene and produce purified cumene as the bottom product.

a. For this problem, consider that the flash contains only propylene and benzene and operates at 3 atm and 100°C. The feed contains 80 mol% benzene and 20 mol% propylene. On a basis of 200 kmol/h feed, determine the flowrates of vapor and liquid leaving the flash.

The feed to the distillation column is 60% liquid at 200 kmol/h containing 48 mol% benzene. It is necessary to produce 99 mol% cumene in the bottom product and 98 mol% benzene in the top product for recycle. The distillation column contains a total condenser and a partial reboiler. The column operates at an average pressure of 3 atm.

b. Determine the flowrates of the distillate and the bottoms.

c. For a reflux ratio of 0.85, calculate all internal flows.

d. Estimate the heat loads (kJ/h) on the condenser and reboiler.

e. Estimate the column efficiency and determine the actual number of trays, the actual number of stages, and the actual feed location. There are 10 equilibrium stages, with feed on Stage 4.

f. Design a sieve tray column for 75% of flooding, with 85% active tray area and with 18-in tray spacing. Specify the column diameter and the actual column height. Ignore the surface tension correction. It is up to you to determine whether to design for the top section or for the bottom section. You should justify your choice.

g. Estimate the pressure drop (kPa) in the column if the trays have 6-in weirs.

h. Design a packed column for 75% of flooding with 2-in, ceramic Raschig rings. It is up to you to determine whether to design for the top section or for the bottom section. You should justify your choice.

i. There is an existing column with the correct number of trays with a 2 m diameter and 85% active area. Will this column work? Explain why or why not. Your answer must be supported with calculations. Could anything be done to make this column work? If so, what are the consequences?

j. The original column is now built and operational, and the feed is 60% liquid. Due to a reactor upset, it is now necessary to accept feed, still 60% liquid, at the same rate but containing 50 mol% benzene, with the remainder cumene. It is still necessary to produce 99 mol% cumene at the same rate from the bottom of this column. You assign your summer intern the job of determining new operating conditions. The answer you get back is to operate at a reflux ratio of 0.75. Will the column operate at these conditions?

Data:

For propylene,

$$\log_{10} p^*(\text{mm Hg}) = 6.77811 - \frac{773.29}{T(°C) + 232}$$

For benzene,

$$\log_{10} p^*(\text{mm Hg}) = 6.90565 - \frac{1211.033}{T(°C) + 220.79}$$

Average column relative volatility = 6

Feed viscosity = 0.3 cP

Benzene density = 800 kg/m^3

Cumene density = 700 kg/m^3

Benzene heat of vaporization = 3.07 × 10^4 kJ/kmol

Cumene heat of vaporization = 3.81 × 10^4 kJ/kmol

Estimated top temperature = 56°C

Estimated bottom temperature = 178°C

Benzene viscosity = 0.4 cP

Cumene viscosity = 0.3 cP

44. A distillation column is fed 500 kmol/h of an equimolar mixture of pentane and hexane. The feed is saturated liquid. It is necessary to produce outlet streams containing 95 mol% pentane and 99 mol% hexane. The distillation column contains a partial reboiler and a total condenser.

   a. Determine the exit flowrates D and B.

   b. If the reflux ratio is one, determine all of the internal molar flowrates in the column.

   c. Estimate the condenser and reboiler heat loads.

   d. In the summer, cooling water is available at 30°C and is returned at 40°C. In the winter, cooling water is available at 25°C and is returned at 35°C. Assuming a 10°C approach in the condenser, based on the cooling water return temperature, determine the seasonal top column pressures.

   e. If the column pressure drop is 15 kPa, determine the seasonal bottom column temperatures.

   f. There are 15 equilibrium stages, with the feed on stage 5. Determine the diameter of a sieve tray column with 24-in tray spacing for 70% of flooding and an active tray area of 88%. Calculate the diameter for both the top and bottom sections. If the column is designed for the larger diameter, how will the other section perform?

   g. Determine the diameter of a packed column with 2-in, ceramic Berl saddles at 70% of flooding. Calculate the diameter for both the top and bottom sections. If the column is designed for the larger diameter, how will the other section perform?

Data:

$$\mu_{feed} = 0.25 \text{ cP } \rho_L = 650 \text{ kg/m}^3$$

$$\lambda_5 = 25{,}720 \text{ kJ/kmol } \lambda_6 = 29{,}632 \text{ kJ/kmol}$$

$$SG_{5,L} = 0.626$$

$$SG_{6,L} = 0.655$$

$$\mu_{5,L} = 0.173 \text{ cP}, \mu_{5,L} = 0.175 \text{ cP}$$

$$\log_{10} P_5^* (\text{mm Hg}) = 6.85296 - \frac{1064.84}{T(°C) + 232.012}$$

$$\log_{10} P_6^* (\text{mm Hg}) = 6.87601 - \frac{1171.17}{T(°C) + 224.408}$$

45. For this problem, the feed is 40 mol% heptane and 60 mol% ethylbenzene. A saturated liquid feed of 200 kmol/h is to be fractionated at atmospheric pressure to give a distillate containing 98 mol% heptane and a bottoms product containing 1 mol% heptane.

   a. Determine the molar flowrates of distillate and bottoms.

   b. Calculate all internal flows for operation at a reflux ratio of 1.5.

   c. Estimate the condenser and reboiler duties, in kW.

   d. Estimate the top and bottom temperatures, in °C.

   e. Design a tray tower with 18-in tray spacing to operate at 75% of flooding. The active tray area is 75%. You should decide whether to design for the top or for the bottom of the column. There are 17 equilibrium stages with the feed on Stage 10.

    f. Design a packed tower with 2-in, ceramic INTALOX™ saddles for 70% of flooding. You should decide whether to design for the top or for the bottom of the column.

Vapor pressure data:

For heptane,

$$\log_{10} p^*(\text{mm Hg}) = 6.90253 - \frac{1267.828}{T(°C) + 216.823}$$

For ethylbenzene,

$$\log_{10} p^*(\text{mm Hg}) = 6.95650 - \frac{1423.543}{T(°C) + 213.091}$$

Other data:

	MW	Liquid Density (kg/m³)	Normal bp (°C)	Latent Heat (kJ/kmol)	Viscosity (cP)
Heptane	100	680	98.4	35,600	0.3
Ethylbenzene	106	867	136.2	35,900	0.35

46. A feed of 500 kmol/h of methanol and ethanol are to be distilled at 2 bar. The feed is 60 mol% methanol, and the feed is 75% liquid and 25% vapor in equilibrium. The desired outlet compositions are 90 mol% methanol in the distillate and 95 mol% ethanol in the bottoms.

    a. Determine the flowrates of the distillate and the bottoms (kmol/h).

    b. Calculate all internal flows for a reflux ratio of 3.

    c. Determine the diameter of a sieve tray tower with 24-in tray spacing at 70% of flooding with an active tray area of 89%. Design for the top and bottom sections. If the larger diameter is used for the entire column, how will the other section perform? There are 19 equilibrium stages with the feed on Stage 7.

Data:

For methanol,

$$\log_{10} p^*(\text{mm Hg}) = 8.07240 - \frac{1574.99}{T(°C) + 238.87}$$

For ethanol,

$$\log_{10} p^*(\text{mm Hg}) = 8.21330 - \frac{1652.050}{T(°C) + 231.480}$$

Methanol heat of vaporization = 40.5 kJ/mol

Ethanol heat of vaporization = 38.3 kJ/mol

Methanol specific gravity = 0.75

Ethanol specific gravity = 0.75

47. CO_2 is to be stripped from water at 20°C and 2 atm using a staged, countercurrent stripper. The liquid flowrate is 100 kmol/h of water with an initial mole fraction of CO_2 of 0.00005.

The inlet air stream contains no CO_2. A 98.4% removal of CO_2 from the water is desired. The Henry's law constant for CO_2 in water at 20°C is 1420 atm.

a. Calculate the outlet mole fraction of CO_2 in water.

b. If there are seven equilibrium stages, calculate V and the outlet mole fraction of CO_2 in the air.

48. An absorber is designed to treat 40 kmol/h of air containing acetone with a mole fraction of 0.02 and reduce the mole fraction to 0.001. Pure water is the solvent, at a rate of 20 kmol/h. The column temperature is 26.7°C, and the pressure is 1 atm. Raoult's law may be assumed to apply, and

$$\ln P^*(\text{atm}) = -\frac{3598.4}{T(\text{K})} + 10.92$$

a. How many equilibrium stages are needed?

b. Assume that a column with five equilibrium stages is built and operational. It is now observed that the outlet mole fraction of acetone in the air is 0.002. Describe and explain at least four possible causes of this situation. Quantify what is happening in the column for each case.

c. Suggest realistic ways to compensate for this problem. An example of a nonrealistic compensation method is to reduce the air feed, which is probably fixed upstream.

d. The column is operating normally, but you are told to expect a 10% increase in air to be treated. How can you change column operation to ensure the desired outlet mole fraction of acetone in air? Be quantitative.

e. The column is operating normally, but you are told to expect an inlet mole fraction of acetone of 0.025. How can you change column operation to ensure the desired outlet mole fraction of acetone in air? Be quantitative.

f. The column is operating normally, but you are told that the outlet mole fraction of acetone must be reduced to 0.00075. How can you change column operation to ensure the desired outlet mole fraction of acetone in air? Be quantitative.

49. Sulfur dioxide (SO_2) is removed from a smelter gas (which may be considered to have the properties of pure air) containing 0.80 mol% SO_2 by absorption into a solution with the properties of pure water in a countercurrent tower. The exit SO_2 concentration in the gas phase must be 0.04 mol%. The tower is designed to operate at exactly 30°C and at 4 bar. At these conditions, the equilibrium relationship is $y = 0.1923x$, where y is the mole fraction of SO_2 in the gas phase, and x is the mole fraction of SO_2 in the water phase. The feed rate of gas is 0.36 kmol/s and 1 mole of aqueous solution is used for every 4 moles of feed gas.

a. What is the ratio of the solvent rate to the minimum solvent rate?

b. For a tray tower, determine the number of equilibrium trays required for this separation. Include fractional trays in your answer.

c. Assume the number of equilibrium stages is 6.5. Due to a temporary upset upstream, it is now necessary to treat gas at the higher feed concentration of 1 mol% SO_2. If the temperature and pressure must remain constant at 30°C and 4 bar, what do you suggest to accommodate the higher concentration in the feed?

50. An air stream flowing at 1500 kmol/h containing benzene vapor is to be scrubbed using 25 kmol/h of hexadecane ($C_{16}H_{34}$) as the solvent. The mole fraction of benzene in the inlet gas is 0.010, and the required outlet mole fraction is 0.00050. The hexadecane is being returned

to the scrubber from a stripper, so the mole fraction of benzene in the feed hexadecane is 0.00010. The column may be assumed to operate at 150 kPa and 35°C. Raoult's law may be assumed to define the benzene equilibrium. Dilute solutions may be assumed.

a. How many equilibrium stages are needed? Keep the decimal places.

b. What is the outlet mole fraction of benzene in the hexadecane? What do you conclude about the assumptions made?

c. The column is operational, and the outlet mole fraction of benzene in the air is measured to be 0.0010. You send an operator to check the conditions in the column, and the temperature, pressure, air flowrate, and hexadecane flowrate are all as per design specifications. Suggest two possible causes for the process upset, and provide exact values for the off-spec conditions, assuming that only one condition at a time is off-spec.

d. Determine the diameter of a sieve tray column for an 18-in tray spacing for operation at 75% of flooding. The fraction of active tray area is 0.88.

e. Determine the diameter of a packed column at 75% of flooding with 2-in, ceramic Raschig rings.

$$\log_{10} P_b^* (\text{mm Hg}) = 6.90565 - \frac{1211.033}{T(°C) + 220.79}$$

$$SG_{hexadecane\ liquid} = 0.77$$

$$SG_{benzene\ liquid} = 0.88$$

$$\mu_{hexadecane\ liquid} = 3.474\ cP$$

51. A wastewater stream at 35,000 kmol/h containing benzene at a mole fraction of 0.001 is to be stripped with air in a column operating at 2 bar and 25°C. The outlet water must contain no more than 0.00004 mole fraction of benzene. The equilibrium between water and air for benzene at 2 bar and 25°C is $y = 150x$, where y is the mole fraction of benzene in air and x is the mole fraction of benzene in water. The air used is maintained in a closed loop. Benzene is recovered by condensation from the air stream; therefore, the air entering the stripper contains 0.005 mole fraction of benzene.

a. As part of the design evaluation for this process, it is necessary to determine whether it is feasible to use an existing tray tower for this process. If the tower contains 9 equilibrium stages, and an air rate of 325 kmol/h is the limit set by 80% of flooding, can the tower be used?

b. You observe that there is some flexibility in the operating conditions of the tower in Part (a). Suggest one set of conditions under which the existing tower could be used.

52. When coal is burned to form synthesis gas (syngas), which contains mostly CO, H_2, H_2S, and CO_2, the H_2S must be removed. The gas is called syngas because H_2 and CO are the building blocks for synthesis of hydrocarbons. One method for removal of H_2S is the Selexol™ solvent, in which the H_2S is highly soluble. The Selexol solvent is the dimethyl ether of polyethylene glycol and may be assumed to have a molecular weight of 300 kg/kmol, $\mu = 2.5 \times 10^{-3}$ kg/m s, and $\rho = 900$ kg/m³.

Consider the removal of H_2S only from an otherwise inert mixture of gases; that is, none of the other gases are soluble in the Selexol solvent. The partition coefficient for H_2S between gas and solvent is given by the relationship

$$\frac{y}{x} = m = 3.6 \times 10^{-4} \left[\frac{460 + T(°F)}{P(\text{atm})} \right]$$

The problem at hand is the removal of H_2S from a gas stream with the properties of air, initially containing 2% H_2S, so that the exit gas only contains 0.05% H_2S. The H_2S-rich Selexol solvent is regenerated in a stripper using pure air. The absorber operates at 100 psig (6.8 atm gauge) and 70°F. The stripper operates at 1 atm absolute and 200°F. The stripper reduces the H_2S content of the Selexol solvent to 0.5%, which is the feed concentration to the absorber. The gas to be treated is flowing at 1000 kmol/h, and it may be assumed that the Selexol solvent circulates at 30,000 kg/h. In the packed-bed absorber, it is known that H_{oV} = 3.5 m. In the packed-bed stripper, it is known that H_{oL} = 0.25 m. It is proposed to use an existing stripper that is packed with a 5 m high section of 1.5-in, ceramic Raschig rings and has a diameter of 1.8 m.

a. Determine the required absorber height. You may assume that the Colburn method is valid.

b. Determine the exit H_2S mole fraction in the Selexol solvent. Comment on this value.

c. For the stripper, determine the amount of air needed.

d. At what percentage of flooding is the stripper operating? Do you recommend operating under these conditions?

e. Determine the pressure drop across the stripper, in kPa.

53. Consider the removal of H_2S only from air. The solvent has a molecular weight of 300, μ = 2.5 × 10^{-3} kg/m/s, σ = 30 dyne/cm^2, and SG = 0.90. The partition coefficient for H_2S between air and solvent is given by the relationship

$$\frac{y}{x} = m = 3.6 \times 10^{-4} \left[\frac{460 + T(°F)}{P(atm)} \right]$$

The problem at hand is the removal of H_2S from an air stream initially containing 0.10% H_2S, so that the exit gas only contains 0.0050% H_2S. The absorber operates at an average pressure of 6 atm and an average temperature of 50°F. The inlet H_2S content of the solvent can be assumed to be zero. The gas to be treated is at 200 lbmol/h, and it may be assumed that the solvent circulates at 2100 lb/h.

a. How many equilibrium stages are needed for this separation?

b. There is a problem with the solvent pump, which must be taken off line soon, and the spare pump can only provide solvent at 5 atm at a flowrate no higher than 105% of the original flowrate. Therefore, the feed gas is to be throttled to 5 atm. Since the heat exchangers for this process already uses refrigerated water, 50°F is the lowest possible temperature. You ask two summer interns to evaluate the situation. One says that everything should be okay. The other says that the absorber will not work. Which intern would you hire?

c. What is the diameter of a sieve-tray column for this absorber with 24-in tray spacing for 70% of flooding with an active area of 90% of the total area?

d. If the trays are 20% efficient, is the height/diameter ratio within typical limits? Explain.

e. Estimate the pressure drop in the actual column if there are 6-in weirs.

f. What is the diameter of a packed column for this absorber with 1.5-in, ceramic Berl saddles at 75% of flooding?

4

Reactors

WHAT YOU WILL LEARN

- The important roles of reaction kinetics, equilibrium, and heat transfer in determining the correct design for chemical reactors
- The hierarchy of configurations for removing heat from exothermic, gas-phase catalytic reactions
- That exothermic, gas-phase reactions may give rise to temperature hot spots in the reactor that can lead to dangerous conditions
- That the performance of existing reactors is complex and requires numerical solutions when temperature effects are taken into account

4.0   INTRODUCTION

As for other chapters in this book, the intention of this chapter is not to replace a textbook on reaction engineering but rather to highlight some basic equations relating to chemical reactor design and then to investigate specific issues relating to equipment design and the performance of such equipment. For many industrially relevant reactions, there is a significant enthalpy change between the reactants and products, and this gives rise to the need to transfer energy (heat) to or from the reaction zone. Moreover, the forms of the reaction kinetics are most often dependent on an inverse exponential of absolute temperature (Arrhenius form). Therefore, most reactions are not isothermal, and the kinetics are strong functions of temperature, which leads to nonlinear behavior that cannot easily be formulated and solved by analytical methods. For many real, industrial processes, sophisticated models must be developed to predict the performance of the reactor accurately. However, in this chapter, only the basic equations are presented and some simple examples (with analytical solutions) are solved. Following this approach, some case studies are covered that investigate more complex situations. Where appropriate, the reader is referred to additional references that cover more complex analyses.

4.1 BASIC RELATIONSHIPS

4.1.1 Kinetics

Consider a basic chemical reaction of the form

$$aA + bB \rightarrow rR + sS \Rightarrow A + \frac{b}{a}B \rightarrow \frac{r}{a}R + \frac{s}{a}S \tag{4.1}$$

where a, b, r, and s are the stoichiometric coefficients for the reaction.

The reaction rate for species i (A, B, R, or S), r_i, is defined as

$$r_i = \frac{1}{V}\frac{dN_i}{dt} = \frac{\text{moles of } i \text{ formed}}{(\text{volume of reactor})(\text{time})} \tag{4.2}$$

The reaction rate is an intensive property. This means that the reaction rate depends only on state variables such as temperature, concentration, and pressure and not on the total mass of material present.

For solid catalyzed reactions, the reaction rate is often defined on the basis of the mass of catalyst present, W:

$$r_i = \frac{1}{W}\frac{dN_i}{dt}\rho_b = \frac{1}{V}\frac{dN_i}{dt} \tag{4.3}$$

where ρ_b is the bulk catalyst density (mass catalyst/volume reactor). The density of solid catalyst is defined as ρ_{cat} (mass catalyst particle/volume catalyst particle). So the bulk density of the catalyst, ρ_b, is defined as

$$\rho_b = (1-\varepsilon)\rho_{cat} \tag{4.4}$$

where ε is the void fraction in the reactor, so $(1-\varepsilon)$ is ratio of the volume of catalyst to the volume of the reactor.

If a reaction comprises an elementary step, the kinetic expression can be obtained directly from the reaction stoichiometry. For example, in Equation (4.1), if the first reaction occurs via an elementary step, the rate expression is

$$-r_A = k_1 c_A^a c_B^b \tag{4.5}$$

For catalytic reactions, the rate expressions are often more complicated, because the balanced equation is not an elementary step. Instead, the rate expression can be obtained by an understanding of the details of the reaction mechanism. The resulting rate expressions are often of the form

$$r_i = \frac{k_i \prod_{i=1}^{n} c_i^{\alpha_i}}{\left[1 + \sum_{j=1}^{m} K_j c_j\right]^\gamma} \tag{4.6}$$

Equation (4.6) describes a form of Langmuir-Hinshelwood (L-H) kinetics. The constants (k_1 and K_j) in Equation (4.6) are catalyst-specific. All the constants in Equation (4.6) must normally be obtained by fitting reaction data.

In heterogeneous catalytic reacting systems, reactions take place on the surface of the catalyst. Most of this surface area is internal to the catalyst pellet or particle. The series of resistances that can govern the rate of catalytic chemical reaction are as follows:

1. Mass film diffusion of reactant from bulk fluid to external surface of catalyst
2. Mass diffusion of reactant from pore mouth to internal surface of catalyst

3. Adsorption of reactant on catalyst surface
4. Chemical reaction on catalyst surface
5. Desorption of product from catalyst surface
6. Mass diffusion of product from internal surface of catalyst to pore mouth
7. Mass diffusion from pore mouth to bulk fluid

Each step offers a resistance to chemical reaction. Reactors often operate in a region where only one or two resistances control the rate. For a good catalyst, the intrinsic rates are so high that internal diffusion resistances are usually controlling. This topic is discussed in more detail in Section 4.1.3.

The temperature dependence of the rate constants in Equations (4.5) and (4.6) is given by the Arrhenius equation:

$$k_i = k_o e^{-\frac{E}{RT}}$$

(4.7)

where k_o is called the pre-exponential factor, and E is the activation energy (units of energy/mol, always positive). Equation (4.7) reflects the significant temperature dependence of the reaction rate. For gas-phase reactions, the concentrations can be expressed or estimated from the ideal gas law, so $c_i = p_i/RT$. This is the origin of the pressure dependence of gas-phase reactions. As pressure increases, so does concentration, and so does the reaction rate. The temperature dependence of the Arrhenius equation usually dominates the opposite temperature effect, that is a decrease in the concentration.

Example 4.1

For a simple first-order reaction, determine the relative change in reaction rate due to a change in temperature from 320°C to 350°C for the following reactions. Assume for Parts (a) and (b) that the concentration is unaffected by temperature.

   a. Elementary first-order reaction rate with an activation energy = 20 kJ/mol
   b. Elementary first-order reaction rate with an activation energy = 80 kJ/mol
   c. By how much would the answer change to Parts (a) and (b) if the effect of a change in temperature on concentration was taken into account?

Solution

Use the ratios of the rates from Equations (4.5) and (4.7), and use subscript 1 to represent the base case at 320°C and subscript 2 to represent the new case at 350°C. After cancelling terms, the following expression is found:

$$\frac{-r_{A,2}}{-r_{A,1}} = \frac{k_0 e^{-E/RT_2} c_{A,2}}{k_0 e^{-E/RT_1} c_{A,1}} = \frac{e^{-E/RT_2}}{e^{-E/RT_1}} = e^{-\frac{E}{R}\left[\frac{1}{T_2} - \frac{1}{T_1}\right]}$$

   a. $(-r_{A,2}/-r_{A,1}) = e^{-\frac{20{,}000}{8.314}\left[\frac{1}{(350+273.2)} - \frac{1}{(320+273.2)}\right]} = 1.216$

   b. $(-r_{A,2}/-r_{A,1}) = e^{-\frac{80{,}000}{8.314}\left[\frac{1}{(350+273.2)} - \frac{1}{(320+273.2)}\right]} = 2.183$

   c. Assuming the ideal gas law,

$$c_A = P/RT \Rightarrow c_{A,2}/c_{A,1} = T_1/T_2 = (320+273.2)/(350+273.2) = 0.9519$$

Therefore, for Parts (a) and (b),

   a. $(-r_{A,2}/-r_{A,1})=(1.216)(0.9519)=1.158$

   b. $(-r_{A,2}/-r_{A,1})=(2.183)(0.9519)=2.088$

Thus, the effect of temperature on the rate constant is significantly greater than the effect on concentration. Note that when the activation energy is high, a small change in temperature can have a dramatic effect on the rate of reaction.

4.1.2   Equilibrium

Thermodynamics provides limits on the conversion obtainable from a chemical reaction. For an equilibrium reaction, the equilibrium conversion may not be exceeded.

    The limitations placed on conversion by thermodynamic equilibrium are best illustrated by Example 4.2.

Example 4.2

Methanol can be produced from syngas by the following reaction:

$$CO+2H_2 \rightleftharpoons CH_3OH$$

For the case when no inerts are present and for a stoichiometric feed, the equilibrium expression has been determined to be

$$K = \frac{X(3-2X)^2}{4(1-X)^3 \, P^2} = 4.8\times10^{-13}\exp(11,458/T) \tag{E4.2}$$

where X is the equilibrium conversion, P is the pressure in atmospheres, and T is the temperature in Kelvin. Construct a plot of equilibrium conversion versus temperature for four different pressures: 15 atm, 30 atm, 50 atm, and 100 atm, and interpret the significance of the results.

Solution

The plot is shown on Figure E4.2. By following any of the four curves from low to high temperature, it is observed that the equilibrium conversion decreases with increasing temperature at constant

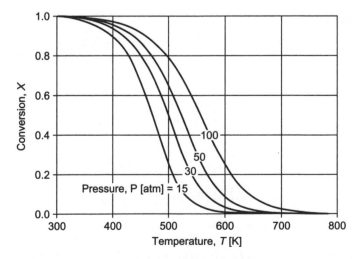

Figure E4.2   Temperature and pressure dependence of conversion for methanol from syngas

pressure. This is a consequence of Le Chatelier's principle, because the methanol formation reaction is exothermic. By following a vertical line from low to high pressure, it is observed that the equilibrium conversion increases with increasing pressure at constant temperature. This is also a consequence of Le Chatelier's principle. Because there are fewer moles on the right-hand side of the reaction, increased conversion is favored at high pressures.

From thermodynamic considerations alone, it appears that this reaction should be run at low temperatures in order to achieve maximum conversion. However, as shown in Example 4.1, the rate of reaction is a strong function of temperature. Therefore, this reaction is usually run at higher temperatures, with low single-pass conversion, in order to take advantage of the faster kinetics. As will be discussed in Section 4.1.4, despite a low single-pass conversion in the reactor, a large overall conversion is still achievable by recycling unused reactants.

4.1.3 Additional Mass Transfer Effects

Up to this point, only the reaction kinetics and the limiting thermodynamics of equilibrium have been considered. These phenomena are often sufficient for quantifying simple, homogeneous reacting systems. However, for heterogeneous systems such as those involving a fluid and catalyst particle, additional effects such as heat and mass transfer within the catalyst may be important. Unlike simple mass transfer, these resistances act both in series and parallel and cannot be combined simply. Figure 4.1 illustrates the processes involved. In order for a reaction to occur, reactant (component A) must first diffuse through an external gas film to reach the catalyst particle's surface. Once at the surface, A reacts but also continues to diffuse into the porous particle and simultaneously reacts inside the particle.

4.1.3.1 No Resistance to External Mass Transfer

First, consider the case when the resistance to external mass transfer is small, and the concentration of gas at the particle's surface is the same as in the bulk, $c_{As} = c_{A0}$. For a first-order reaction with no change in the number of moles, the reaction rate is expressed in terms of the rate at the surface multiplied by an effectiveness factor, η. The derivation of the catalyst effectiveness is similar to the fin effectiveness for heat transfer that was covered in Chapter 2.

$$-r'''_{A,ov}[\text{mol/m}^3\text{catalyst/s}] = k'''[\text{m}^3\text{reactor/m}^3\text{catalyst/s}]c_{A0}[\text{mol/m}^3\text{reactor}]\eta \qquad (4.8)$$

where the triple prime (''') signifies a quantity based on a unit volume of catalyst, and

$$k'''V_{catalyst} = kV_{reactor} \Rightarrow k''' = k\left[\frac{V_{reactor}}{V_{catalyst}}\right] = (1-\varepsilon)k \qquad (4.9)$$

where ε is the voidage of the catalyst bed defined as the volume of the void spaces between the particles divided by the total volume of gas and particles, and the effectiveness factor is given by

$$\eta = \frac{\tanh(M_T)}{M_T} \qquad (4.10)$$

where

$$M_T = L\sqrt{\frac{k'''}{D_{cat}}} \qquad (4.11)$$

The term M_T is the Thiele modulus, which is a ratio of the rate of surface reaction to the rate at which reactant diffuses through the catalyst; L is the characteristic length of the catalyst particle ($L = D_p/6$ for a sphere and $L = D_p/4$ for a cylinder); and D_{cat} is the effective diffusion coefficient

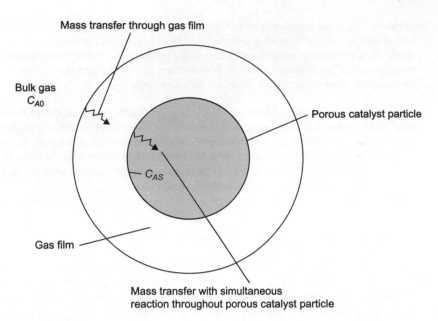

Figure 4.1   Processes occurring in a heterogeneous catalyst reaction system

within the porous catalyst particle that takes account of the pore structure in the catalyst (tortuosity) and the contributions of both Fickian and Knudsen diffusion.

The expression in Equation (4.10) is strictly correct for flat-plate geometry but works very well for cylinders and spheres if the appropriate length dimension, L, is used. Equation (4.10) is plotted in Figure 4.2.

From Figure 4.2, when the value of M_T is <0.4, the catalyst effectiveness factor is 1, and the interior of the particle is bathed in gas with the same concentration as at the surface, c_{As}. For this region, the resistance to pore diffusion is negligible. For values of $M_T > 4.0$, the concentration of gas within the porous catalyst drops rapidly from the surface concentration. For this condition, there is strong pore resistance, and the effectiveness factor is essentially equal to the inverse of the Thiele modulus. In the intermediate regime, the effect of pore diffusion resistance increases with increasing M_T. Expressions for the Thiele modulus for reaction kinetics different from simple first order can be found in the common texts on reaction engineering.

4.1.3.2   Including External Resistance to Mass Transfer

To account for mass transfer effects, an external mass transfer coefficient, k_m, must be introduced, thus

$$N_A[\text{mol/m}^2/\text{s}] = k_m[\text{m/s}](c_{A0} - c_{As})[\text{mol/m}^3] \tag{4.12}$$

where N_A is the flux of reactant A transported from the bulk to the surface of the catalyst. N_A can be related to the reaction rate in Equation (4.8) as follows:

$$N_A[\text{mol/m}_{\text{ext}}^2/\text{s}] = -r_A'''[\text{mol/m}^3\text{catalyst/s}]\, a\,[\text{m}^3\text{catalyst/m}_{\text{ext}}^2] \tag{4.13}$$

where a is the ratio of the volume of catalyst to the external surface area or L, the characteristic length.

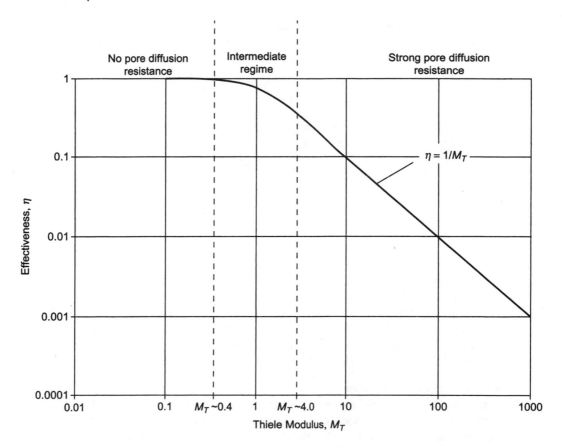

Figure 4.2 Catalyst effectiveness as a function of Thiele modulus

Equating the terms in Equation 4.13 in terms of concentration gives

$$k_m(c_{A0} - c_{As}) = k'''c_{As}\eta L \Rightarrow c_{As} = \frac{k_m}{k_m + k'''\eta L}c_{A0}$$

$$c_{As} = \frac{1}{1 + \dfrac{k'''\eta L}{k_m}}c_{A0}$$

$$-r'''_{A,ov} = k'''c_{As}\eta = \frac{\eta}{1 + \dfrac{\eta k'''L}{k_m}}k'''c_{A0} = \Omega k'''c_{A0} \tag{4.14}$$

where Ω is the overall effectiveness including internal and external mass transfer,

$$\Omega = \frac{\eta}{1 + \dfrac{\eta k'''L}{k_m}} \tag{4.15}$$

For the case when the surface reaction is so fast that the external mass transfer resistance becomes limiting ($k_m << k'''$), the second term in the denominator becomes very large compared to 1. Combination with Equation (4.11), then, yields

$$\Omega = \frac{\eta}{1+\dfrac{\eta k''' L}{k_m}} \cong \frac{k_m}{k''' L} = \frac{k_m L}{D_{cat}} \frac{1}{M_T^2}$$ (4.16)

For this condition, the concentration of reactant A at the surface of the particle $c_{As} \to 0$, and the effective reaction rate is given by

$$-r_{A,ov}''' = \frac{k_m(c_{A0} - \cancel{c}_{As})}{L} = \frac{k_m c_{A0}}{L}$$ (4.17)

For more complex reaction kinetics such as Langmuir-Hinshelwood, analytical solutions for the overall effectiveness are not available. However, in practice, over quite a wide range of operating conditions, it is usually possible to approximate the reaction kinetics by a first-order process, and the development given here can be applied over that range of conditions.

4.1.3.3   Temperature Effects
For highly exothermic reactions, significant temperature variations may occur within the catalyst particle and across the external gas film surrounding the particle. In extreme cases, the particles may be much hotter in the center than the surrounding gas, causing catalyst sintering and rapid catalyst decay. The analysis of such systems is not covered here, but the interested reader is referred to Levenspiel (1989) for a practical method of determining the temperature effects in catalyst systems.

4.1.3.4   Choosing a Catalyst
In practice, it does not make sense to operate far to the right of Figure 4.2 (large M_T), because so much of the internal catalyst surface area is unavailable for reaction. For very active catalysts where k''' is very high, it does makes sense to reduce the size of the catalyst particle in order to reduce the value of M_T or to dilute the active material within the particle, that is, reduce the loading of active material. Small catalyst particles also tend to be isothermal with little external mass transfer resistance. However, at some point, the pressure drop through a packed bed of small particles becomes excessive and, unless fluidized bed operation can be used, this sets a practical limit to the catalyst particle size.

Example 4.3

Consider the reaction of methanol to dimethyl ether (DME) via the gas-phase catalytic reaction using an amorphous alumina catalyst treated with 10.2% silica.

$$2CH_3OH \rightleftharpoons (CH_3)_2O + H_2O$$

There are no significant side reactions at temperatures less than 400°C, and the equilibrium conversion at these temperatures is greater than 99%. At temperatures greater than 250°C, the rate equation is given by Bondiera and Naccache (1991) as

$$-r_{MeOH}''' = k_o \exp\left[-\frac{E}{RT}\right] c_{MeOH}$$

where $-r_{MeOH}'''$ is the rate of reaction of methanol in mol/m³catalyst/s, $k_0 = 3.475 \times 10^6$ m³reactor/m³catalyst/s, $E = 84.06$ kJ/mol, and c_{MeOH} = concentration of methanol in mol/m³reactor. A typical feed to a packed bed reactor is 98% methanol at a pressure of approximately 14 atm. This reaction is exothermic and the temperature is expected to rise from 250°C to 350°C across the reactor bed with an expected conversion of methanol across the reactor of 80%.

For this system, determine if there are any pore diffusion effects at the

  a. Reactor inlet
  b. Reactor outlet
  c. For Part (b) determine the overall effectiveness including external mass transfer, and comment on the influence of external mass transfer to the overall reaction rate.

The internal diffusion coefficient for the porous catalyst is estimated to be 8×10^{-8} m²/s, and the minimum catalyst particle size to avoid excessive pressure drop is estimated to be 3 mm diameter spheres. The mass transfer coefficient for methanol at 350°C and 14 atm pressure for this size of catalyst is approximately $k_m = 2.4 \times 10^{-3}$ m/s.

Solution

  a. At the reactor inlet, the temperature is 250°C = 523.2 K, and

$$k = k_0 \exp\left[-\frac{E}{RT}\right] = 3.475 \times 10^6 \exp\left[-\frac{84,060}{(8.314)(523.2)}\right] = 0.01407$$

From Equation 4.11, the Thiele modulus is

$$M_T = L\sqrt{\frac{k'''}{D_{cat}}} = \frac{(0.003)}{6}\sqrt{\frac{0.01407}{8 \times 10^{-8}}} = 0.21$$

and

$$\eta = \frac{\tanh(M_T)}{M_T} = \frac{\tanh(0.21)}{0.21} = 0.986$$

So approximately 1.4% of the total surface area in the catalyst is not "used" because of the reduction in reactant concentration due to diffusion, and therefore, pore diffusion resistance is negligible at the front end of the reactor.

  b. At the reactor outlet, the temperature is 350°C = 623.2 K, and

$$k = k_0 \exp\left[-\frac{E}{RT}\right] = 3.475 \times 10^6 \exp\left[-\frac{84,060}{(8.314)(623.2)}\right] = 0.3127$$

From Equation 4.11, the Thiele modulus is

$$M_T = L\sqrt{\frac{k'''}{D_{cat}}} = \frac{(0.003)}{6}\sqrt{\frac{0.3127}{8 \times 10^{-8}}} = 0.989$$

and

$$\eta = \frac{\tanh(M_T)}{M_T} = \frac{\tanh(0.989)}{0.989} = 0.765$$

This location in the reactor is in the intermediate pore diffusion resistance region.

  c. At the reactor outlet, compare the overall catalyst effectiveness Ω to the internal effectiveness η using Equation (4.15):

$$\Omega = \frac{\eta}{1 + \frac{\eta k''' L}{k_m}} = \frac{0.765}{1 + \frac{(0.765)(0.3127)(0.003)}{(0.002410)(6)}} = 0.729$$

From this result, it can be seen that the overall effectiveness decreases by only 4.7% (from 0.765 to 0.729); therefore, external mass transfer does not play an important role at these conditions.

4.1.4   Mass Balances

Consider a stream (F_0 [mole/s]) containing several chemical species i ($1 \to n$) entering a control volume (CV) of volume V [m³]. The conditions within the control volume are such that some of the species j will react to form products. In order to keep the formulation simple, it is assumed that the products are a subset of $1 \to n$. In general, the rate at which a given species is created may be expressed by a reaction rate per unit volume as r_i [mole/s/m³]. The stream leaving the control volume (F [mole/s]) will have a different composition than the stream entering because some species will have reacted, and if the system is not operating at steady state, then the total number of moles within the control volume (N) will change with time. The situation described here is shown in Figure 4.3. Writing a material balance for species i gives the following relationship:

$$F_{i0} + \int^V r_i dV - F_i = \frac{dN_i}{dt} \tag{4.18}$$

In words, Equation 4.18 means input of component i + reaction of component i – output of component i = accumulation of component i. So, this is nothing more than a straightforward material balance. The integral term on the left-hand side takes account of the changing conditions within the control volume. It should be noted that if multiple input and output streams existed, then Equation (4.18) should be expanded to include all the input and output streams. The overall mole balance would then be given by Equation (4.19).

$$\sum_{i=1}^{n} F_{i0} + \sum_{i=1}^{n} \int^V r_i dV - \sum_{i=1}^{n} F_i = \sum_{i=1}^{n} \frac{dN_i}{dt} = \frac{dN}{dt} \tag{4.19}$$

At this point, it is convenient to introduce some terms that are commonly used to describe the behavior of reactors. In general, there is more than one reaction taking place in any reactor, and only a subset of those reactions produce the desired product. The other reactions produce undesired products (that have no value or that have to be removed at a cost) or by-products (products with some value that may be sold).

$$\text{Single-pass Conversion, } X = \frac{\text{Reactant consumed in reactor}}{\text{Reactant fed to reactor}} \tag{4.20}$$

$$\text{Overall Conversion, } X_{ov} = \frac{\text{Reactant consumed in process}}{\text{Reactant fed to process}} \tag{4.21}$$

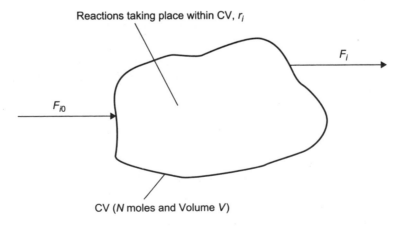

Figure 4.3   Material flows in a reacting system

If the process does not involve separation and recycle of unused reactants, then these two measures of conversion are identical. However, in the majority of cases, recycle of unused reactant does take place, and this leads to a higher overall conversion than the single-pass conversion.

$$\text{Yield, } Y = \frac{\text{Moles of limiting reactant reacted to produce desired product}}{\text{Moles of limiting reactant reacted}} \tag{4.22}$$

The yield is a measure of how effectively the limiting reactant is converted to the desired product. The yield is affected by the selectivity, defined as follows:

$$\text{Selectivity, } S = \frac{\text{Rate of production of desired product}}{\text{Rate of production of undesired products (including by-products)}} \tag{4.23}$$

It should be noted that the yield has an upper bound of 1, but selectivity is unbounded. In optimizing chemical processes with an economic objective function, the economics of separating and recycling reactants plays an important role. Therefore, in real processes, the optimal configuration for a reactor is rarely (never) one that maximizes any of the four parameters discussed previously. However, high values of the parameters defined in Equations (4.21), (4.22), and (4.23) are usually desirable.

4.1.5   Energy Balances

In a similar way to the derivation of Equation (4.19), the general formulation for the energy balance for a reacting flow system can be written as

$$\dot{Q} - \dot{W}_s + \sum_{i=1}^{n} F_{io} h_{io} - \sum_{i=1}^{n} F_i h_i = \sum_{i=1}^{n} \left(\frac{dU_i}{dt} \right)_{cv} \tag{4.24}$$

where $\dot{Q}$ is the rate of heat input into the CV, $\dot{W}_s$ is the rate of shaft work done by the CV on the surroundings, h is the specific molar enthalpy of the stream, and U is the internal energy of the control volume. In most cases, the work term is negligible, but if a stirrer is required to mix reactants, then this work would be included in the energy balance. In Equation (4.24), work is defined as negative when done on the system. For steady-state operations, the right-hand side of Equation (4.24) (and Equations [4.18] and [4.19]) is zero. The enthalpy terms in Equation (4.24) are usually broken up into terms that contain the heat of reaction and the change in enthalpy of a stream due to reaction. To illustrate this concept, the reaction given in Equation (4.1) is used:

$$aA + bB \rightarrow rR + sS \Rightarrow A + \frac{b}{a}B \rightarrow \frac{r}{a}R + \frac{s}{a}S \tag{4.1}$$

One additional parameter is defined as

$$\Theta_i = \frac{\text{moles of species } i \text{ entering the reactor}}{\text{moles of limiting reactant A entering the reactor}} \tag{4.25}$$

Using the stoichiometry and the definition of Θ_i given in Equation (4.25), the molar flows of each chemical species may be related to the conversion of A, X, as

$$F_i = F_{Ao}(\Theta_i + v_i X) \tag{4.26}$$

where v_i is the stoichiometric coefficient for reactant i using a basis of unity for reactant A.

For the four species in the reaction, the following relationships hold:

$$F_A = F_{Ao}(1-X)$$

$$F_B = F_{Ao}(\Theta_B - \frac{b}{a}X)$$

$$F_R = F_{Ao}(\Theta_R + \frac{r}{a}X)$$

$$F_S = F_{Ao}(\Theta_S + \frac{s}{a}X)$$

By substituting these relationships into Equation (4.24) and simplifying, the following relationship for steady-state operations with all reactants entering the reactor at the same temperature may be derived (Fogler, 2006):

$$\dot{Q} - \dot{W}_s + F_{Ao}\sum_i \Theta_i \hat{c}_{pi}[T-T_0] - F_{Ao}X[\Delta H_R(T_R) + \Delta \hat{c}_{pi}(T-T_R)]$$

$$= \dot{Q} - \dot{W}_s + F_{Ao}\sum_i \Theta_i \hat{c}_{pi}[T-T_0] - F_{Ao}X[\Delta H_R(T)] = 0 \tag{4.27}$$

where $\hat{c}_{pi}$ is the molar specific heat capacities of species i, T_o is the inlet temperature, T is the outlet temperature, T_R is a reference temperature and ΔH_R is the heat of reaction evaluated at the reference temperature (T_R). In the first form of Equation (4.27), the heat of reaction is specified at some reference temperature (often 25°C). In the second form of Equation (4.27), the reference temperature (T_R) is taken as the outlet temperature (T) of the reactor, and a more compact version of the energy balance is obtained.

4.1.6 Reactor Models

In order to determine the size (usually volume of reactor or mass of catalyst) required for a certain amount of material to react to produce a specified amount of product, certain assumptions must be made about the way material flows through the reactor. When there is more than one phase present (heterogeneous systems), the flow patterns for each phase may be very complicated and simple models are often not accurate. Even for the case of a single phase (homogeneous systems), only idealized flow patterns give rise to analytical solutions. However, despite these limitations, it is still useful to review the idealized flow models for reactors, because they often represent limiting conditions for reactor behavior.

4.1.6.1 Continuous Stirred Tank Reactors

A continuous stirred tank reactor (CSTR) refers to a system in which the fluid (almost always liquid or a mixture of liquid and gas or solid) is mixed so well that there are no concentration differences (gradients) anywhere in the vessel. A CSTR is illustrated in Figure 4.4(a). Because the contents of the reactor are well mixed, it is also assumed that the temperature is constant everywhere and thus the system is isothermal. Therefore, the outlet is identical to the conditions in the reactor. By analogy, since the concentration and temperature everywhere in the reactor are the same, the rate of reaction is the same everywhere in the reactor. Using these assumptions, for a single-input and single-output reactor, operating at steady state, Equation (4.18) gives

$$F_{i0} + \int^V r_i dV - F_i = 0 \Rightarrow V = \frac{F_{i0} - F_i}{-r_i|_{exit}} \tag{4.28}$$

where the reaction rate is evaluated at the conditions (concentration) in the reactor, which for a stirred tank are identical to the exit conditions. For reactant A in terms of conversion, Equation (4.28) can be written as

$$V = \frac{F_{i0} - F_i}{-r_i\big|_{exit}} \Rightarrow V = \frac{XF_{A0}}{-r_A\big|_{exit}} \tag{4.29}$$

Equation (4.29) is the design equation for a CSTR.

4.1.6.2   Plug Flow Reactor (PFR)

In a plug flow reactor (PFR), all the fluid is assumed to move in plug or piston flow. Therefore, the velocity profile is flat, and each element of fluid stays in the reactor for the same amount of time. This situation is close to that of a single fluid flowing in a pipe under turbulent conditions, where the velocity profile is relatively flat across the cross-section of the pipe. A PFR is illustrated in Figure 4.4(b). The concentration of reactants (and products) changes along the length of the reactor, and for steady-state operation, Equation (4.18) gives

$$F_{i0} + \int^V r_i dV - F_i = 0 \tag{4.30}$$

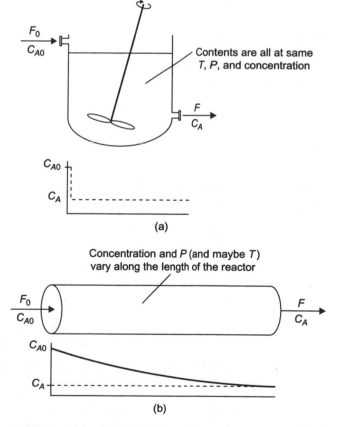

Figure 4.4   Reactor flow patterns (a) continuous stirred tank reactor and (b) plug flow reactor

The differential form of Equation (4.30) for reactant A is

$$\frac{dF_A}{dV} = -r_A \tag{4.31}$$

This leads to the design equation for plug flow reactors as

$$\frac{dF_A}{dV} = -r_A \text{ and } dF_A = F_{A0}dX$$

$$\therefore \frac{F_{A0}dX}{dV} = -r_A \text{ and}$$

$$V = F_{A0}\int_0^X \frac{dX}{-r_A} \tag{4.32}$$

Equation (4.32) is the design equation for a PFR. Integration of this equation requires knowledge of both the temperature and concentration dependence of the rate expression. For isothermal systems with simple rate laws, analytical expressions are available. However, for nonisothermal reactors, integration of Equation (4.32) must be performed simultaneously with the appropriate energy balances and, in general, this requires the numerical solution of coupled ordinary (or partial) differential equations.

Example 4.4

Consider a simple liquid phase reaction in which reactant A is converted to product B. The reaction is second order in A and the kinetics are given by the equation $-r_A$ (reaction rate) $= \text{mol/L/s} = 0.05c_A^2 [\text{mol/L}]^2$. A stream of material A of concentration 10 mol/L is fed to an isothermal reactor at a rate of 8 L/s. Determine the volume of the reactor to obtain 95% conversion of A assuming

    a. A CSTR is used.

    b. A PFR is used.

Solution

    a. For a CSTR, use Equation (4.29) and note that the concentration in the CSTR is the same as at the outlet conditions $\Rightarrow$

$$c_A = (1-X)c_{A0} = (1-0.95)(10) = 0.5 \text{ [mol/L] and } F_{A0} = (10)(8) = 80 \text{ mol/s}$$

$$V = \frac{XF_{A0}}{-r_A\big|_{exit}} = \frac{(0.95)(80)}{(0.05)(0.5)^2} = 6080 \text{ L}$$

    b. For a PFR, use Equation (4.32):

$$V = F_{A0}\int_0^X \frac{dX}{-r_A} = 80\int_0^{0.95} \frac{dX}{(0.05)c_A^2} = \frac{80}{c_{A0}^2}\int_0^{0.95} \frac{dX}{(0.05)(1-X)^2}$$

$$V = \frac{80}{(10)^2}\frac{1}{(0.05)}\frac{1}{(1-X)}\bigg|_0^{0.95} = \frac{80}{5}\left(\frac{1}{0.05} - \frac{1}{1}\right) = 304 \text{ L}$$

This result shows that the PFR requires significantly less volume (1/20) than the CSTR to achieve the same conversion. This result is true for all reactions of an elementary form. This result should not be surprising either, since in the CSTR, the reaction rate everywhere in the reactor is evaluated at the exit concentration of A. However, in the PFR, the concentration everywhere in the reactor is greater than at the exit condition, and hence the reaction rate is also greater than at the exit concentration.

4.1.6.3   Batch Reactor

As the name implies, a batch reactor operation consists of loading reactants into a reactor, allowing them to react for a period of time t, and then removing the unused reactants and products. The design equation for a batch reactor may be derived from Equation (4.18), under the assumption that the conditions in the reactor are the same everywhere at any given point in time. This implies that the reaction rate, r_i, is a function of time but not a function of position. Therefore, for reactant A,

$$F_{i0} + \int^V r_i dV - F_i = \frac{dN_i}{dt} \Rightarrow \frac{dN_A}{dt} = r_A V \qquad (4.33)$$

When there is negligible volume change in the reacting mixture, V is constant, and Equation (4.33) may be rewritten in terms of concentrations:

$$-\frac{dc_A}{dt} = -r_A \text{ or } c_{A0}\frac{dX}{dt} = -r_A \qquad (4.34)$$

Either Equation (4.33) or Equation (4.34) may be considered the design relationship for a batch reactor. The independent variable is the time spent in the reactor, while the volume dictates how much material is charged to the reactor in a given batch. It should be noted that the batch reactor is an unsteady-state operation, unlike the CSTR and PFR discussed previously.

Example 4.5

A mixture of 5 moles of reactant A are introduced into a batch reactor by dissolving them in 100 L of solvent. The reaction of A to product B occurs by a first-order reaction of the form $-r_A(\text{reaction rate}) = \text{mol/L/min} = 0.036 c_A[\text{mol/L}]$. Determine the time required to react 90% of A to give B. You may assume that there is no change in density of the reaction mixture during the reaction process.

Solution

$c_{A0} = 5/100 = 0.05$ mol/L
Conversion = 0.9
Using Equation (4.34),

$$c_{A0}\frac{dX}{dt} = -r_A \Rightarrow c_{A0}\frac{dX}{dt} = 0.036 c_A = 0.036 c_{A0}(1-X) \Rightarrow \frac{1}{(1-X)}\frac{dX}{dt} = 0.036$$

$$\int_0^{0.9}\frac{dX}{(1-X)} = 0.036\int_0^t dt \Rightarrow -\ln(1-X)\big|_0^{0.9} = 0.036 t$$

$$t = \frac{\ln(1-0)-\ln(1-0.9)}{0.036} = 63.96 \text{ min}$$

Time for reaction is ~64 minutes.

4.1.6.4   Other Reactor Models (Laminar Flow)

The PFR and CSTR models represent two idealized flow patterns within reactors. In reality, reactors often have flow patterns that are somewhere between these two models. However, in some cases, especially in heterogeneous systems such as fluidized beds and slurry reactors, the flow patterns of the fluids are very complex and lie outside the bounds provided by these two ideal cases. Needless to say for these complex flow patterns, complex modeling techniques and experimental verification are required to obtain the size of the reactor.

An example of a reactor model that is more complex than either a CSTR or a PFR, but for which analytical solutions exist, is the laminar flow reactor, which is illustrated in Figure 4.5.

As the name suggests, the flow of liquid in the reactor (usually a long pipe) is in the laminar flow regime, meaning that the velocity profile is parabolic. If the flow regime was turbulent, then a PFR would probably be a good representation of the reactor, since the velocity profile for turbulent flow is quite flat. However, with a parabolic flow profile, elements of fluid at the center of the reactor move at twice the average velocity, and those elements close to the wall move much slower than the average. Thus, there is a wide distribution of times that fluid elements spend in the reactor, so the amount of conversion that each of these elements achieves differs widely. The formulation for the conversion of reactant A in such a reactor requires a knowledge of the residence time distribution (RTD) in the reactor, which is outside the scope of this text, but the interested reader is referred to Levenspiel (1999) and Fogler (2006) for fuller coverage of this and a variety of other topics in reactor design. The conversion within a laminar flow reactor is simply stated here without proof and is given by

$$(1-X)=\frac{c_A}{c_{A0}}=\int_{\bar{t}/2}^{\infty}\left(\frac{c_A}{c_{A0}}\right)_t E(t)\,dt \qquad (4.35)$$

where $E(t)$ is the RTD of fluid elements in the vessel or reactor, $\bar{t}$ is the average time spent in the reactor, and the first term in the integral is the expression for the ratio of the concentration in a batch reactor to the initial concentration after time t. For laminar flow, $E(t)$ is given by

$$E(t)=\frac{\bar{t}^2}{2t^3} \quad \text{for } t\geq\frac{\bar{t}}{2} \quad \text{and} \quad E(t)=0 \text{ otherwise} \qquad (4.36)$$

The lower bound of $\bar{t}/2$ for the integral in Equation 4.35 is needed because the centerline velocity for laminar flow is twice the average velocity, so the least amount of time a fluid element can stay in the reactor is $\bar{t}/2$. For known kinetics and a given reactor size V, $\bar{t} = V/F$, (c_A/c_{A0}) are known, and Equation (4.35) may be solved to find the conversion X. In effect, Equation (4.35) is the design equation for a laminar flow reactor, but to find the reactor volume V (equivalent to finding $\bar{t}$) requires an iterative solution.

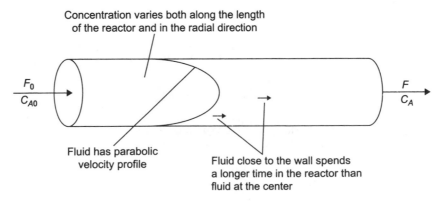

Figure 4.5   Laminar flow reactor

Example 4.6

Using the results from Example 4.4, Part (b), determine the conversion in the reactor if the flow pattern is laminar and not plug flow.

Solution

From Equation (4.34) for a batch reactor,

$$-\frac{dc_A}{dt} = -r_A = -kc_A^2 \implies \int_{c_A}^{c_{A0}} \frac{dc_A}{c_A^2} = kt$$

$$-1\left(\frac{1}{c_{A0}} - \frac{1}{c_A}\right) = kt \implies \left(\frac{c_A}{c_{A0}}\right)_t = \frac{1}{1 + kc_{A0}t}$$

From Equation (4.35),

$$(1-X) = \int_{\bar{t}/2}^{\infty} \left(\frac{c_A}{c_{A0}}\right)_t E(t)\, dt = \int_{\bar{t}/2}^{\infty} \frac{1}{1 + kc_{A0}t} \frac{\bar{t}^2}{2t^3}\, dt$$

After integration and simplification, this gives

$$(1-X) = 1 - kc_{A0}\bar{t} + \frac{(kc_{A0}\bar{t})^2}{2}\ln\left(1 + \frac{2}{kc_{A0}\bar{t}}\right) \tag{E4.6}$$

For this problem,

$$\bar{t} = \frac{V}{v} = \frac{304}{8} = 38 \text{ s}, \ c_{A0} = 10 \text{ mol/liter, and } k = 0.05 \text{ liter/mol/s}$$

$$\therefore kc_{A0}\bar{t} = 19$$

Substituting in Equation (E4.6) gives

$$(1-X) = 1 - kc_{A0}\bar{t} + \frac{(kc_{A0}\bar{t})^2}{2}\ln\left(1 + \frac{2}{kc_{A0}\bar{t}}\right) = 1 - 19 + \frac{(19)^2}{2}\ln\left(1 + \frac{2}{19}\right) = 0.0651$$

$$X = 0.935$$

This conversion is a little less than that for a PFR with the same volume ($X = 0.95$). The main reason is that some of the fluid elements pass through the reactor in a time less than the average time, and these elements have a lower conversion than those elements that spend more time in the reactor. The net result is that when the conversion is averaged, a lower overall conversion compared to plug flow is obtained.

Another important practical reactor configuration is the fluidized bed that is often used for highly exothermic reactions because of its favorable, near isothermal, operation. The flow patterns in this type of reactor are very complex and difficult to characterize. Fluidized bed reactors are covered separately in Section 4.2.3.

4.1.6.5   Selectivity and Yield for Parallel and Series Reactions
Consider the following general reaction scheme:

$$aA + bB \xrightarrow[\text{rxn 1}]{k_1} pP \xrightarrow[\text{rxn 2}]{k_2} uU \tag{4.37}$$

$$\beta B \xrightarrow[\text{rxn 3}]{k_3} vV$$

Equation (4.37) shows three reactions involving five species A, B, P, U, and V, with stoichiometric coefficients a, b, p, u, β, and v, respectively. It is assumed that P is the desired product and that U and V are unwanted by-products. The reaction scheme in Equation (4.37) can be used to illustrate the effects commonly observed in many industrial reaction systems.

1. A **single** reaction produces desired product. Here, $k_2 = k_3 = 0$, and only the first reaction proceeds. An example is the catalytic dealkylation of toluene to benzene, where the catalyst suppresses the side reactions:

$$\underset{\text{toluene}}{C_7H_8} + H_2 \rightarrow \underset{\text{benzene}}{C_6H_6} + CH_4 \tag{4.38}$$

2. **Parallel** (competing) reactions produce desired products and unwanted by-products. Here $k_2 = 0$, and no U is formed. Species B reacts to form either P or V. In the phthalic anhydride reaction sequence, the reaction of o-xylene to form either phthalic anhydride, maleic anhydride, or combustion products is an example of a parallel reaction (R-1, R-2, and R-3).

$$\text{R-1} \quad \underset{\text{o-xylene}}{C_8H_{10}} + 3O_2 \rightarrow \underset{\substack{\text{phthalic} \\ \text{anhydride}}}{C_8H_4O_3} + 3H_2O$$

$$\text{R-2} \quad \underset{\text{o-xylene}}{C_8H_{10}} + \frac{15}{2}O_2 \rightarrow \underset{\substack{\text{maleic} \\ \text{anhydride}}}{C_4H_2O_3} + 4H_2O + 4CO_2$$

$$\text{R-3} \quad \underset{\text{o-xylene}}{C_8H_{10}} + \frac{21}{2}O_2 \rightarrow 5H_2O + 8CO_2 \tag{4.39}$$

$$\text{R-4} \quad \underset{\substack{\text{phthalic} \\ \text{anhydride}}}{C_8H_4O_3} + \frac{15}{2}O_2 \rightarrow 2H_2O + 8CO_2$$

$$\text{R-6} \quad \underset{\substack{\text{maleic} \\ \text{anhydride}}}{C_4H_2O_3} + 3O_2 \rightarrow H_2O + 4CO_2$$

3. **Series** (sequential) reactions produce desired products and unwanted by-products. Here $k_3 = 0$, and no V is formed. Species A reacts to form desired product P, which further reacts to form unwanted by-product U. In the phthalic anhydride reaction sequence, shown previously, the reaction of o-xylene to phthalic anhydride to combustion products (R-1 and R-4) is an example of a series reaction.

4. **Series** and **parallel** reactions produce desired products and unwanted by products. Here, all three reactions in Equation (4.37) occur to form desired product P and unwanted by-products U and V. The entire phthalic anhydride sequence (R-1 through R-5) is an example of series and parallel reactions.

Example 4.7

Consider the reaction scheme given in Equation (4.37) where P is the desired product, with both U and V as undesired by-products. Here it is assumed that the reactions occur in the gas phase. Assume that Equation (4.37) represents elementary steps, that $a = b = p = u = v = 1$, that $\beta = 2$, and that the activation energies for the reactions are as follows: $E_1 > E_2 > E_3$.

   a. For the case where $k_2 = 0$, what conditions maximize the selectivity for P?
   b. For the case where $k_3 = 0$, what conditions maximize the selectivity for P?

Solution

a. For this case, from Equation (4.23), the selectivity, S, is written as

$$S = \frac{r_P}{r_V} = \frac{k_1 c_A c_B}{k_3 c_B^2} = \frac{k_1 c_A}{k_3 c_B} \tag{E4.7a}$$

There are several ways to maximize the selectivity. Increasing c_A/c_B increases the selectivity. This means that excess A is needed and that B is the limiting reactant. Many reactions are operated with one reactant in excess. The reason is usually to improve selectivity, as shown here. Because, for gas-phase reactions, pressure affects all concentrations equally, it is seen for this case that pressure does not affect the selectivity. Temperature has its most significant effect on the rate constant. Because the activation energy for rxn 1 is larger than that for rxn 3, k_1 is more strongly affected by temperature changes than is k_3. Therefore, increasing the temperature increases the selectivity. In summary, higher temperatures and excess A maximize the selectivity for P.

b. For this case, the selectivity is written as

$$S = \frac{r_P}{r_U} = \frac{k_1 c_A c_B - k_2 c_P}{k_2 c_P} = \frac{k_1 c_A c_B}{k_2 c_P} - 1 \tag{E4.7b}$$

Because the concentration and hence pressure appears to the second power in the numerator and to the first power in the denominator, increasing the pressure increases the selectivity. Because the activation energy for rxn 1 is larger, increasing the temperature increases the selectivity. Increasing both reactant concentrations increases the selectivity, but increasing the concentration of Component P decreases the selectivity. The question is how the concentration of Component P can be kept to a minimum. The answer is to run the reaction at low conversions (small reactor volumes for a given feed rate). Quantitatively, the selectivity in Equation (E4.7b) is maximized by running at conditions where the concentration of P is very low. Thus, for a series reaction with the desired product intermediate in the series, a low conversion maximizes the intermediate product and minimizes the undesired product. This is illustrated by the concentration profiles obtained by assuming that these reactions take place in a PFR, as shown in Figure E4.7. It is seen that the ratio of P/U is at a maximum at low reactor volumes, which corresponds to low conversion. Therefore, increasing the temperature or the pressure or both increases the selectivity for Component P. Running at low conversion probably does more to

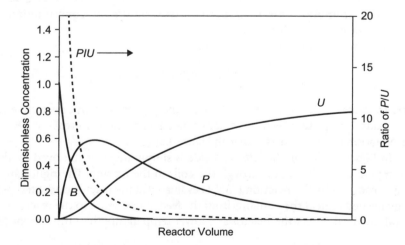

Figure E4.7   Concentration profiles for a series reaction ($A \rightarrow P \rightarrow U$) in a plug flow reactor

increase the selectivity than can be accomplished by manipulating temperature and pressure alone. However, there is a trade-off between selectivity and overall profitability, because low conversion per pass means very large recycles, more difficult separations, and larger equipment.

4.2   EQUIPMENT DESIGN FOR NONISOTHERMAL CONDITIONS

In the previous section, design equations were introduced that allow the calculation of the volume needed for a few ideal reactor types. In practice, the assumption of isothermal behavior is often not valid, and more rigorous sizing calculations need to be carried out. In this section, some of the basics of real reactor behavior are covered, and the approaches needed to perform preliminary sizing estimates are described.

4.2.1   Nonisothermal Continuous Stirred Tank Reactor

The design of a CSTR for an exothermic reaction may require the removal of heat from the reactor through an internal or external heat exchanger. For most practical situations, CSTRs are only used for liquid-phase reactions, and the discussion here focuses on these types of reactions. When the heat of reaction is small or when there is a difference between the feed temperature and the reaction temperature, adiabatic operation may be possible. This is true, in part, because the mass of the inventory in the reactor is usually quite high, and a significant amount of cooling can be absorbed in sensible heat by heating the contents from the inlet temperature to the reaction temperature. However, as reaction temperatures increase, so do the reaction rates, and this may lead to excessive temperatures when operating adiabatically, so such conditions require some form of additional heat transfer or cooling.

Some configurations for heat removal in CSTRs are shown in Figure 4.6. The use of an external heating loop, Figure 4.6(a), allows easier maintenance of heat transfer equipment and gives more precise control of reactor temperature, as both the circulation rate and coolant rate may be used as control variables. However, the use of cooling coils within the CSTR (Figure 4.6[b]) is also a viable method. When the temperature of the reactor is high enough, partial vaporization of the reactor contents and their subsequent condensation and recycle provides another means of cooling (Figure 4.6[c]), which can be modeled in an identical way as for the case shown in Figure 4.6(a).

The equations describing the situations in Figure 4.6 are different depending on which configuration is modeled. Only the equations describing Figure 4.6(b) are given here. Note that the well-mixed assumptions mean that both the temperature and concentration in the reactor are the same as at the exit conditions.

An energy balance on the reactor gives

$$\dot{m}_c c_{p,c}(T_2 - T_1) = V r_A(c_A, T)(-\Delta H_r) - \dot{m}_p c_{p,p}(T - T_{in}) \qquad (4.40)$$

where the subscripts c and p refer to coolant and process streams, respectively. In words, Equation (4.40) reads, Energy removed by coolant = Energy generated in reactor − Energy required to heat feed. Notation for this case is shown in Figure 4.6(d).

In Equation (4.40), the left-hand side is simply the heat removed by the coolant in changing temperature as it passes through the coils in the reactor. The right-hand side represents the energy produced by the reaction ($-\Delta H_r$ is evaluated at the reactor outlet temperature, T) minus the energy needed to heat the contents from the feed temperature to the reactor temperature. Since, in general, the feed enters the reactor at a lower temperature than the reactor temperature, the two

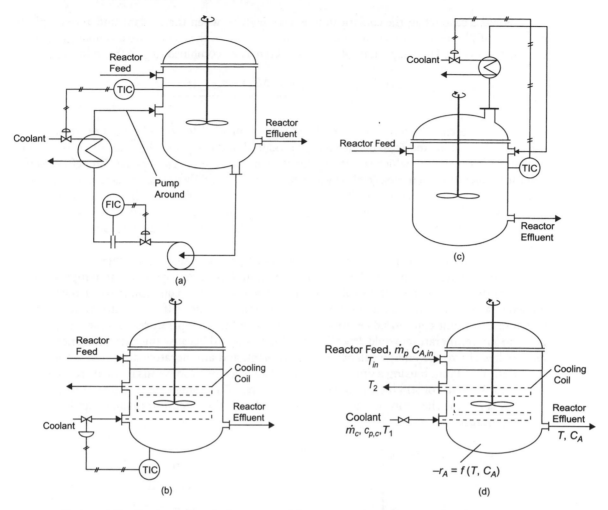

Figure 4.6 Configurations for heat removal from CSTRs: (a) pump around with external heat exchange, (b) internal cooling coils, (c) partial vaporization and subsequent condensation of reactor contents in an external heat exchanger; (d) notation for configuration (b) used in model development

terms on the right-hand side are of opposite sign. The condition for adiabatic operation is then also described by Equation (4.40) with the left-hand side set to zero. For the case of heat removal, the heat exchanger design equation must also be solved in conjunction with the energy balance. This requires the following to be true:

$$\dot{m}_c c_{p,c}(T_2 - T_1) = UA \frac{(T - T_1) - (T - T_2)}{\ln \frac{(T - T_1)}{(T - T_2)}} = UA \frac{(T_2 - T_1)}{\ln \frac{(T - T_1)}{(T - T_2)}} \tag{4.41}$$

For a given reactor volume, heat exchanger area, process throughput, process stream inlet temperature, and coolant flowrate and inlet temperature, the only unknowns are the reactor temperature and coolant outlet temperature, and these can be solved using Equations (4.40) and (4.41).

For the case when the coolant flow is very high or when the coolant undergoes a phase change (boils), then the coolant temperatures into and out of the reactor are essentially the same, that is, $T_1 \cong T_2 = T_c$. Then, Equations (4.40) and (4.41) may be combined to give the following:

$$\underbrace{UA(T-T_C)}_{Q_C}=\underbrace{Vr_A(c_A,T)(-\Delta H_r)-\dot{m}_p c_{p,p}(T-T_{in})}_{Q_R} \qquad (4.42)$$

where Q_C represents the term on the left-hand side of Equation (4.42) and is the heat removed by the cooling medium and Q_R is the net energy produced in the reactor. The value of Q_R is a strong function of conversion, which is a strong function of the reactor temperature. Therefore, as the reactor temperature increases, Q_R slowly increases and then rapidly rises as the conversion starts to "take off" and then slows down as the conversion approaches 100%. The T-Q diagram for the reactor, shown in Figure 4.7, shows the relationships for Q_C and Q_R as a function of temperature, and the point(s) of intersection of the two curves represents a mathematical solution(s) to Equation (4.42).

From Figure 4.7, it is clear that for low coolant temperatures (T_{C1}), there is only a single low-temperature solution (Point a) to Equation (4.42). As the coolant temperature is increased (to T_{C2}), there are three solutions (Points b, c, and d) to Equation (4.42). However, the intermediate-temperature solution (Point c) is unstable. Therefore, if the reactor were designed to operate at Point c, it would be impossible to control the temperature, because any positive deviation in temperature would drive the solution to Point d because the energy balance for the reactor lies above the energy balance for the coolant, and any negative deviation would drive the solution to Point b using a similar reasoning. However, both solutions b and d are stable and can be controlled. For still higher coolant temperatures (T_{C3}), a single high-temperature solution (Point e) is found. Therefore, it is important not to design the reactor to maintain an intermediate temperature.

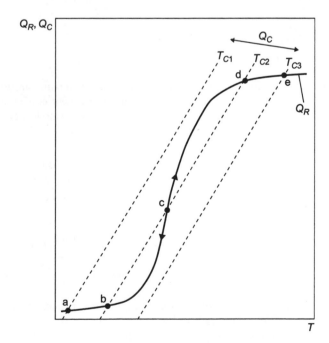

Figure 4.7   Graphical solution of Equation (4.42)

Example 4.8

An exothermic, liquid-phase reaction is to take place in a stirred tank reactor. The feed to the reactor (F) is pure A at 500 mol/min with a concentration of 10 mol/L. The volume of the reactor is 100 L, and the feed temperature is 30°C. A reaction takes place in which A is converted to B (A → B) by a first-order reaction:

$$-r_A = kc_A = 3e^{-\frac{1650}{T[K]}}c_A\,[mol/m^3/s]$$

The heat of reaction ($-\Delta H_r$) is 75,240 J/mol, and it is approximately constant from 30°C to 200°C. Other data for the process and coolant are as follows:

Data

$$c_{p,p} = 4150\,J/kg/°C,\,c_{p,c} = 4180\,J/kg/°C,\,T_{C,2} = 30°C,\,T_{C,1} = 20°C,\,U = 1000\,W/m^2/°C,\,\rho_p = 1050\,kg/m^3,$$

$$\text{and } m_p = \frac{F}{c_{A0}}\rho_p = \frac{(500/60)}{(10\times1000)}(1050) = 0.875\,kg/s$$

For this system, calculate the following:

a. If the reactor is to be run adiabatically (no coolant), at what temperature and conversion will the reactor run?

b. If it is desired to obtain 80% conversion, at what temperature must the reaction run?

c. For the case in Part (b), what must be the flow of coolant fed to the cooling coil in the reactor?

d. What heat transfer area is required for Part (c)?

Solution

a. Using Equation (4.40) and setting the left-hand side of the equation to 0, the following expression must be solved for the unknown variable T:

$$0 = Vr_A(c_A,T)(-\Delta H_r) - \dot{m}_p c_{p,p}(T - T_{in}) \qquad (E4.8a)$$

The integrated form for the design equation for a stirred tank reactor can be written in terms of the space time,

$$c_A = \frac{c_{A0}}{1+k\tau} \qquad (E4.8b)$$

where

$$\tau = \frac{c_{A0}V}{F} = \frac{(10)(100)}{(500)/(60)} = 120\,sec \qquad (E4.8c)$$

and

$$r_A = kc_A = \frac{kc_{A0}}{1+k\tau} \qquad (E4.8d)$$

Substituting Equations (E4.8c) and (E4.8d) into the right-hand side of Equation (E4.8a) along with the other parameters gives

$$0 = \frac{(100)}{1000}3e^{-\frac{1650}{T+273.15}}\frac{(10\times1000)}{1+3e^{-\frac{1650}{T+273.15}}(120)}(75,240) - (0.875)(4150)(T-30) \qquad (E4.8e)$$

Simplifying gives

$$22.572 \times 10^7 \frac{e^{-\frac{1650}{T+273.15}}}{1+360e^{-\frac{1650}{T+273.15}}} - 3631.3(T-30) = 0 \qquad \text{(E4.8f)}$$

where T is in °C. Solving Equation (E4.8f) gives

$$T = 186.9°C \text{ and } c_A = \frac{(10 \times 1000)}{1+360e^{-\frac{1650}{T+273.15}}} = 911.4 \text{ mol/m}^3 \text{ giving } x = 1 - \frac{c_A}{c_{A0}} = 1 - \frac{911.4}{10000} = 0.9089$$

b. For 80% conversion,

$$c_A = 2000 = \frac{c_{A0}}{1+k\tau} \Rightarrow k = \frac{1}{120}\left[\frac{10,000}{2000}-1\right] = 0.03333 = 3e^{-\frac{1650}{T+273.15}} \Rightarrow T = 93.5°C$$

c. At $T = 93.5°C$, the right-hand side of Equation (4.40) is 2.711×10^5 W.

An energy balance on the cooling medium gives

$$Q_C = \dot{m}_c c_{p,c}(T_2 - T_1) = 2.711 \times 10^5 \Rightarrow \dot{m}_c = \frac{(2.711 \times 10^5)}{(4180)(30-20)} = 6.486 \text{ kg/s}$$

d. For the design of the heat transfer surface,

$$UA\Delta T_{lm} = Q_C$$

$$\Delta T_{lm} = \frac{(93.5-20)-(93.5-30)}{\ln\frac{(93.5-20)}{(93.5-30)}} = 68.38°C \text{ and } A = \frac{Q_C}{U\Delta T_{lm}} = \frac{(2.711 \times 10^5)}{(1000)(68.38)} = 3.96 \text{ m}^2$$

The performance aspects of the reactor in Example 4.8 are considered in Section 4.3.2.

4.2.2   Nonisothermal Plug Flow Reactor

4.2.2.1   Hierarchy for Exothermic Reactions

The design of a reactor for a set of exothermic reactions must include consideration of the extent of reaction and the amount of heat generated. For reactions with low to moderate heats of reaction, the reactor may be designed without the need to add heat transfer surfaces within the reactor. Such configurations are inherently simpler, less expensive, and easier to operate. As the heat generated by the reactions increases, some form of internal heat transfer must be included in the reactor design. Since most industrially relevant reactions include the need for a catalyst and often these reactions are performed in the gas phase, the emphasis given here is on heterogeneous gas-phase reactions. A diagram illustrating the hierarchy of reactor design for increasingly exothermic reactions is shown in Figure 4.8. It should be noted that the direction of gas flow is always downward through a packed bed to avoid the possibility of unwanted fluidization. For fluidized beds, the direction of gas flow is always upward.

In Figure 4.8(a), the reactions of interest are only slightly exothermic, and a simple packed bed of catalyst is used. The gas temperature increases across the bed of catalyst, and this increase should be relatively small (≤50°C), which limits the amount of conversion within a single reactor. For higher conversions or more exothermic reactions, a series of packed beds with either intermediate cooling or the addition of supplemental cool feed material (cold

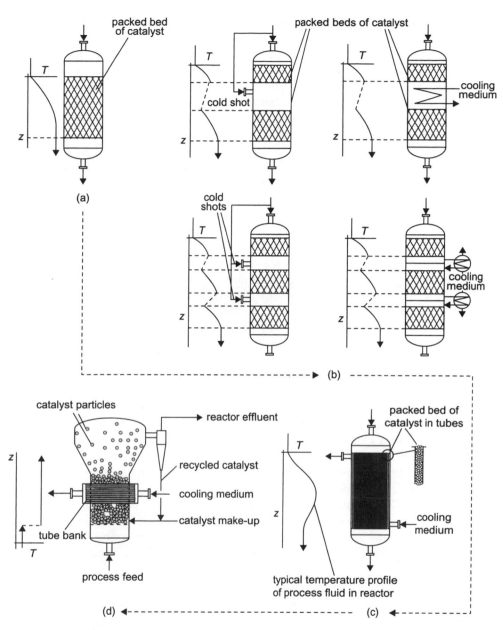

Figure 4.8   Reactor configurations for increasingly exothermic reactions (shown by the direction of the dotted line): (a) single packed bed, (b) multiple packed beds with intermediate cooling or the additions of cold shots, (c) shell-and-tube arrangements, and (d) fluidized beds

shots) can be used, as shown in Figure 4.8(b). In practice, not more than three or four packed beds in series are normally used. In Figure 4.8(b), internal and external intermediate cooling coils are shown. These coils or heat exchangers serve the same function of cooling the gas prior to its being fed to the next catalyst bed. This differs from the type of heat exchange that is required in Figures 4.8(c) and 4.8(d), where heat must be removed at the same location at which reactions are taking place. Therefore, for more exothermic reactions, the use of some form of integrated heat exchanger is employed, as shown in Figure 4.8(c). A common configuration for highly exothermic catalyzed reactions is to load the catalyst into tubes through which the process gas flows. The heat generated by the reactions is then transferred out of the catalyst, through the reacting process gas and the walls of the tube, and into a cooling medium on the shell side of the reactor. The cooling medium may be a liquid such as Dowtherm™, a molten salt, or boiler feed water that is turned into steam. The temperature profiles along the length of the tube (and possibly across the tube diameter) are not linear and are discussed in more detail in the following section.

For extremely exothermic reactions, such as the partial combustion of certain hydrocarbons (e.g., the production of acrylic acid, maleic anhydride, and phthalic anhydride), the control of the temperature within the tubes of a shell-and-tube type reactor becomes difficult or leads to excessively large reactor volumes due to the need for significant dilution of the catalyst. For such cases, the use of a fluidized bed with internal heat removal may be required. Fluidized beds are usually more complicated than other forms of reactors, they need a solids handling and recycle system, and they often require the continuous make up of catalyst. They are generally more expensive than other reactor types, but they are capable of near isothermal operation and provide very stable operating conditions for highly exothermic reactions. The behavior and fluid flow patterns within a fluidized bed depend on the ratio of the superficial velocity to the minimum fluidizing velocity (u_s/u_{mf}) and the density and size of the particles being fluidized. The modes of operation of these reactors depend on the value of the preceding parameters and, for increasing gas flowrate, vary between bubbling bed, turbulent bed, fast fluidization, and transport reactor. An example of a bubbling fluidized bed with heat removal is shown in Figure 4.8(d). It should be noted that the solids act like a well-stirred tank and remain at a constant temperature. The inlet gas stream is usually cooler than the reactor temperature but undergoes a very rapid temperature change when entering the bed and essentially passes through the bed at isothermal conditions. However, it should be noted that the flow pattern for the gas is generally not plug flow, and for the case shown in Figure 4.8(d), gas bubbles are formed that do not contact the catalyst efficiently. Therefore, the flow pattern of gas is quite complex and not well described by a simple flow model. Thus, models for fluidized beds are complex and often require experimental verification to determine all the parameters needed in the model. Usually the choice of a fluidized bed is a necessity because of extremely high heats of reaction that require very good temperature control and may often require the use of very small catalyst particles or catalyst that must be regenerated often. For such cases, the pressure drop in a packed bed and the heat transfer surface area requirements would be prohibitively high. Fluidized beds are covered in more detail in Section 4.2.3.

4.2.2.2 Hierarchy for Endothermic Reactions
The same configurations shown in Figure 4.8 can be used for endothermic reactions. If the reaction(s) is very endothermic, for example, the catalytic reformation of naphtha or the dehydrogenation of ethylbenzene to form styrene, then as the reaction proceeds, the temperature drops, the rate decreases, and the reaction may quench out. In such cases, some form of heat transfer (heat transferred to the reactant stream from a hot source) may be required. The normal configurations for endothermic reactions are shown in Figures 4.8(a) and 4.8(b). If heat addition is required to prevent the quenching of the reactions or to reduce the size of the reactor then the normal approach is to use a series of staged packed beds with intermediate heating. The use of "hot shots" to the feed of sequential packed beds may also be used but is less common. The degree of endothermicity of commercial reactions is generally less than the exothermicity for partial oxidation reactions, and as a result, the configurations shown in Figures 4.8(c) and 4.8(d) are not usually employed for endothermic reactions.

4.2.2.3   Understanding Reactor Concentration and Temperature Profiles

In Figures 4.8(a) and 4.8(b), the exothermicity of the reactions taking place is generally low enough that a given catalyst particle may be assumed to be isothermal. This means that there are no radial temperature gradients within a catalyst particle and that temperature gradients across the diameter of the reactor are small. Thus, it is justified to model such reactors using a simple one-dimensional model in which temperature and concentration vary only in the z-direction, that is, along the length of the reactor. These are the models that process simulators typically employ in their standard libraries. For more sophisticated (higher-dimensional models), custom written operations are usually needed. For more exothermic reactions, such as shown in Figure 4.8(c), these assumptions must be carefully reevaluated and checked. When reactions are extremely exothermic, radial temperature variations within the catalyst pellet and within the tube carrying the catalyst are possible. A typical situation for a single catalyst tube in a shell-and-tube type reactor is shown in Figure 4.9.

From the illustration in Figure 4.9, it should be clear that temperature and composition may vary with both the location within the catalyst pellet and with the location of the pellet in the tube. Clearly, a simple one-dimensional model is incapable of determining these radial variations and modeling these types of reactors accurately; therefore, some form of coupled set of partial differential equations (for both material and energy balances) will be required.

The potential complexities of temperature and concentration profiles in the tubes of a shell-and-tube reactor were identified in the previous discussion. In many cases, it is possible to minimize these radial variations and reduce the analysis to one in which temperature and concentration variations need to be considered for only the z-direction. The discussion of these reactors starts with a qualitative consideration of the profiles that are shown in Figure 4.10.

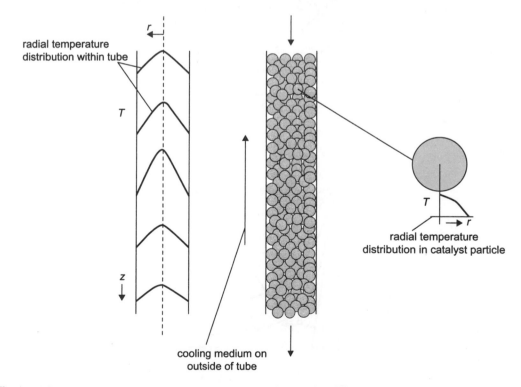

Figure 4.9   How temperature variations may occur within catalyst filled tube and catalyst pellet for highly exothermic reactions

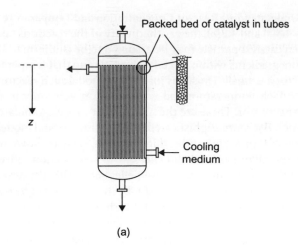

(a)

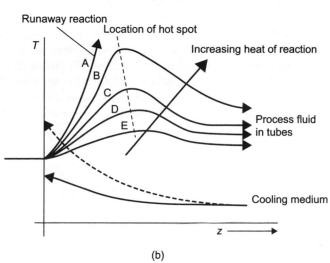

(b)

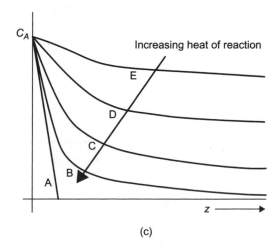

(c)

Figure 4.10   Temperature and concentration profiles in a shell-and-tube reactor for increasing heat of reaction

In Figure 4.10, the direction of the cooling medium is shown to be countercurrent to the flow of process gas in the tubes. This does not have to be the case, and cocurrent arrangements may be advantageous in some circumstances, especially when a hot spot needs to be reduced at the entrance to a reactor by using a large temperature driving force. Also shown (as a dotted line) is an alternative curve for the cooling medium that actually crosses over the temperature curve for the reacting process fluid. Such behavior is obviously impossible in simple recuperating heat exchangers, discussed in Chapter 2. However, unlike the simple heat exchangers covered previously, heat is generated within the process fluid because of the chemical reactions that take place. Therefore, it is possible for the cooling medium to be heated above the temperature of the process fluid at a certain location in the reactor. This situation does not have to occur and is usually undesirable for controlling the reactor, but it does not violate any physical laws if it does take place.

Returning to the temperature profiles in Figure 4.10(b), as fluid enters the reactor, it is exposed to catalyst in the tubes and starts to react. Simultaneously, heat is transferred from the process gas, through the tube wall into the cooling medium. Therefore, there are two opposing terms in the energy balance: generation through chemical reaction and heat exchange to the coolant. At the entrance of the reactor, the concentration of reactants is high and the heat generated by the chemical reaction is higher than the transfer of heat to the coolant. As a result, the process gas heats up. Because the rate constant is a strong function of temperature, the increase in reaction rate due to the increase in temperature is relatively large and causes the process gas temperature to increase as it flows down through the tubes. Considering Curve B in Figures 4.10(b) and 4.10(c), initially the reaction rate and temperature increase quickly. Therefore, the concentration of reactant A decreases quickly, and this decrease opposes the increase in reaction rate due to the increasing temperature. Simultaneously, the temperature driving force for heat transfer between the process gas and the cooling medium increases. Therefore, as the gas moves down the tube, at some point the rate of heat generation by the reaction is exactly equal to the rate of heat removal through the tube wall. At this point, a maximum temperature is seen in the tube (often called the *hot spot*). As the process gas moves further down the tube, the rate of heat transfer to the cooling medium is greater than the rate of heat generation due to chemical reaction, and the gas decreases in temperature. Since reaction is still taking place, the concentration continues to decrease until the gas leaves the reactor. Curves C, D, and E represent cases in which the heat of reaction is smaller than for Curve B. For these cases, the same description of the processes telling how the gas temperature and concentration profiles change along the length of the reactor applies, except that the magnitude of the hot spot temperature decreases and its location shifts further down the reactor.

Consider now the case given in Curve A, for which the heat of reaction is greater than for Curve B. The same initial processes as described for Curve B still take place, but now the rate of heat generation due to the chemical reactions is too great for the cooling medium to counterbalance the temperature increase. Therefore, the temperature continues to increase rapidly, and a condition is reached at which the reactions "run away," that is, the temperature increases unbounded. This is a condition that must be avoided at all costs, because material failure within the reactor may very well occur, and some form of catastrophic event (explosion or vessel rupture) may happen. In addition, the catalyst will undergo sintering, the active material comprising the catalyst may evaporate or will certainly agglomerate, and virtually all catalytic activity will be destroyed.

The sequence of curves shown in Figures 4.10(b) and 4.10(c) are labeled to show the effect of increasing heat of reaction (from E to A), but the same set of curves could be obtained in other ways. For example, a change in cooling medium flow, a change in inlet temperature for the cooling medium, or a change in process gas inlet temperature could all cause similar effects. In the next section, the equations describing these types of reactors and the importance of the heat transfer process are discussed.

4.2.2.4 The Role of Heat Transfer in Reactor Design

From the preceding section, it should be clear that, for shell-and-tube reactors, the role of heat transfer is critically important. It was also pointed out that, in general, radial temperature profiles within the catalyst pellets and also in the tubes holding the catalyst pellets will exist. However, Levenspiel (1989) points out that for commercial heterogeneous (gas-solid) catalyzed reactions, temperature variations within the catalyst pellet are usually not significant. Froment, Bischoff, and De Wilde (2011) note that radial temperature variations within the catalyst tubes may be present for some commercial reactions. In such cases, a two-dimensional model of the reactor is required to capture correctly the heat transfer and reaction processes, but this level of analysis is not covered here, and interested readers are referred to Froment et al. (2011) for a detailed discussion on the different approaches to modeling packed bed reactors with and without heat exchange.

In the discussion given here, a one-dimensional approach to solving the equations for packed beds with heat exchange is covered. In the one-dimensional approach, it is assumed that temperature and concentration are only functions of the axial position in the reactor and that radial variations within the tubes and between tubes at the same axial position do not exist. With these assumptions, it is only necessary to consider the processes occurring in a single packed tube of catalyst, shown in Figure 4.11.

Figure 4.11 shows a slice of the reactor tube and accompanying cooling medium. Heat is generated in the catalyst filled tube due to reaction, and some of this heat is removed by the cooling medium (shown flowing countercurrently to the process gas flow) by heat transfer through the wall of the tube. The amount of heat transferred through the tube wall is a function of the temperature difference between the process gas and the coolant at the same axial location in the reactor. This transfer of heat will, in general, not equal the amount of heat generated in the tube; therefore, the gas leaving the slice of the bed will not be equal to the temperature entering. The following assumptions are made to simplify the analysis:

1. The overall heat transfer coefficient between the tube wall and cooling medium, U, is constant over the length of the reactor.
2. The heat capacities for process gas and coolant do not change appreciably over the length of the tubes.
3. The heat of reaction does not change appreciably over the length of the tubes.

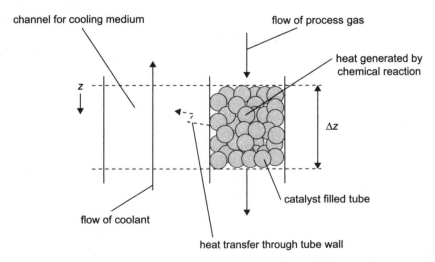

Figure 4.11   Processes taking place in a shell-and-tube reactor

The equations describing these processes for a reaction involving species A are given below:

$$\dot{m}_p c_{p,p} T_p \Big|_z + \frac{\pi D_{tube}^2}{4} \Delta z (1-\varepsilon)(-r_A''')(-\Delta H_r) = \dot{m}_p c_{p,p} T_p \Big|_{z+\Delta z} + \pi D_{tube} \Delta z U (T_p - T_c) \quad (4.43)$$

The terms in Equation (4.43) are explained by the following word equation:

Energy flowing in with the process gas + Energy generated by chemical reaction = Energy flowing out with process gas + Energy transferred to cooling medium through the tube wall

Simplifying terms, dividing by Δz, and taking the limit, the following differential equation is obtained:

$$-G_p c_{p,p} \frac{dT_p}{dz} = -(1-\varepsilon)(-r_A''')(-\Delta H_r) + \frac{4U}{D_{tube}}(T_p - T_c) \quad (4.44)$$

where the subscripts p and c refer to the process gas and coolant, respectively, the term G is the superficial mass velocity, that is, $G_p = 4\dot{m}_p / \pi D_{tube}^2$, and the rate $(-r_A''')$ is given per unit volume of catalyst. The process mass flowrate in the above relationship for G_p is the mass flow per tube and not the total mass flow rate of the process gas. In order to integrate Equation (4.44), an expression for the rate of reaction of Component A must be known and, in addition, a mole balance on A is required to keep track of the composition and/or partial pressure of A as the gas moves down the tube.

A mole balance on A is as follows:

$$-\frac{dF_A}{dz} = -r_A''' \frac{\pi D_{tube}^2 (1-\varepsilon)}{4} \quad (4.45)$$

To obtain the concentration of A or the partial pressure, an overall mole balance based on the stoichiometry of the reaction is required and the ideal gas law (or another appropriate equation of state). Therefore,

$$F = F_{Ao} \sum_{all\ i} \left(\Theta_i + \frac{v_i}{v_a} X \right) \quad (4.46)$$

where $X = 1 - F_A/F_{A0}$ and

$$c_A = \frac{F_A}{F} \frac{P}{RT} \quad \text{and} \quad p_A = \frac{F_A}{F} P \quad (4.47)$$

Cooling Medium
An energy balance on the cooling medium gives,

$$\dot{m}_c c_{p,c} T_c \Big|_{z+\Delta z} = \dot{m}_c c_{p,c} T_c \Big|_z + \pi D_{tube} \Delta z U (T_p - T_c) \quad (4.48)$$

Simplifying terms and dividing by Δz and taking the limit, the following differential equation is obtained:

$$G_c c_{p,c} \frac{dT_c}{dz} = \frac{4U}{D_{tube}}(T_p - T_c) \quad (4.49)$$

For cocurrent flow of the cooling medium and process gas, a negative sign would be added to the term on the left-hand side of Equation (4.49).

Pressure of Process Gas

From Chapter 1, the pressure drop for the process gas flowing through the packed bed of catalyst in the tubes is given by the differential form of the Ergun equation:

$$\frac{dP}{dz} = \frac{150\mu G_p}{\rho_g D_p^2}\frac{(1-\varepsilon)^2}{\varepsilon^3} + \frac{1.75 G_p^2}{\rho_g D_p}\frac{(1-\varepsilon)}{\varepsilon^3} \tag{4.50}$$

The set of equations ([4.44]–[4.47], [4.49], and [4.50]) must be solved simultaneously to obtain the concentration and temperature profiles in a shell-and-tube reactor. For the case illustrated in Figure 4.11, the set of equations is a split boundary problem, because the conditions of the entering process gas are known at $z = 0$, but the inlet conditions for the cooling medium are known at the end of the reactor at $z = L$. For cocurrent flow of the cooling medium and the process gas, the conditions of both streams are known at the beginning of the reactor, which makes the solution more straightforward.

Overall Heat Transfer Coefficient

The overall heat transfer coefficient U used in Equations (4.44) and (4.49) can be calculated in the same way described in Chapter 2. However, for the heat transfer from a gas flowing inside a catalyst packed bed, no correlation was given in Chapter 2. In all cases, the limiting heat transfer coefficient will be on the tube side of the reactor, since the shell-side heat transfer coefficients are expected to be one to two orders of magnitude greater than the tube-side coefficient for a liquid cooling medium, the resistance of the metal tube wall will be small, and fouling should also be small. A large body of work is available in the literature on effective tube-wall heat transfer coefficients for gas flowing in packed beds. The majority of this work has been for air at ambient conditions. According to Li and Finlayson (1977), the following heat transfer correlations may be used for particle-air systems:

Spherical Catalyst Particles (diameter = D_p) in Tubes (diameter = D_{tube})

$$\frac{h_{w,eff} D_{tube}}{k_g} = 2.03\,\text{Re}^{0.8}\exp\left[-\frac{6D_p}{D_{tube}}\right] \quad (20 \le \text{Re} \le 7600) \text{ and } \left(3.33 \le \frac{D_{tube}}{D_p} \le 20\right) \tag{4.51}$$

Cylindrical Catalyst Particles (diameter = D_p) in Tubes (diameter = D_{tube})

$$\frac{h_{w,eff} D_{tube}}{k_g} = 1.26\,\text{Re}^{0.95}\exp\left[-\frac{6D_p}{D_{tube}}\right] \quad (20 \le \text{Re} \le 800) \text{ and } \left(5.0 \le \frac{D_{tube}}{D_p} \le 33.3\right) \tag{4.52}$$

where Re is the particle Reynolds number, $\text{Re} = \dfrac{G_g D_p}{\mu}$.

The preceding correlations are insensitive to the material properties of the solid particles, and in practice, this has been shown to be an oversimplification. A more comprehensive and complicated approach is given by Dixon and Cresswell (1979). They define an effective radial thermal conductivity in the bed that is a function of both the properties of the solid and the gas and the gas flowrate, and then they define a wall Biot number that correlates with the particle Reynolds number. In addition, an overview of heat and mass transfer processes in packed beds is given by Wakao and Kaguei (1982). The interested reader is referred to these works for more details on this subject. For the sake of simplicity, Equations (4.51) and (4.52) are used in this text to calculate the overall heat transfer coefficient for shell- and catalyst-packed-tube reactors.

Example 4.9 illustrates the use of the equations developed in this section to solve a shell-and-tube reactor problem.

Example 4.9

The synthesis of phthalic anhydride from o-xylene is considered. The following parameters and pseudo-first-order reaction rate may be used for an approximate preliminary design of a shell-and-tube reactor.

$$C_8H_{10}+3O_2 \rightarrow C_8H_4O_3+3H_2O$$

Auto ignition temperature of o-xylene = 465°C, UEL = 0.9% at room temperature

Data (data taken from Froment et al. [2011])

Tube length, $L > 1$ m, D_{tube} = 0.0254 m, catalyst is V_2O_5 on promoted silica gel, diameter of catalyst particle, D_p = 3mm, catalyst bulk density, ρ_{bulk} = 1300 kg/m³, bed voidage, ε is 0.45, the acceptable operating range for the catalyst is 335°C to 415°C, reactor pressure ~ 1 atm. A typical gas mass velocity in the tubes, G_g, is 4680 kg/m²/h (1.3 kg/m²/s), average specific heat of the gas, c_{pg}, is 1107 J/kg/K. The properties of gas at 350°C are $\mu = 30.78 \times 10^{-6}$ kg/m/s and k_g = 0.0469 W/m/K. Assume physical properties are constant over the temperatures used in the simulation. To find the overall heat transfer coefficient, Equation (4.51) is used.

$$\text{Re} = \frac{G_g D_p}{\mu} = \frac{(1.3)(0.003)}{(30.78 \times 10^{-6})} = 127$$

and

$$\frac{D_{tube}}{D_p} = \frac{0.0254}{0.003} = 8.467,$$

which are in the correct range to use Equation (4.51),

$$\frac{h_{w,eff} D_{tube}}{k_g} = (2.03)(127)^{0.8} \exp\left[-\frac{6}{8.467}\right] = 48.17$$

$$h_{w,eff} = (48.17)\frac{(0.0469)}{(0.0254)} = 88.9 \text{ W/m}^2/\text{K}$$

$$\text{and } U \cong h_{w,efff} = 88.9 \text{ W/m}^2/\text{K}$$

The reaction rate can be approximated by a pseudo-first-order equation:

$$-r_A''' \text{ (reaction rate)} = \text{mol/h/m}^3\text{catalyst} = \rho_{bulk}kp_{B,0}p_A$$

where Component A is o-xylene, $k = 4.122 \times 10^8 e^{-13,636/T[K]}$ mol/h/bar²/kg (catalyst), and Component B is oxygen with an initial partial pressure, $p_{B,0}$ = 0.21 bar. The average heat of reaction for this temperature range $-\Delta H_r$ is 1150 kJ/mol.

For illustration purposes, it is assumed that cooling takes place by heat exchange with a cooling medium flowing through the shell side at 335°C that vaporizes at constant pressure and temperature. With this assumption, the shell-side temperature, T_{cool}, is fixed at 335°C.

Determine the temperature profiles in the tube for partial pressures of o-xylene in the feed ranging from 0.01 to 0.02. Assume that the effect of changing pressure is small; that is, assume constant pressure conditions and a process gas inlet temperature of 335°C.

Solution

The energy balance Equation (4.44) can be written:

$$\frac{dT_p}{dz} = \frac{(1-\varepsilon)(-r_A''')(-\Delta H_r)}{G_p c_{p,p}} - \frac{4U}{G_p c_{p,p} D_{tube}}(T_p - T_{cool})$$

$$\frac{dT_p}{dz} = \frac{(0.55)(1300)(4.122\times10^8 e^{-13,636/T_p})(0.21)p_A(1.150\times10^6)}{(1.3)(1107)}$$

$$- \frac{4(88.9)}{(1.3)(1107)(0.0254)}(T_p - 608.2)$$

$$\frac{dT_p}{dz} = 49.46\times10^{10} e^{-13,636/T_p} p_A - 9.73(T_p - 608.2) \tag{E4.9a}$$

The material balance is given by Equation (4.45):

$$-\frac{dF_A}{dz} = -r_A'''\frac{\pi d_{tube}^2(1-\varepsilon)}{4} = (4.122\times10^8 e^{-13,636/T_p})(1300)(0.21)p_A\frac{\pi(0.0254)^2(1-0.45)}{4}$$

$$\frac{dF_A}{dz} = -31.36\times10^6 e^{-13,636/T_p} p_A \tag{E4.9b}$$

In order to relate F_A and p_A, the stoichiometry of the reaction needs to be considered. Assume that the inlet gas to the reactor contains y mole of o-xylene per 100 moles of air. Using Equation 4.46, a mole balance for any conversion is

$$F = F_{Ao}\sum_{all\ i}\left(\Theta_i + \frac{v_i}{v_a}X\right) = F_{Ao}\left[1 - X + \frac{21}{y} - \frac{3}{1}X + X + \frac{3}{1}X + \frac{79}{y}\right] = F_{Ao}\left[1 + \frac{100}{y}\right] \tag{E4.9c}$$

$$\underset{OX}{\longleftrightarrow}\quad\underset{O_2}{\longleftrightarrow}\quad\underset{PA}{\diagdown}\;\underset{H_2O}{\diagdown}\;\underset{N_2}{\diagdown}$$

The result in Equation E4.9c is obvious from the stoichiometry, as there is no net change in the number of moles during the reaction. The relationship between p_A and F_A in Equation (4.47) is also used to close the two equations. Finally, Equation (4.49) is not required because T_{cool} does not change with z. By integrating these equations, the plots shown in Figure E4.9 are obtained. It can be seen that for partial pressures above approximately 0.0165 bar, the temperature bump or spike starts to increase significantly, and with a partial pressure of 0.0175, the reaction becomes uncontrollable resulting in a runaway. The conversions of o-xylene are also shown on this figure, and as expected, the conversions increase rapidly when runaway occurs.

When the cooling medium does not change phase, the temperature profile will change because T_{cool} changes down the length of the reactor. Examples for such cases are given in the problems at the end of this chapter.

A note about process safety: The lower explosion limit (LEL) for o-xylene in ambient air is about 0.90 mol%, and this drops to about 0.66 mol% at the temperature of the reactor inlet used in this example (335°C). The inlet concentrations shown in Figure E4.7 are all above the LEL (and well below the upper explosion limit), so these are dangerous conditions at which to operate the reactor. Special design considerations must be employed if operation in the flammable region is unavoidable. These include **containment, explosion venting,** and **explosion suppression.** The idea behind containment is that the reactor is built to withstand the explosion pressure if it were to occur. Explosion venting allows the rapid release of vapor if an explosion were to occur, while explosion suppression would include the removal of all sources of ignition, suppression of flames, and flame propagation. All of these techniques require significant additional capital investment and should be carefully analyzed and other alternatives considered prior to the final design. For more information on these topics, the interested reader is referred to Mannan (2014).

It should also be noted that as the temperature increases, the total combustion of the organic compounds will start to take place, giving water and carbon dioxide as products. The heat of combustion for o-xylene is 4375 kJ/mol, which is 3.8 times the heat of reaction to produce phthalic anhydride. Therefore, when the reaction starts to run away, the heat generated will be extremely large and the temperature profiles would be even steeper than those shown in Figure E4.9.

Example 4.9 considered the system to be at a constant pressure of approximately 1 atm. The pressure drop can be calculated using the Ergun equation. To account for the changes in gas properties along the length of the reactor, the differential form of the Ergun Equation, Equation (4.50), can be solved simultaneously with Equations (E4.9a) and (E4.9b). However, if the pressure does not change very much, then the integral form of the Ergun equation can be used. This is shown in Example 4.10.

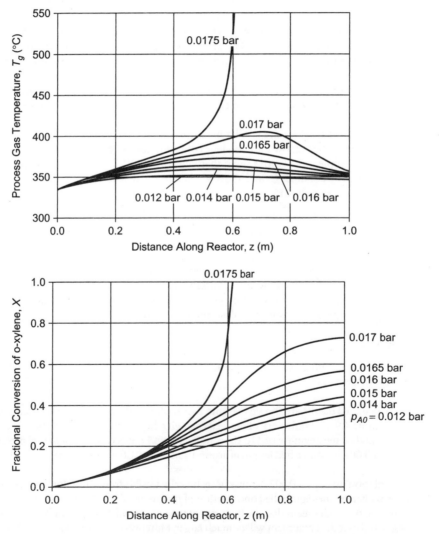

Figure E4.9   Temperature and conversion profiles in the shell-and-tube reactor from Example 4.9

Example 4.10

For the reactor given in Example 4.9, determine the pressure drop across the reactor. Assume that the gas has the properties of air at the inlet conditions.

Solution

For air, the density at 1 atm and 335°C is 0.5778 kg/m³ and the viscosity is 30.85 × 10⁻⁶ kg/m/s. Integrating Equation (4.50) and assuming the fluid properties remain unchanged gives

$$-\frac{\Delta P}{L} = \frac{150\mu G_p}{\rho_g D_p^2}\frac{(1-\varepsilon)^2}{\varepsilon^3} + \frac{1.75G_p^2}{\rho_g D_p}\frac{(1-\varepsilon)}{\varepsilon^3}$$

$$\therefore -\frac{\Delta P}{L} = \frac{150(30.85\times10^{-6})(1.3)}{(0.5778)(0.003)^2}\frac{(1-0.45)^2}{(0.45)^3} + \frac{1.75(1.3)^2}{(0.5778)(0.003)}\frac{(1-0.45)}{(0.45)^3}$$

$$\therefore -\frac{\Delta P}{L} = 3840+10,298 = 14,138 \text{ Pa/m} \Rightarrow \Delta P = 14.14 \text{ kPa}$$

(E4.10a)

This pressure drop is significant when compared to the absolute pressure of the system (~101.3 kPa). An air blower would be required to push the air through the reactor, and the other process equipment and the reactor would operate at a pressure slightly above ambient.

The temperature profiles shown in Example 4.9 are very sensitive to the overall heat transfer coefficient, and this and alternative operating conditions are considered in the problems at the end of this chapter.

In reviewing the temperature profiles in Example 4.9, it is tempting to consider that perhaps an adiabatic bed operating at low inlet o-xylene concentrations might be a possible alternative configuration. For example, calculations show that when p_A = 0.01 bar, the temperature across the reactor only changes by about 12°C or 13°C, so would it be possible to operate without cooling? This case is considered in Example 4.11.

Example 4.11

Repeat Example 4.9 for the case when catalyst is placed in a packed bed without internal cooling, as shown in Figure 4.8(a). Determine the temperature and conversion profiles within such a packed bed for inlet partial pressures of o-xylene from 0.008 to 0.002 bar. Assume that the maximum allowable temperature to avoid catalyst sintering is 410°C.

Solution

The same equations given in Example 4.9 are used, except the value of the overall heat transfer coefficient, U, is set to 0.

The calculated results are shown in Figure E4.11. Clearly, the cooling of the process gas is significant in Example 4.9, and only very dilute feeds can give high conversions and still not exceed the maximum catalyst temperature of 410°C. From Figure E4.11, it is clear that a partial pressure of o-xylene around 0.002 bar and a bed length of around 3 m will satisfy these criteria.

The solution using p_A = 0.002 bar may be technically feasible from a reaction standpoint, but the feed is approximately one-eighth the concentration of that required in a shell-and-tube reactor (0.002 bar vs. 0.0165 bar), which means that the front end of the process will have to handle eight times the flowrate of gas and that the equipment will be much larger. Moreover, when the phthalic anhydride is separated from the reactor effluent stream, the separation will be more difficult (lower partial pressure driving force) and more expensive (larger equipment and more utilities). Therefore, it is easy to see why a shell-and-tube reactor is favored in practice for this process.

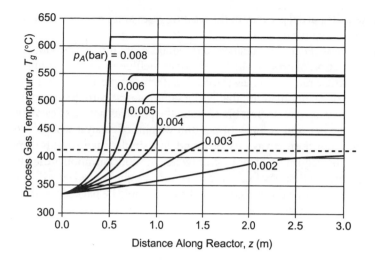

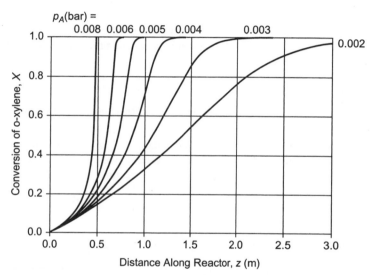

Figure E4.11   Temperature and conversion profiles for adiabatic packed bed reactor for Example 4.11

4.2.2.5   Matching Volume and Heat Transfer Area

Often when designing a shell-and-tube reactor using a simulator, it is necessary to specify the heat exchanger configuration (cocurrent, countercurrent, number of tubes, and tube diameter) prior to simulation. However, in practice, it is the desired conversion of limiting reactant that is usually known (or a range of conversions that might be investigated is known), and the amount of catalyst and the heat exchanger surface area must be determined. Because the catalyst is contained within the tubes, the ratio of the heat transfer area to the volume of catalyst is always equal to $\pi D_{tube} L / (\pi D_{tube}^2 L / 4) = 4 / D_{tube}$. This constrains the problem and may lead to many infeasible solutions. For example, for pressure drop considerations, it might be necessary to use a large catalyst particle and hence fairly large tubes. Note that it is common practice for the ratio of D_{tube}/D_p to be ≥10. With a given cooling medium, it may be impossible to find a configuration that will prevent a runaway condition in the reactor. One solution is to dilute the feed stream to the reactor by adding an inert gas (e.g., steam or nitrogen) that would reduce the partial pressure of the reacting gases and slow the reaction rate. This solution has the disadvantage of requiring that extra material flow through the process, which will increase equipment size, increase heating/cooling utilities, and require more difficult product separation, because of the lower concentration of products in the reactor exit (effluent) stream. An alternative method is to "turn-down" the reaction rate by diluting the catalyst. For example, the catalyst could be mixed with similar sized particles of carrier material; that is, the same catalyst particles without the active catalytic material. Alternatively, it may be possible to dope the particles with less catalytic material. In both cases, the reaction rate per unit volume (or per unit weight) of catalyst would be decreased, and a solution avoiding temperature runaway may be found. In a similar manner, it is sometimes effective to dilute the catalyst at the beginning of the reactor (where the concentration is the highest) to avoid a runaway condition. Thus, the first few feet of a packed tube might be packed with a mixture of active catalyst and inert material, while the remainder of the tube would be packed with just active catalyst. Example 4.12 illustrates the concept of catalyst dilution.

Example 4.12

Consider the reactor in Example 4.9 using an inlet o-xylene partial pressure of 0.0175 bar. Without dilution, this feed concentration leads to a runaway condition in the reactor. Explore the possibility of diluting the catalyst.

Solution

The reaction rate used in Example 4.9 was $-r_A'''$ (reaction rate) $= \text{mol/h/m}^3 \text{ catalyst} = \rho_{bulk} k p_{B,0} p_A$, where

$k = 4.122 \times 10^8 e^{-13,636/T_p[K]}$ [mol/h/bar^2/kg catalyst]. The reaction rate is expressed in terms of per unit volume of catalyst, so if the catalyst were mixed with, say, 5% of inert material, then the effective rate of reaction of this mixture would be 95% ($f = 0.95$) of the rate given previously, and the value of k would be f times that of the undiluted catalyst, that is, $k_{95\%-dilute} = (0.95)(4.122 \times 10^8 e^{-13,636/T_p[K]}) = 3.916 \times 10^6 e^{-13,636/T_p[K]}$.

The results for different dilution rates (f) are shown in Figure E4.12.

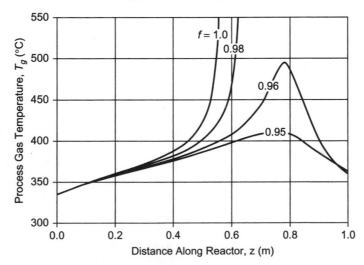

Figure E4.12   Process gas temperature and conversion profiles for different levels of catalyst dilution (f) (Continued)

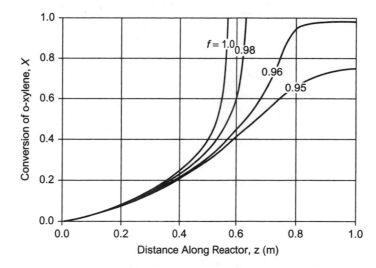

Figure E4.12   (Continued) Process gas temperature and conversion profiles for different levels of catalyst dilution (f)

With a dilution factor of only 5%, the process gas temperature can be kept below the catalyst sintering temperature of 410°C with close to 80% conversion in a 1 m length tube. For a 4 m long tube, the conversion increases to approximately 90%.

4.2.3   Fluidized Bed Reactor

The term *fluidized bed reactor* is used to describe a wide variety of gas-solid (and sometimes liquid-solid) reactor configurations in which the upward flow of fluid (gas) through a bed of catalyst particles exceeds the minimum fluidizing velocity. The type of fluidization is determined by the ratio of the superficial velocity to the minimum fluidizing velocity (u_s/u_{mf}) and the size of the particles. Several flow regime maps describing fluidized flow behavior have been published, and the one that is attributed to Grace (1986) is probably the most widely used. A simplified version of this flow map was presented in Chapter 1, Figure 1.11. Pictorial representations of the four most common flow regimes are shown in Figure 4.12 and are discussed here:

- **Bubbling fluidized bed (Figure 4.12[a]):** For relatively low values of u_s/u_{mf} and small particles, the system behaves as a bubbling fluidized bed. According to the two-phase theory of fluidization first introduced by Toomey and Johnstone (1952), the bed can be separated into two distinct phases: the emulsion phase and the bubble phase. The majority of the bed containing the solids is at or slightly above the minimum fluidizing velocity and comprises the emulsion phase. The excess gas, above that required for minimum fluidization, passes through the bed in the form of distinct bubbles. The two-phase theory was extended by the pioneering work of Davidson and Harrison (1963) who proposed the first accurate hydrodynamic model of bubbling fluidized beds. The exchange of the gas between the emulsion and the bubble phases means that at times the gas will see regions of high catalyst loading (emulsion) and at times it will see low catalyst loading (bubbles). If the particles are small, then Davidson and Harrison showed that a bubble is surrounded by a cloud of circulating gas that stabilizes the bubbles that rise through the emulsion and also limits the exchange of gas between the two regions or phases. When this is the case, it leads to significant bypassing of gas through the bed with

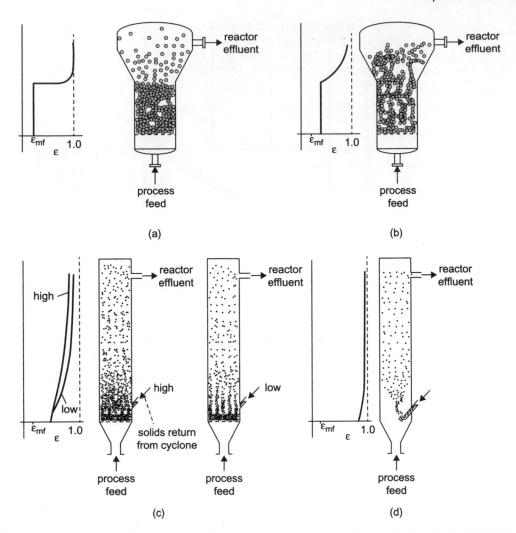

Figure 4.12 Common flow regimes and corresponding voidage profiles for fluidized beds: (a) bubbling bed, (b) turbulent bed, (c) fast fluidized or circulating bed (showing high and low solids circulation rates), and (d) transport (pneumatic conveying) reactor (Adapted from Kunii and Levenspiel [1991])

a corresponding reduction in conversion compared to pure plug flow. Conversely, when the particles are large, the gas bubbles tend have very thin clouds and gas from the emulsion tends to "short-circuit" through the bubbles and back into the emulsion phase.

- **Turbulent fluidization (Figure 4.12[b]):** For higher values of u_s/u_{mf} and all but the largest particles, the system behaves as a turbulent fluidized bed. In this regime, the gas no longer flows through the bed as distinct bubbles. Turbulent bed behavior is characterized by elongated voids of gas and streamers or strands of solid particles moving about the bed in a rather violent and haphazard fashion, making it difficult to identify individual phases. The bed surface also becomes somewhat indistinct. Pressure fluctuations across the bed are large, and the general form of the bed is unstructured and chaotic. Due to the violent nature of the fluidization, solids are flung upward into the freeboard region of the bed and tend to overflow with the gas stream. For this reason, internal cyclones are usually employed to help return the catalyst particles to the reactor.

- **Fast fluidization (Figure 4.12[c]):** As the value of u_s/u_{mf} continues to increase, solids start to be transported from the top of the bed and out of the reactor. For such conditions, much of the reaction takes place above the dense bed formed at the bottom of the equipment. Reactors operating in this fast fluidization regime are constructed with a long riser configuration that discharges into an external cyclone. Solids disengage from the gas in the cyclone and are circulated back to the base of the reactor where they are re-injected into the bed and get transported back up through the riser and into the cyclone. In this way, solids circulate in a loop and sometimes these reactors are referred to as *circulating fluid-bed reactors*. The height of the dense bed at the bottom of the riser is a function of the solids circulation rate and increases with increasing circulation rate. The solids moving up in the riser form clusters and may move toward the wall and fall downward before being re-entrained. The net movement of solids is up, but significant solids recirculation exists in the riser.
- **Transport (pneumatic conveying) reactors (Figure 4.12[d]):** When the gas velocity exceeds about 20 times the terminal velocity for small particles and the mass loading of solids-to-gas is also about 20:1, then particles move upward in a lean state with little interaction with each other. Thus, both solids and gas move essentially in plug flow upward through the reactor.

The type of fluidized bed reactor chosen for a given process depends on many factors, including the flowrate of gas and the size of catalyst particles. For example, in a bubbling bed, bubbles drag catalyst particles upward and provide good solids circulation that leads to near isothermal conditions in the bed. The disadvantages are that relatively low values of u_s/u_{mf} are needed, and to obtain such velocities, it may lead to beds that have large diameters and small heights. If higher gas velocities are required, circulating fluidized beds or transport reactors are normally used but do not act isothermally, although they still provide stable temperature control. A good review of the practical application of fluidized beds to a wide number of processes is given by Kunii and Levenspiel (1991).

4.2.3.1   Reactor Models

As stated previously, reactor models for fluidized beds are generally complex and vary widely depending on the flow regime. The simplest models are *phenomenological* in nature, whereby a picture of the flow patterns of gas and solids is assumed (based on experimental observations), and then a model is developed around this picture of the reactor. At the complex end of the modeling spectrum, *rigorous hydrodynamic* models for each phase are postulated along with appropriate closure equations that describe the interactions between the phases. These models are then integrated with the material and energy balances along the length of the reactor to give concentration and temperature profiles. For details of the different models, the interested reader is referred to Kunii and Levenspiel (1991) for phenomenological models and Gidaspow (1994) for rigorous computational models of gas-solid hydrodynamics.

One very simple model is presented here to illustrate the approach used with phenomenological models. The model is the two-phase model given by Froment et al. (2011) that is based on the work of May (1959) and van Deemter (1961), which may be used as a first approximation for a bubbling fluidized bed reactor. The model is illustrated in Figure 4.13, where the fluidized bed is split into an emulsion region and a bubble region. The incoming flow is split between the two regions with the gas required for minimum fluidization going into the emulsion region and the remainder of the feed gas going to the bubble region. There is an interchange of gas between the two regions that is characterized by k_I (m^3gas/m^3total reactor/s).

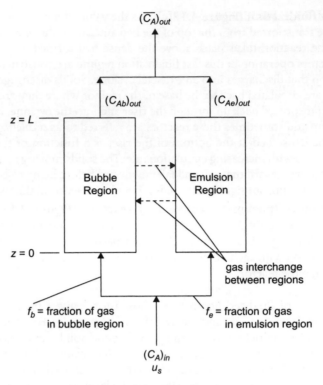

Figure 4.13   Representation of two-phase model for bubbling fluidized beds

The equations describing the change in concentration in each phase are given as

$$f_b u_b \frac{dc_{Ab}}{dz} = k_I(c_{Ae} - c_{Ab}) - r_A \rho_b f_b \tag{4.53}$$

$$f_e u_e \frac{dc_{Ae}}{dz} = k_I(c_{Ab} - c_{Ae}) + f_e D_e \frac{d^2 c_{Ae}}{dz^2} - r_A \rho_e(1 - f_b) \tag{4.54}$$

and the exit gas concentration is given by the average of the gas leaving both regions as

$$u_0(\bar{c}_A)_{out} = f_b u_b(c_{Ab})_{out} + f_e u_e(c_{Ae})_{out} \tag{4.55}$$

where u_e is the interstitial velocity in the emulsion phase (u_{mf}/ε_{mf}), u_b is the rise velocity of the bubbles, f_b is the fraction of the bed volume taken up by bubbles, and f_e is the fraction of the bed volume taken up by the gas in the emulsion. k_I is an interchange coefficient between the two phases and is expressed in m^3/(m^3 total reactor volume)/s. Values of k_I must be found through experimentation using tracer techniques and are strong functions of the system parameters (particle size, fluidizing velocity, etc.). The term D_e is a dispersion parameter to describe the flow of gas in the emulsion phase. Expressions for D_e exist in the literature but are defined for only narrow ranges of operating and bed parameters, such as particle size, bed height, and u_s/u_{mf}. However, if plug flow or mixed flow of the gas are assumed in the emulsion phase, then D_e equals 0 or ∞, respectively. Van Swaaij and Zuiderweg (1972) concluded that the simple two-phase model gave an adequate description of the mass transfer and reaction taking place in the ozone decomposition reaction occurring in a bubbling fluidized bed of silica. A problem based on this system is given at the end of this chapter.

4.3   PERFORMANCE PROBLEMS

4.3.1   Ratios for Simple Cases

Most reactions and reactors of industrial importance do not operate isothermally and cannot be described with simple reactor models (PFR or CSTR). Therefore, analytical solutions for reactor problems in the real world are rare. Nevertheless, when changes in the process occur, then simple models may often be used to determine the direction of the expected trends and methods to mitigate (if needed) the process change that has occurred. Example 4.13 illustrates this point.

Example 4.13

Consider the reaction in Example 4.4, which described the isothermal liquid-phase conversion of feed A to product B in a plug flow reactor. The reactor volume was found for a given temperature, and a conversion of 95% to be 304 L. The following data were provided:

$-r_A$ (reaction rate) $= \text{mol/lit/s} = 0.05 c_A^2 [\text{mol/lit}]^2$, $C_{A0} = 10$ mol/L is fed to an isothermal reactor $F_{A0} = 80$ mol/s, and the design equation (Equation [4.32]) was

$$V = F_{A0} \int_0^X \frac{dX}{-r_A} = 80 \int_0^{0.95} \frac{dX}{(0.05)c_A^2} = \frac{80}{c_{A0}^2} \int_0^{0.95} \frac{dX}{(0.05)(1-X)^2}$$

$$V = \frac{80}{(10)^2} \frac{1}{(0.05)} \frac{1}{(1-X)} \Big|_0^{0.95} = \frac{80}{5} \left(\frac{1}{0.05} - \frac{1}{1} \right) = 304 \text{ L}$$

Now consider the case when the reactor is up and running, but the throughput to the reactor is increased by 20%. It may be assumed that heat generation within the reactor is negligible, that is, the heat of reaction is small. For this reactor, consider the following:

a. What will the new conversion be (assuming the temperature remains constant)?

b. If the original temperature of the reactor was 150°C, what should the temperature in the reactor be adjusted to so that the same conversion is achieved with the new throughput? The activation energy for this reaction is $E = 20$ kJ/mol.

Solution

a. The form of the reaction rate does not change, so the integrated form of the design equation is

$$V = \frac{F_{A0}}{k c_{A0}^2} \int_0^X \frac{dX}{(1-X)^2} = \frac{F_{A0}}{k c_{A0}^2} \left[\frac{1}{(1-X)} - \frac{1}{1} \right] = \frac{F_{A0}}{k c_{A0}^2} \left[\frac{X}{1-X} \right] \qquad \text{(E4.13a)}$$

Using ratios with subscript 1 representing the base case and 2 representing the new case and noting the $k_1 = k_2$ and $C_{A0,1} = C_{A0,2}$ gives

$$\frac{V_1}{V_2} = \frac{F_{A0,1}}{k_1 c_{A0,1}^2} \left[\frac{X_1}{1-X_1} \right] \frac{k_2 c_{A0,2}^2}{F_{A0,2}} \left[\frac{1-X_2}{X_2} \right] \qquad \text{(E4.13b)}$$

Solving Equation (E4.13b),

$$1 = \frac{F_{A0,1}}{F_{A0,2}} \frac{X_1}{1-X_1} \frac{1-X_2}{X_2} = \frac{1}{1.2} \left[\frac{0.95}{1-0.95} \right] \left[\frac{1-X_2}{X_2} \right] \Rightarrow X_2 = 0.9406 \text{ or } 94.06\%$$

b. In order to keep the conversion at 95%, the ratio k_2/k_1 must $= F_{A0,2}/F_{A0,1} = 1.2$. Therefore,

$$\frac{k_2}{k_1} = 1.2 = \frac{k_0 e^{-E/RT_2}}{k_0 e^{-E/RT_1}} \Rightarrow \frac{-E}{RT_2} - \frac{-E}{RT_1} = \ln(1.2) = 0.1823$$

$$\frac{1}{T_1} - \frac{1}{T_2} = \frac{R}{E}(0.1823) \Rightarrow \frac{1}{(150+273.15)} - \frac{1}{T_2} = \frac{8.314}{20,000}(0.1823)$$

$$\Rightarrow T_2 = 437.17\,K = 164.0\,°C$$

There are two takeaways from Example 4.13. First, when the feed flowrate to the reactor increases and conditions of temperature (and pressure) remain constant, then the conversion goes down. This should be intuitively obvious, since the time spent by the reactants in the reactor (space time) is decreased giving less time for A to react to form B. Second, by increasing the reactor temperature (from 150°C to 164°C), the conversion can be maintained at 95% with the new throughput. Again this result should agree with intuition in that a shorter time in the reactor may be compensated by a higher rate of reaction caused by a higher reactor/reaction temperature.

4.3.2 More Complex Examples

As stated previously, when the operation of the reactor involves significant temperature changes, then the energy balance must be considered in conjunction with the material balance, and the solution of such problems involves the simultaneous solution of two (or more) algebraic (CSTR) or ordinary differential (PFR) equations. To illustrate a more complicated (nonisothermal) problem, Example 4.8 is revisited.

Example 4.14   Performance of Reactor in Example 4.8

The problem statement for Example 4.8 was given as follows: An exothermic, liquid-phase reaction is to take place in a stirred tank reactor. The feed to the reactor (F) is pure A at 500 mol/min with a concentration of 10 mol/L. The volume of the reactor is 100 L, and the feed temperature is 30°C. A reaction takes place in which A is converted to B (A → B) by a first-order reaction:

$$-r_A = kc_A = 3e^{-\frac{1650}{T[K]}} c_A \,[\text{mol}/\text{m}^3/\text{s}] \tag{E4.14a}$$

The heat of reaction ($-\Delta H_r$) is 75,240 J/mol, and it is approximately constant from 30–200°C. Other data for the process and coolant are as follows:

Data

$c_{p,p} = 4150\,\text{J/kg/°C}$, $c_{p,c} = 4180\,\text{J/kg/°C}$, $T_{C,2} = 30°C$, $T_{C,1} = 20°C$, $U = 1000\,\text{W/m}^2/°C$, $\rho_p = 1050\,\text{kg/m}^3$, and

$$m_p = \frac{F}{c_{A0}}\rho_p = \frac{(500/60)}{(10\times1000)}(1050) = 0.875\,\text{kg/s}$$

In Example 4.8, the conversion was controlled at 80% by maintaining the reactor temperature at 93.5°C using cooling water entering the reactor at 20°C and leaving at 30°C. The mass flowrate of cooling water was calculated to be 6.486 kg/s, and the required area for heat transfer was 3.96 m².

The reactor designed in this example was implemented and runs as predicted. However, it has been decided to increase the flow of reactant to 600 mol/min keeping the inlet concentration at 10 mol/L. What must the cooling water's flowrate and exit temperature be in order to maintain the conversion at 80%?

Solution

With the new flowrate, the time spent in the reactor will decrease, so the reactor temperature must be increased to maintain the desired conversion. Referring to Part (b) of Example 4.8,

$$c_A = 2000 = \frac{c_{A0}}{1 + k\tau} \tag{E4.14b}$$

but now the value of τ has changed to

$$\tau = \frac{c_{A0}V}{F} = \frac{(10)(100)}{(600)/(60)} = 100 \text{ sec} \tag{E4.14c}$$

$$\therefore c_A = 2000 = \frac{c_{A0}}{1 + k\tau} \Rightarrow k = \frac{1}{100}\left[\frac{10,000}{2000} - 1\right] = 0.040 = 3e^{-\frac{1650}{T + 273.15}} \Rightarrow T = 109.0\,^{\circ}\text{C}$$

The energy balance on the reactor and the heat exchanger design equations (Equations [4.40] and [4.41]) must now be solved simultaneously to determine the new coolant flowrate and exit temperature. Using ratios with subscript 1 representing the design case and 2 representing the new case gives

$$\frac{\left[\dot{m}_c c_{p,c}(T_2 - T_1)\right]_2}{\left[\dot{m}_c c_{p,c}(T_2 - T_1)\right]_1} = \frac{\left[Vr_A(c_A, T)(-\Delta H_r) - \dot{m}_p c_{p,p}(T - T_{in})\right]_2}{\left[Vr_A(c_A, T)(-\Delta H_r) - \dot{m}_p c_{p,p}(T - T_{in})\right]_1} \tag{E4.14d}$$

$$\frac{(T_2 - 20)\left[\dot{m}_c\right]_2}{(10)\left[\dot{m}_c\right]_1} = \frac{\left[Vkc_A(-\Delta H_r) - \dot{m}_p c_{p,p}(T - T_{in})\right]_2}{\left[Vkc_A(-\Delta H_r) - \dot{m}_p c_{p,p}(T - T_{in})\right]_1} \tag{E4.14e}$$

$$\therefore M_c \frac{(T_2 - 20)}{(10)} =$$

$$\frac{(100)(3e^{\frac{-1650}{109 + 273.15}})(2000)(75,240) - (600/500)(0.875)(4150)(109 - 20)}{(100)(3e^{\frac{-1650}{93.5 + 273.15}})(2000)(75,240) - (0.875)(4150)(93.5 - 20)} = 1.20 \tag{E4.14f}$$

$$M_c \frac{(T_2 - 20)}{10} = 1.20 \tag{E4.14g}$$

$$\frac{\left[\dot{m}_c c_{p,c}(T_2 - T_1)\right]_2}{\left[\dot{m}_c c_{p,c}(T_2 - T_1)\right]_1} = \frac{\left[UA\Delta T_{lm}\right]_2}{\left[UA\Delta T_{lm}\right]_1}$$

$$\therefore M_c \frac{(T_2 - 20)}{10} = \frac{U_2}{U_1}\frac{1}{68.38}\frac{T_2 - 20}{\ln\dfrac{89}{109 - T_2}} \Rightarrow M_c = \frac{U_2}{U_1}\frac{10}{68.38}\frac{1}{\ln\dfrac{89}{109 - T_2}} \tag{E4.14h}$$

To evaluate how the overall heat transfer coefficients change, the relative contributions to the inside and outside coefficients must be determined. For the sake of this example, it is assumed that the resistances are equal for the base-case design. The inside (cooling water flow) resistance will change by $Re^{0.8}$, but the outside resistance will remain approximately constant assuming that the stirring speed does not change. Therefore,

$$\frac{U_2}{U_1} = \frac{\left[\dfrac{1}{2000 M_c^{0.8}} + \dfrac{1}{2000}\right]^{-1}}{1000} \tag{E4.14i}$$

By substituting this equation into Equation (E4.14h) gives

$$M_c = \frac{1}{6838} \frac{\left[\dfrac{1}{2000 M_c^{0.8}} + \dfrac{1}{2000} \right]^{-1}}{\ln \dfrac{89}{109 - T_2}} \tag{E4.14j}$$

Solving equations (E4.14g) and (E4.14j) simultaneously gives

$$M_c = 0.9791 \ (\dot{m}_c = (6.486 \times 0.9791) = 6.35 \text{ kg/s}) \text{ and } T_2 = 32.25°C$$

Summary

By decreasing the cooling water flow to 6.35 kg/s, the exit temperature increases to 32.25°C and the reactor temperature increases to 109°C (from 93.5°C). This allows the flow to the reactor to increase by 20% and still maintains 80% conversion.

Example 4.14 illustrates the need to solve the material balances, energy balances, and the design equation simultaneously when considering the behavior of reactors with heat exchange (or nonisothermal) behavior. In the case considered, the reactor was a CSTR and the equations are relatively simple to solve, since they are algebraic. For a PFR, the energy balance and material balances are in the forms of differential equations, and their solution is more complicated, but the same principles apply. Performance problems involving nonisothermal PFRs are considered in the problems at the end of this chapter.

WHAT YOU SHOULD HAVE LEARNED

- Reaction kinetics, equilibrium, and heat transfer all play important roles in determining the correct design for chemical reactors.
- There is a hierarchy of reactor configurations for removing heat from gas-phase exothermic (and endothermic) reactions.
- Exothermic, gas-phase reactions may give rise to temperature hot spots in the reactor that can lead to dangerous conditions.
- Simultaneous removal of heat from gas-catalyzed, reacting systems occurring in shell-and-tube reactors is necessary for highly exothermic reactions. Design of such systems often requires the solution of simultaneous coupled ordinary differential equations (ODEs).
- Simultaneous removal of heat from liquid-phase reactions occurring in stirred tank reactors may be necessary. The design of such systems requires the solution of a set of coupled, nonlinear algebraic equations.
- The role of mass transfer in determining reaction rates for heterogeneous (gas-solid catalyzed) systems is important and may dominate the kinetics at high temperatures.
- The performance of simple reactor systems without temperature effects may be accomplished by comparing ratios.

NOTATION

Symbol	Definition	SI Units
$a, b, c,$	stoichiometric coefficients	
A	area	m^2
c	concentration	mol/m^3
c_p	specific heat capacity	$J/kg/K$
D	diffusion or dispersion coefficient	m^2/s
D	diameter	m
E	activation energy	J/mol
$E(t)$	residence time distribution in reactor	s^{-1}
f	fraction of stream	
F	molar flowrate	mol/s
G	mass velocity	$kg/m^2/s$
h	specific enthalpy of stream	J/mol
K, k	kinetic constants in Langmuir-Hinshelwood reaction rate	various
k_0	pre-exponential factor	various
L	characteristic length of a catalyst particle	m
$\dot{m}$	mass flowrate	kg/s
M_T	Thiele modulus	
N	number of moles	mol
P, p	pressure and partial pressure	bar
$\dot{Q}$	rate of heat input into a control volume	W
r	rate of reaction	$mol/m^3/s$
R	universal gas constant	$J/mol/K$
Re	Reynolds number	
S	selectivity	
t	time	s
$\bar{t}$	average time spent in the reactor	s

Symbol	Definition	SI Units
T	temperature	K
u	velocity	m/s
U	internal energy of system or overall heat transfer coefficient	J or W/m^2/K
V	volume of reactor	m^3
W_s	rate of shaft work done by the control volume	W
W	weight of catalyst	kg
X	conversion of reactant	
Y	yield	
z	distance along reactor	m

GREEK SYMBOLS

ΔH_R	heat of reaction	J/mol
ε	voidage of packed (or fluidized) bed	
Θ	stoichiometry parameter defined in Equation (4.25)	
ρ	density	kg/m^3
η	catalyst effectiveness factor	
Ω	overall catalyst effectiveness (including internal and external resistances)	
ν	stoichiometric coefficient	

SUBSCRIPTS AND SUPERSCRIPTS

α, γ	powers in Langmuir-Hinshelwood reaction kinetics
a, b	powers in simple rate laws
A, B, R, S	components A, B, R, and S
b	bulk
b	bubble phase in a fluidized bed
B, R, S	relating to component species B, R, S
c	coolant
cat	catalyst
cv	control volume

e	emulsion phase of a fluidized bed
eff	effective
p	process gas
i, j	components i and j
in	inlet condition for process stream
int	internal, initial or outside
m	mass transfer
out	evaluated at exit of reactor
ov	overall
p	particle or process
r	reaction
R	reference
rxn	reaction
s	surface or superficial
$tube$	tube in reactor
w	wall
z	at point z in reactor
$'''$	signifies reaction rate per unit mass of catalyst
$1,2$	inlet and outlet conditions for coolant

REFERENCES

Bondiera, J., and C. Naccache. 1991. "Kinetics of Methanol Dehydration in Dealuminated H-mordenite: Model with Acid and Basic Active Centres." *Applied Catalysis* 69: 139–148.

Davidson, J. F., and D. Harrison. 1963. *Fluidised Particles.* New York: Cambridge University Press.

Dixon, A. G., and D. L. Cresswell. 1979. "Theoretical Prediction of Effective Heat Transfer Parameters in Packed Beds." *AIChE J.* 25: 663–676.

Fogler, H. S. 2006. *Elements of Chemical Reaction Engineering,* 4th ed. Upper Saddle River, NJ: Prentice Hall.

Froment, G. F., K. B. Bischoff, and J. De Wilde. 2011. *Chemical Reactor Analysis and Design,* 3rd ed. New York: Wiley.

Gidaspow, D. 1994. *Multiphase Flow and Fluidization.* London: Academic Press.

Grace, J. R. 1986. "Contacting Modes and Behavior Classification of Gas—Solid and Other Two-Phase Suspensions." *Can J Chem Eng* 64: 353–363.

Kunii, D., and O. Levenspiel. 1991. *Fluidization Engineering*, 2nd ed. Boston: Butterworth-Heinemann.

Levenspiel, O. 1989. *The Chemical Reactor Omnibook.* Corvallis, OR: Distributed by OSU Book Stores.

Levenspiel, O. 1999. *Chemical Reaction Engineering*, 3rd ed. New York: Wiley.

Li, C.-H., and B. A. Finlayson. 1977. "Heat Transfer in Packed Beds—A Reevaluation." *Chem Eng Sci* 12: 1055–1066.

Mannan, S. 2014. *Lees' Process Safety Essentials: Hazard Identification, Assessment and Control.* Amsterdam: Elsevier.

May, M. G. 1959. "Fluidized Bed Reactor Studies." *Chem Eng Progr* 55: 49.

Toomey, R. D., and H. F Johnstone. 1952. "Gas Fluidization of Solid Particles." *Chem Eng Prog* 48(5): 220–226.

Van Deemter, J. J. 1961. "Mixing and Contacting in Gas-Solid Fluidized Beds." *Chem Eng Sci* 13: 143–154.

Van Swaaij, W. P. M., and F. J. Zuiderweg. 1972. "Investigation of Ozone Decomposition in Fluidized Beds on the Basis of a Two-Phase Model." In *Proc. 5th Eur. and 2nd Int. Symp. Chem. React. Eng.*, B9-25-36. Amsterdam: Elsevier.

Wakao, N., and S. Kaguei. 1982. *Heat and Mass Transfer in Packed Beds.* New York: Gordon and Breach.

PROBLEMS

Short Answer Problems

1. A reaction between two components occurs in the gas phase and produces a single third component according to the following reaction:

$$A+B \rightleftharpoons C$$

   a. Write the form of the reaction kinetics for the forward and reverse reactions assuming the reactions are elementary in form.

   b. If the reaction in the forward direction is highly exothermic, and the reaction is at equilibrium at a given temperature T_1, what is the effect on equilibrium of increasing the temperature?

2. Comment on the following statement: "For reactions in equilibrium, if the forward reaction is exothermic, then an increase in temperature reduces the conversion, which means that the forward reaction rate decreases with temperature."

3. Comment on the following statement: "For highly exothermic reactions, some form of heat exchange should be integrated within the reactor."

4. Comment on the following statement: "For an exothermic reaction, if the temperature is increased at constant pressure with a fixed reactor size, the conversion decreases."

5. Sketch the T-Q diagram for an endothermic reaction in a PFR for

   a. Countercurrent flow of HTM (heat transfer medium)

   b. Cocurrent flow of HTM

   What are the consequences on reactor size of choosing each configuration? Explain your answer by examining and discussing trends.

Problems to Solve

6. It is known that for a certain third-order, gas-phase reaction, the rate of reaction doubles when the temperature goes from 250°C to 270°C. The form of the rate equation is

$$-r_A = k_o e^{-\left(\frac{E}{RT}\right)} c_A c_B^2$$

If the rate of reaction is 10 moles/m³/s at the base conditions, which are 20 atm pressure and 250°C, and an equal molar flow of A and B are in the feed (no inerts), answer the following questions:

   a. Compared to the base case, by how much does the rate of reaction (at inlet conditions) change if the temperature is increased to 260°C?

   b. Compared to the base case, by how much does the rate of reaction (at inlet conditions) change if the pressure is increased by 15%?

7. Consider the following reaction[a] for the production of benzene via the homogeneous thermal dealkylation of toluene in the temperature range 700°C to 950°C:

$$C_7H_8 + H_2 \rightarrow C_6H_6 + CH_4$$

   toluene   benzene

$$-r_{tol} = 3.0 \times 10^{10} e^{-\frac{25,614}{T[K]}} c_{tol} c_{H_2}^{0.5} \text{ [mol/m}^3 \text{ of reactor/s]}$$

where the concentrations of reactants are in mol/liter reactor and the temperature is in K. The heat of reaction is −52.0 kJ/mol (at 900°C) and −50.5 kJ/mol (at 700°C), and the average specific heats of the reactor feed and effluent are 3.3216 kJ/kg/K (at 950°C) and 3.042 kJ/kg/K (700°C).

   a. If feed enters the reactor at 700°C and 25 bar, solve the material and energy balances for a plug flow reactor to determine the volume of reactor needed to give 90% conversion of toluene. The feed to the reactor comprises 80 kmol/h of toluene and 320 kmol/h of hydrogen. Plot the conversion and temperature profiles in the reactor.

   b. Compare your results for Part (a) with the results from a commercial simulator. Use the Soave-Redlich-Kwong and latent heat methods for the vapor-liquid equilibrium and enthalpy packages, respectively.

8. For the solution to Problem 4.7(a), and assuming that the reactor is packed with inert ceramic spheres, 5 mm in diameter with a voidage of 0.45 and the reactor has a *Length/Diameter* ratio of 8:1 determine the pressure drop across the reactor. For this problem use an average viscosity of the process gas of 26.8×10^{-6} kg/m/s and an average gas density calculated from the ideal gas law. Comment on the accuracy of using these average property values to determine the pressure drop across the reactor.

9. Consider the production of cumene (c) from the catalytic alkylation reaction of benzene (b) using propylene (p) – Reaction 1. A second undesirable reaction (Reaction 2) between propylene and cumene occurs to produce p-diisopropyl benzene (DIPB):

a   Zimmerman, C. C., and R. York, "Thermal Demethylation of Toluene," *Ind. Eng. Chem. Proc. Des. Dev.* **3**: 254–258 (1964).

$$C_3H_6 + C_6H_6 \xrightarrow{k_1} C_9H_{12}$$

$$\text{p} \qquad \text{b} \qquad \text{c}$$

$$C_3H_6 + C_9H_{12} \xrightarrow{k_2} C_{12}H_{18}$$

$$\text{p} \qquad \text{c} \qquad \text{DIPB}$$

where $r_1 = k_1 c_p c_b$ and $r_2 = k_2 c_p c_c$ and the rates of reaction are given in mol/L/s and the rate constants are $k_1 = 2.8 \times 10^7 \exp\left(-\dfrac{12{,}530}{T[K]}\right)$ and $k_2 = 2.32 \times 10^9 \exp\left(-\dfrac{17{,}650}{T[K]}\right)$. These reaction are quite exothermic, and typical reactor temperatures are in the range 350°C to 410°C. The heats of reaction per mol of product in this temperature range are −100.6 kJ/mol and 121.3 kJ/mol for Reactions 1 and 2, respectively. The heat capacities for the reactant and product streams may be taken to be the same and equal to 2.44 kJ/kg/K in this range. For this system, do the following:

a. Determine the conversion of an equimolar feed of propylene and benzene (feed conditions are 350°C, 3 MPa, and 100 mol/s) if the reactor is run as an adiabatic packed bed with a maximum outlet temperature of 425°C.

b. Based on the results from Part (a), is it feasible to obtain an overall conversion of 80% of the benzene using a series of staged packed beds with cooling between stages to bring the temperature back to 350°C at the inlet of the next reactor stage? Note: In practice, no more than four stages would be used.

10. Reexamine Problem 9 for the following cases:

a. Design a packed bed reactor with intercooling (shell-and-tube design) using a cooling medium with a phase change (T_{boil} = 350°C) to obtain 80% overall conversion of benzene. Assume that the catalyst tubes are 2-in in diameter and 3 m in length packed with 5 mm diameter spherical catalyst particles of density, 2350 kg/m^3, and voidage = 0.5. The properties of the process gas may be taken as constant with the viscosity as 17×10^{-6} kg/m/s and the thermal conductivity as 0.0525 W/m/K. For this case, you need to determine the number of tubes needed to give the desired conversion of benzene. For your design, the hot spot in the reactor must not exceed the maximum value of 425°C.

b. Repeat Part (a) above using a process simulator.

11. For the stirred tank reactor in Example 4.8, determine whether the chosen operating/design point (T = 93.5, X = 80%, $T_{C,1}$ = 20°C, $T_{C,2}$ = 30°C, and A = 3.96 m^2) represents a stable operating condition.

12. For the case of methanol synthesis discussed in Example 4.2, determine the appropriate expression for the equilibrium constant $K(T)$ for the case when inerts are present in the feed along with the stoichiometric amounts of CO and hydrogen. What is the effect of adding 50% inerts to the feed for the case when the temperature is 500 K and the pressure is 100 bar?

13. Using the data from Example 4.3, determine the volume of catalyst needed to convert 80% of a feed of pure methanol (at a rate of 12,500 kg/h) at 14 atm and 250°C in an adiabatic packed bed reactor for the following conditions:

a. When pore diffusion effects are ignored.

b. Taking account of pore diffusion—note that the value of M_T and the catalyst effectiveness should be evaluated along the length of the reactor.

Hint: You may assume that the following gas properties (except gas density) are constant along the length of the reactor: c_p = 2.10 kJ/kg/K, μ = 1.85 × 10^{-5} kg/m/s, k = 0.049 W/m/K, ρ_{in} = 10.93 kg/m^3, $-\Delta H_r$ = 12,000 J/mol, D_p = 3 mm, and voidage, ε = 0.45.

14. In Example 4.9, the temperature and conversion profiles along the length of a plug flow reactor were generated for the production of phthalic anhydride from o-xylene assuming an overall heat transfer coefficient of 88.9 W/m^2/K. The accuracy of heat transfer correlations are known to vary significantly. Assuming that the reactor was designed for an inlet partial pressure of o-xylene of 0.0165 bar, determine the following:

    a. The temperature and conversion profiles for a catalyst filled reactor tube 1 m long when the heat transfer coefficient is 10% greater than predicted by the correlation.

    b. The temperature and conversion profiles for a catalyst filled reactor tube 1 m long when the heat transfer coefficient is 10% lower than predicted by the correlation.

    c. Based on these results, how would you determine the actual reactor design needed to avoid a temperature runaway?

15. In Example 4.9, the temperature and conversion profiles along the length of a plug flow reactor were generated for the production of phthalic anhydride from o-xylene assuming a tube diameter of 1-in (25.4 mm). Repeat the problem for an inlet partial pressure of o-xylene of 0.0165 bar and a 2-in tube. Compare your results to those for the 1-in tube and discuss the differences.

16. In Example 4.9, the temperature and conversion profiles along the length of a plug flow reactor were generated for the production of phthalic anhydride from o-xylene assuming a particle size of 3 mm. Repeat this problem for an inlet partial pressure of o-xylene of 0.0165 bar using the following catalyst particle sizes, and determine the pressure drop across the bed. You may assume the same average gas properties as given in the example.

    a. 5 mm

    b. 2 mm

17. In Example 4.9, the secondary combustion reaction of o-xylene was ignored. If this reaction can be approximated by the following reaction rate, rework the example for inlet partial pressures of o-xylene (Component A) of 0.015, 0.016, 0.0170, and 0.0175 bar.

$$C_8H_{10} + \frac{21}{2}O_2 \rightarrow 8CO_2 + 5H_2O$$

o-xylene

$$-r_A'''[\text{mol/h/m}^3\text{-cat}] = 2 \times 10^{12} e^{-25,000/T[K]} p_A[\text{bar}]$$

$$\Delta H_{comb} = 4375 \text{ kJ/mol}$$

Do the results change significantly from those given in Example 4.9?

Are your conclusions surprising, and could you have determined these results by simply comparing the basic kinetic equations?

18. One technique to control the generation of a hot spot at the beginning of a reactor is to load the front end of the reactor with a diluted catalyst and then to load the remainder of the reactor with the undiluted catalyst. For the phthalic anhydride example given in Examples 4.9, 4.10, and 4.11, it has been decided to load the front end of the reactor tubes with a catalyst containing 5% inactive material (f = 0.95) and the remainder with fully active catalyst (f = 1.0). For a 1 m length, 25.4 mm diameter tube, what is the maximum

conversion of o-xylene ($p_{A,in}$ = 0.0175 bar) that can be achieved without exceeding the maximum catalyst temperature of 410°C?

19. In the production of cumene from propylene, the following elementary, vapor-phase, irreversible reaction takes place:

$$C_3H_6 + C_6H_6 \xrightarrow{k_1} C_9H_{12}$$
$$\text{propylene} \quad \text{benzene} \quad \text{cumene}$$

The reaction rate is given by

$$r_1 = k_1 c_p c_b \text{ mol}/(\text{g cat})/\text{s and } k_1 = 3.5 \times 10^4 \exp\left(\frac{-12{,}530}{T(K)}\right)$$

The feed to a fluidized bed reactor consists of an equal ratio of benzene and propylene. The reaction takes place in a fluidized bed reactor operating at 300°C and 3 MPa pressure. For this problem, the fluidized bed may be assumed to be a constant-temperature CSTR reactor.

At design conditions, you may assume that side reactions do not take place to any great extent and the conversion is 68%.

It is desired to scale up production by 25% and all flows to the reactor will increase by 25% at the same feed concentration. Determine the following:

a. What is the single-pass conversion if the process conditions and amount of catalyst remain unchanged?

b. What percentage change in catalyst would be required to achieve the scale-up assuming that the pressure, temperature, and conversion were held constant?

c. Estimate how much the temperature would have to be changed (without changes in catalyst amount, operating pressure, or conversion) to achieve the desired scale-up.

d. By how much would the pressure have to be changed (without changes in catalyst amount, operating temperature, or conversion) to achieve the desired scale-up?

20. Consider a liquid-phase reaction occurring in a constant-volume, isothermal, batch reactor.

a. For a first-order decomposition, what is the ratio of the time to reach 75% conversion to the time to reach 50% conversion?

b. For a first-order decomposition, what is the ratio of the time to reach 90% conversion to the time to reach 50% conversion?

c. Repeat Parts (a) and (b) for a second-order reaction between reactants initially in equimolar quantities.

d. Explain the results for Parts (a), (b), and (c).

21. In an isothermal batch reactor, a first-order, irreversible, liquid-phase reaction, A → B, occurs. The initial concentration is C_{Ao}.

a. By what percentage must the reaction time be increased to raise the conversion of A from 0.9 to 0.95?

b. If the initial conversion is 0.8, what will the final conversion be if the reaction time is increased by 50%?

22. Repeat Problem 21 for a catalytic/enzymatic reaction with a rate expression of the form

$$-r_A = \frac{k_1 c_A}{1 + k_2 c_A}$$

where c_{A0} = 10 mol/L, k_1 = 3.5 s^{-1}, and k_2 = 0.45 L/mol

23. It is known that for a certain second-order, elementary, gas-phase reaction (first order in A, first order in B), the rate of reaction increases by 50% when the temperature goes from 250°C to 280°C. In a laboratory setup, a plug flow reactor operates at essentially isothermal conditions (250°C) and at an operating pressure of 20 atm. For a fixed inlet flowrate (equal molar flows of A and B in the feed with no inerts), the measured conversion of A is 57%.

$$A+B \rightarrow C+D$$

  a. Compared to the base case, by how much does the conversion change if the temperature is increased to 265°C?

  b. Compared to the base case, by how much does the conversion change if the pressure is increased by 20%?

24. Consider a liquid-phase reaction occurring in a constant-volume, isothermal, CSTR.

  a. For a first-order decomposition, what is the ratio of the volume of reactor needed for 75% conversion to the volume needed for 50% conversion?

  b. For a first-order decomposition, what is the ratio of the volume of reactor needed for 90% conversion to the volume needed for 50% conversion?

  c. Repeat Parts (a) and (b) for a second-order reaction between reactants initially in equimolar quantities.

  d. Explain the results of Parts (a), (b), and (c).

25. Consider the dehydration of isopropyl alcohol (IPA) to yield acetone and hydrogen:

$$(CH_3)_2CHOH \overset{k_1}{\rightarrow} (CH_3)_2CO + H_2$$
$$\text{IPA} \qquad\qquad \text{acetone}$$

This reaction is endothermic, with a heat of reaction of 57.2 kJ/mol. The reaction is kinetically controlled and occurs in the vapor phase over a catalyst. The reaction kinetics for this reaction are first order with respect to the concentration of alcohol and can be estimated from the following equation:[b,c]

$$-r_{IPA} = k_0 \exp\left[-\frac{E_a}{RT[\text{K}]}\right] c_{IPA} \text{ kmol/m}^3\text{(catalyst)/s}$$

where $E_a = 72.38$ MJ/kmol, $k_0 = 1.931 \times 10^5$ m^3(gas)/m^3(reactor)/s, and c_{IPA} kmol/m^3(gas)

The feed to the reactor is 87 wt% IPA in water at 240°C and 2 bar. Spherical catalyst particles (5 mm in diameter) are placed in 50 mm ID tubes of length 6 m. The heat transfer coefficient on the tube side of the reactor is limiting. A heat transfer medium (HTM) is available as part of a heating loop that supplies the HTM to the reactor shell at 400°C.

Properties of the heat transfer fluid are

$$c_p = 2800 \text{ J/kg/°C}, \rho = 680 \text{ kg/m}^3, k = 0.078 \text{ W/m/K}, \mu = 0.00012 \text{ kg/m/s}$$

Properties of the process gas are

$$c_{p,p} = 2200 \text{ J/kg/°C}, k_p = 0.055 \text{ W/m/K}, \mu_p = 1.48 \times 10^{-5} \text{ kg/m/s}$$

b McKetta, J. J., and W. A. Cunningham (Ed.), *Encyclopedia of Chemical Processing and Design*, vol. 1 (New York: Marcel Dekker, 1976), 314–362.

c Sheely, C. Q., *Kinetics of Catalytic Dehydrogenation of Isopropanol*, Ph.D. Thesis, University of Illinois, 1963.

Properties of catalyst are

$$\text{Voidage} = 0.45,\ \rho_{cat} = 2400\ \text{kg/m}^3$$

a. Determine the conversion of alcohol to acetone assuming that the HTM flows cocurrently with the process gas. Assume that the HTM flows at a rate of 200,000 kg/h, the inlet flowrate to the reactor is 8,000 kg/h of IPA and water, and the reactor has a total of 1600 tubes arranged on a square pitch with the tube centers 75 mm apart.

b. Sketch the temperature concentration profiles in a single reactor tube.

26. For the setup in Problem 25, do the following,

a. Reconfigure the utility flow to be countercurrent with the process gas and determine the conversion of IPA.

b. Sketch the temperature concentration profiles in a single reactor tube.

27. For the setup in Problem 25, do the following,

a. Determine the change in IPA conversion if the flowrate to the reactor increases by 25% but the utility flow does not change.

b. Determine the change in utility flowrate that must accommodate the change in feed gas rate in order to maintain the same conversion of IPA as in Problem 25.

28. Determine the pressure drop across the reactor tubes for Problem 25.

29. The catalytic decomposition of ozone in fluidized beds of silica particles impregnated with Fe_2O_3 were studied by Van Swaaij and Zuiderweg.[d] The decomposition reaction for a given temperature and catalyst loading was shown to follow a simple first-order model:

$$2O_3 \xrightarrow[Fe_2O_3]{k_1} 3O_2 \ \text{ and } \ k_1 = 1\left[s^{-1}\right]$$

These researchers report the following experimental results:

ε (voidage in emulsion phase) = 0.45

$H_{bed} = 2.4$ m

$u_s = 0.15$ m/s (superficial velocity)

$u_{mf} = 0.01$ m/s

$u_{bubble} = 1.62$ m/s

For the case where the interchange coefficient between the bubble and emulsion phase is zero (i.e., gas in bubbles does not interact with the catalyst), determine the conversion of pure ozone feed when the emulsion phase is completely well mixed.

30. Repeat Problem 29 for the case when the gas flows through the emulsion phase in plug flow.

31. For the laminar flow reactor problem given in Example 4.6, determine the volume of the reactor needed to obtain a conversion of 95%.

32. For the laminar flow reactor problem given in Example 4.6, determine the flowrate of feed needed to increase the conversion to 95%.

33. For the laminar flow reactor problem given in Example 4.6, determine conversion in the reactor when the flowrate of feed is increased by 25%.

d   van Swaaij, W. P. M., and F. J. Zuiderweg, "Investigation of Ozone Decomposition in Fluidised Beds on the Basis of a Two-Phase Model," *Chem. React. Eng., Proc. Eur. Symp.* B, B9-25-36, 1972.

CHAPTER
5
Other Equipment

WHAT YOU WILL LEARN

- The basic equations governing the design of pressure vessels
- How to estimate the size and mass of pressure vessels made from common materials
- How to determine the length and diameter of knockout drums, including vapor-liquid (V-L), liquid-liquid (L-L), and L-L-V separators
- How to estimate the pressure drop, flooding condition, and droplet size distribution for mist eliminators
- How to predict the performance of phase separators
- The key design parameters for the design of liquid-liquid separators
- The principles of operation of steam ejectors
- How to specify single- and multistage steam ejectors
- How to predict the performance of steam ejectors

5.0   INTRODUCTION

The purpose of this chapter is to introduce the concepts necessary for a process engineer to perform preliminary analyses and obtain preliminary designs of pressure vessels, simple phase-separators or knockout drums, and steam ejectors. Pressure vessels are used nearly everywhere in the process industry to contain fluids safely at pressures above atmospheric. Knockout drums are used extensively in the process industry to allow different phases to separate prior to feeding to downstream processing equipment. The basic principle is to provide enough volume and residence time to allow a dispersed phase to separate from a continuous phase, usually under the influence of gravity. Steam ejectors are widely used to maintain vacuum conditions in process equipment through the expansion of low- to medium-pressure steam (or air) through a nozzle and diffuser (ejector).

5.1   PRESSURE VESSELS

A pressure vessel is defined as any container that has a pressure differential between the inside and the outside (Chattopadhyay, 2005). The American Society of Mechanical Engineers (ASME) Boiler and Pressure Vessel Code, Section VIII (2015) specifies design methods for vessels over a range of pressures from 100 kPa to 30 MPa (~1–300 atm). By contrast, the American Petroleum Institute (API) defines storage tanks as tanks that are designed with internal pressures to accommodate no more than that generated by the static head of the fluid in the tank. In this chapter, the focus is on the design of pressure vessels covered by the ASME code.

The design of pressure vessels is a very involved process that requires complex analysis of stress and strain relationships in all directions within the shell of the vessel. These stress patterns may be very complex, requiring sophisticated software to calculate the forces in the shell. Moreover, when the vessel undergoes repeated cycling of pressure and/or temperature, the history of this cycling has an effect on the vessel integrity. All these detailed calculations and analyses must be performed by mechanical engineers with specific expertise. However, for the purposes of this book, a simplified approach to vessel sizing is given. Such an analysis should provide the chemical or process engineer with enough information to determine the approximate shell thickness and the thicknesses of the heads or end caps. With such information, an estimate of the weight of the vessel and an approximate capital cost can be made.

5.1.1   Material Properties

Vessels for chemical processes are nearly always made from metal. However, for some small-scale plants and some pharmaceutical and food processing plants, polymer tanks may be used. The discussion here is restricted to metal vessels. There are many mechanical properties that effect the design of a pressure vessel, including tensile strength, ductility, corrosion resistance, and fracture toughness. Here, tensile strength and corrosion resistance are emphasized. In Table 5.1, the maximum allowable tensile stress (S_{max}) for some common carbon, low-alloy, and high-alloy steels are given; for their compositions and other properties, the ASME Boiler and Pressure Vessel Code, Section II—Material Properties should be consulted. In Table 5.2, the values for S_{max} for some other common metals are given.

Corrosion resistance of metals is a complex subject and, again, beyond the scope of the current text. However, Table 5.3, from Sandler and Luckiewicz (1987), provides a cursory summary of material suitability for some common chemicals and processes.

5.1.2   Basic Design Equations

As noted by Chattopadhyay (2005), there are two basic philosophies that can be applied to the design of pressure vessels: design by rule and design by analysis. The approach used here is the design-by-rule philosophy; in other words, a simple equation-based approach is adopted.

5.1.2.1   Cylindrical Shells
For moderately thick shells ($t \leq 0.25D_i$), Section III, Division 1, of the ASME Boiler and Pressure Vessel Code gives the following formula for the thickness of a shell (t_{shell}) to withstand an internal pressure (P in gauge pressure) in a vessel of internal diameter D_i:

$$t_{shell} = \frac{PD_i}{2SE - 1.2P} \tag{5.1}$$

where S is the allowable stress taken to be equal to S_{max} from Tables 5.1 and 5.2, and E is a welded joint efficiency. The need for the term E arises because vessels are most often constructed by

Table 5.1   Maximum Allowable Tensile Stress (MPa) of Carbon and Low- and High-Alloy Steels Carbon Steel

T [°C] (°F)	SA515-55	SA515-70	SA516-55	SA516-70	SA285-A	SA285-B	SA285-C
−29 (−20)	94.4	120.6	94.4	120.6	77.2	86.1	94.4
343 (650)	94.4	120.6	94.4	120.6	77.2	86.1	94.4
371 (700)	91.0	114.4	91.0	114.4	75.8	83.4	91.0
427 (800)	70.3	82.7	70.3	82.7	62.0	66.2	70.3
482 (900)	44.8	44.8	44.8	44.8	44.8	44.8	44.8
538 (1000)	17.2	17.2	17.2	17.2			

Low-Alloy Steel			
T [°C] (°F)	SA202-A	SA202-B	SA387
−29 (−20)	128.9	146.1	103.4
343 (650)	128.9	146.1	103.4
371 (700)	122.0	136.4	103.4
427 (800)	86.8	88.2	103.4
482 (900)	44.8	44.8	90.3
538 (1000)	17.2	17.2	19.3
593 (1100)			28.9
649 (1200)			11.0

High-Alloy Steel				
T [°C] (°F)	SA-240	SA-241	SA-242	SA-243	SA-244
Stainless steel grade	**304**	**304L**	**310S**	**316**	**410**
−29 (−20)	128.9	107.5	128.9	128.9	111.6
38 (100)	128.9	107.5	128.9	128.9	111.6
93 (200)	107.5	91.7	116.5	110.9	106.1
204 (400)	88.9	68.9	102.7	91.7	99.2
371 (700)	75.8	64.1	87.5	77.9	90.3
482 (900)	69.6		79.9	74.4	71.7
538 (1000)	66.8		67.5	73.0	44.1
593 (1100)	60.6		34.5	71.0	20.0
649 (1200)	41.3		17.2	51.0	6.9
704 (1300)	25.5		4.8	28.3	
760 (1400)	15.8		2.1	15.2	
816 (1500)	9.6		1.4	11.7	

Source: Excerpted from Perry and Chilton (1973).

Table 5.2   Maximum Allowable Tensile Stress (MPa) of Other Alloys and Nonferrous Metals

T [°C] (°F)	Hastelloy® C-276	Nickel 200	Monel® 400	Inconel® 625	Inconel 825	Incoloy® 800
	SB-575	SB-160	SB-127	SB-443	SB-424	SB-409
−29 (−20)	172.3	68.9	128.2	189.5	146.1	111.6
38 (100)	172.3	68.9	128.2	189.5	146.1	111.6
93 (200)	172.3	68.9	113.0	189.5	146.1	111.6
204 (400)	148.2	68.9	102.0	184.7	146.1	111.6
316 (600)	129.6	68.9	101.3	175.0	146.1	110.3
371 (700)	122.7		101.3	172.3	144.7	108.2
427 (800)	117.8		97.9	169.5	143.3	105.4
482 (900)	114.4		55.1	165.4	141.3	102.0
538 (1000)	113.7			163.3	135.8	99.2
593 (1100)	103.4			161.3		79.9
621 (1150)	84.1			144.7		64.1
649 (1200)	67.5			91.0		51.0
704 (1300)						40.7
760 (1400)						20.7
816 (1500)						13.1
871 (1600)						8.3
899 (1650)						6.8

T [°C] (°F)	Titanium	Titanium		Aluminum 1060	Aluminum 1060	Aluminum 1060	Aluminum 1060
	ASME Grade 2	ASME Grade 9	Temper	0	H112	H112	H112
			Thickness (inch)	0.051–3.0	0.25–0.499	0.5–1.0	1.001–3.0
38 (100)	86.1	155.1		11.0	18.6	17.2	15.2
66 (150)	82.7	155.1		11.0	17.9	16.5	14.5
93 (200)	75.1	149.5		11.0	16.5	14.5	13.1
121 (250)	68.2	143.3		9.6	13.8	13.1	11.7
149 (300)	62.0	136.4		8.3	12.4	11.0	9.6
177 (350)	57.9	128.2		6.9	11.0	9.6	6.9
204 (400)	53.1	121.3		5.5	6.9	6.9	5.5
260 (500)	45.5	108.9					
316 (600)	39.3	104.1					

Source: Data excerpted from ASME Boiler and Pressure Vessel Code, 1995, Section II, Parts A–D.

Table 5.3 Corrosion Characteristics for Some Materials of Construction

Chemical Component	Carbon Steel	304 Stainless Steel	316 Stainless Steel	Aluminum	Copper	Brass	Monel	Hastelloy C	Titanium	TFE	Graphite
Acetaldehyde	N		A			C		A	A	A	A
Acetic acid, glacial	N		A	A	A	C	B	A	A	A	A
Acetic acid, 20%	N	A	A	A	A	C	B	A	A	A	A
Acetic anhydride	N	A	B	A	A	C		A	A	A	A
Acetone	A	A	A	A	A	A	A	A	A	A	A
Ammonia, 10%	C	A	A	C	N	N	N	A	A	A	A
Aniline	A	A	A	N	N	N	A	A	A	A	A
Aqua regia	N	N	N	N	N	N	N	C	A	A	
Benzaldehyde		A	A	A	A	A	A	A	A	A	A
Benzene	A	A	A	A	A	A	A	A	A	A	A
Benzoic acid		C	A					A	B	A	A
Furfural	A	C	C	A	A	A	A	A	A	A	A
Gasoline	C	A	A	A	A	A	A	A	N	A	A
Heptane	A	A	A	A	A	A	A	A	A	A	A
Hexane		A	A	A			A	A	A	A	A
HCl, 0–25%	N	N	N	N	C	N	C	C	C	A	A
HCl, 25–37%	N	N	N	N	C	N	C	C	C	A	A
HF, 30%	N	B	B	N	N	N	A	A	N	A	A
HF, 60%	N	B	B	N	N	N	A	A	N	A	A
H2O2, 30%	C	C	A	C	C	N	C	A	A	A	A
H2O2, 90%	C	C	A	C	C	N	C	A	A	A	A
H2S, aqueous	C	C	A	A	N	N	N	A	A	A	A
Maleic acid		A	A	A	A	A	A	A	A	A	A
Methanol	A	A	A	A	A		A	A	A	A	A

(continued)

Table 5.3 Corrosion Characteristics for Some Materials of Construction *(Continued)*

Chemical Component	Carbon Steel	304 Stainless Steel	316 Stainless Steel	Aluminum	Copper	Brass	Monel	Hastelloy C	Titanium	TFE	Graphite
Methyl chloride		A	A	N			A	A	A	A	A
Methyl ethyl ketone	A	A	A	A	A		A	A	A	A	A
Methylene chloride		A	A		N		N	A	A	A	A
Naphthalene		A	A	A			A	A	A	A	A
Nitric acid, 10%	N	A	A	B	N		N	A	A	A	A
Nitric acid, 50%	N	C	C	B	N		N	A	A	A	N
Oleic acid	C	A	A	A	C		A	A	A	A	A
Oxalic acid	C	C	B	C	C		A	A	A	A	A
Phenol	N	C	C	B	N		A	A	A	A	A
Phosphoric acid, 0–50%	C	C	C	N	C		C	A	B	A	A
Phosphoric acid, 51–100%	C	C	C	N	C		C	A	B	A	A
Propyl alcohol		A	A	A	A			A	A	A	A
Sodium hydroxide, 20%	A	A	A	N	C	N	A	A	A	A	A
Sodium hydroxide, 50%	A	A	A	N	C	N	A	A	A	A	A
Stearic acid		A	A	A	A		B	A	A	A	A
Sulfuric acid, 0–10%	N	N	N	N	N		C	A	B	A	A
Sulfuric acid, 10–75%	N	N	N	N	N		C	A	C	A	A
Sulfuric acid, 75–100%	N	N	N	N	N		C	C	N	A	A
Tartaric acid		A	A	A	A		C	A	A	A	A
Toluene	A	A	A	A	A		A	A	A	A	A
Urea		A	A	A			A	A	A	A	A
Xylene		A	A					A	A	A	A

A = acceptable; B = acceptable up to 30°C; C = caution, use under limited conditions; N = not recommended; no entry = information is not available. (Reproduced from Sandler and Luckiewicz, *Practical Process Engineering, a Working Approach to Plant Design*, with permission of XIMIX, Inc., Philadelphia, 1987.)

Source: Reproduced by permission from Sandler and Luckiewicz (1987).

welding curved plates of metal together, and the welded joint generally will not be as strong as the metal plate. Values of E, a dimensionless quantity, range from 1.0 (100%) for double-welded butt joints that are fully radiographed to about 0.6 (60%) for single-welded butt joints without backing strips and without radiographing. The types of welds are illustrated in Figure 5.1, and suggested efficiencies from Perry and Chilton (1973) are given in Table 5.4.

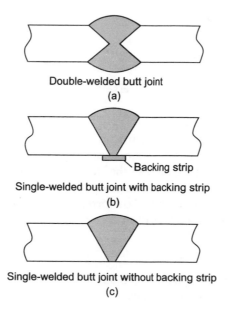

Double-welded butt joint
(a)

Single-welded butt joint with backing strip
(b)

Single-welded butt joint without backing strip
(c)

Figure 5.1   Different types of welds (Used by permission from Perry and Chilton [1973])

Table 5.4   Efficiencies of welds to be used in Equation (5.1)

Type of Weld	Efficiency, E
Double-Welded Butt Joints	
Fully radiographed	1.0
Spot radiographed examination	0.85
No radiographed examination	0.70
Single-Welded Butt Joints with Backing Strip	
Fully radiographed	0.90
Spot radiographed examination	0.80
No radiographed examination	0.65
Single-Welded Butt Joints without Backing Strip	
No radiographed examination	0.60
Source: Data excerpted from Perry and Chilton (1973).	

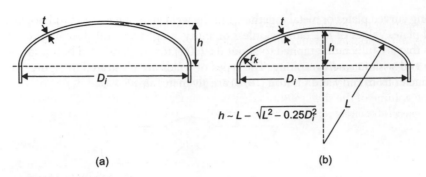

Figure 5.2   Examples of vessel heads: (a) elliptical, (b) dished or torispherical

5.1.2.2   Corrosion Allowance

When specifying a material of construction for a vessel, information such as shown in Table 5.3 should be consulted. Even though a material may be designated as acceptable for a given service, some corrosion will likely occur over the lifetime of the vessel, which might be 50 years or more. It is standard practice that additional thickness above that calculated by Equation (5.1) is provided as a safety margin for material removal by corrosion. Therefore, Equation (5.1) is modified to include an additional thickness for a corrosion allowance, t_{CA}, as

$$t_{shell} = \frac{PD_i}{2SE - 1.2P} + t_{CA} \tag{5.2}$$

where typical values of t_{CA} are between 3.175 and 6.35 mm (1/8–1/4 in). In addition, the minimum thickness for any process vessel set by the ASME code is 4.763 mm (3/16 in).

5.1.2.3   Heads

A variety of heads (end sections) for vessels is possible, the most common being flanged elliptical and dished (torispherical) heads. Flat flanged, conical, hemispherical and toriconical heads are also used for special applications. Equations (5.3) and (5.4) are for elliptical and torispherical (dished) heads, respectively. The geometries and geometric parameters of these heads are illustrated in Figure 5.2 .

$$t_{head} = \frac{PD_i K}{2SE - 0.2P} + t_{CA} \quad \text{elliptical} \tag{5.3}$$

$$t_{head} = \frac{PLM}{2SE - 0.2P} + t_{CA} \quad \text{dished} \tag{5.4}$$

where D_i is the internal diameter of the vessel to which the head is attached, K is a geometric factor for elliptical heads given by $K = \frac{1}{6}\left[2 + \left(\frac{D_i}{2h}\right)^2\right]$ and for the case when $h = D_i/4$ (a 2:1 elliptical head), $K = 1$. The parameter M is called the *stress intensity factor* for dished heads and is given by $M = \frac{1}{4}\left[3 + \sqrt{\frac{L}{2r_k}}\right]$, where L is the radius of the head and r_k is the knuckle radius, as shown in Figure 5.2. A standard design known as an *ASME head* uses values of $L = D_i$ and $r_k = 0.03D_i$, which gives values of $h \cong 0.134\, D_i$ and $M = 1.77$.

5.1.2.4   Nozzles

Equation (5.2) may be used to assess the thickness of the wall of an inlet or outlet nozzle, where the value of D_i is that of the inlet/outlet pipe. A common practice when installing nozzles is to provide reinforcement for the shell around the nozzle by welding additional material onto the shell in the vicinity of the nozzle. Details of this practice are given by Chattopadhyay (2005).

5.1.2.5   Mass of Vessels and Heads

The mass of a cylindrical shell (m_{shell}) of inside diameter D_i and length (tangent-to-tangent) z is simply the volume of the shell multiplied by the density of the material of construction (ρ_m). Here the term *tangent-to-tangent* refers to the distance of the straight portion of the cylindrical shell.

$$m_{shell} = \pi t_{shell}(D_i + t_{shell})z\rho_m \tag{5.5}$$

The approximate formulae for elliptical (2:1) and dished (ASME) heads are

$$m_{elliptical,2:1} = \frac{\pi}{4}D_i^2\left[1+\left(\frac{2h}{D_i}\right)^2\left(2-\frac{2h}{D_i}\right)\right]t_{head}\rho_m = 1.375\frac{\pi}{4}D_i^2 t_{head}\rho_m a \tag{5.6}$$

$$m_{dished,ASME} = \frac{\pi}{4}D_i^2\left[1+\left(\frac{2h}{D_i}\right)^2\left(2-\frac{2h}{D_i}\right)\right]t_{head}\rho_m = 1.124\frac{\pi}{4}D_i^2 t_{head}\rho_m \tag{5.7}$$

For nozzles, manways, skirts, and other miscellaneous items, an additional 10% to 20% should be added to the mass of the shell and heads to give the approximate weight of the vessel. The densities of some common metals used in the construction of pressure vessels are given in Table 5.5.

The equations and methods for sizing pressure vessels are illustrated in Example 5.1.

Table 5.5   **Densities of Common Metals Used in Pressure Vessel Construction**

Metal	Density (kg/m³)
Carbon steel (all grades)	7,850
Low-alloy steels (all grades)	7,800
High-alloy steels (all grades)	
SS-304, SS-310, SS-316	8,000
SS-410	7,800
Hastelloy C-276	8,890
Nickel 200	8,894
Monel 400	8,830
Inconel 625	8,446
Inconel 825	8,141
Incoloy 800	7,949
Titanium Gr 2	4,510
Titanium Gr 9	4,480
Aluminum 1060	2,705

Example 5.1

A vessel is to be designed to withstand an internal pressure of 10 MPag (1450 psig) at 250°C. The tangent-to-tangent length (z) of the vessel is 15 m, and the inside diameter (D_i) is 1.52 m. The vessel is designed to have 2:1 elliptical heads and will be made of grade 316 stainless steel. Estimate the thickness of the shell and heads and the mass of the vessel. Use a corrosion allowance of 3.175 mm (1/8-in) and a joint efficiency of 0.85.

Solution

From Table 5.1 for SS 316,

S (204°C) = 91.7 MPa

S (371°C) = 77.9 MPa

Interpolating gives $S\ (250°C) = \dfrac{(250-204)}{(371-204)}(77.9-91.7)+91.7 = 87.9$ (MPa)

Determine the thickness of the shell using Equation (5.2):

$$t_{shell} = \frac{PD_i}{2SE-1.2P}+t_{CA} = \frac{(10\times10^6)(1.52)}{(2)(87.9\times10^6)(0.85)-(1.2)(10\times10^6)}+0.003175 = 0.1138 \text{ m}$$

$$t < 0.25D_i = 0.38 \text{ m}$$

From Equation (5.5), using a density of 8000 kg/m³,

$$m_{shell} = \pi t_{shell}(D_i+t_{shell})z\rho_m = \pi(0.1138)(1.52+0.1138)(15)(8{,}000) = 70{,}092 \text{ kg}$$

From Equation (5.3) with K = 1.0 for a 2:1 elliptical head,

$$t_{head} = \frac{PD_iK}{2SE-0.2P}+t_{CA} = \frac{(10\times10^6)(1.52)(1.0)}{(2)(87.9\times10^6)(0.85)-(0.2)(10\times10^6)}+0.003175 = 0.1063 \text{ m}$$

From Equation (5.6) for a 2:1 elliptical head,

$$m_{elliptical,2:1} = 1.375\frac{\pi}{4}D_i^2 t_{head}\rho_m = (1.375)\frac{\pi}{4}(1.52)^2(0.1063)(8{,}000) = 2{,}122 \text{ kg}$$

Taking account of the additional nozzles, and so on, using a factor of 15% additional mass, the total mass of the vessel is estimated to be

$$m_{vessel} \approx [70{,}092+(2)(2{,}122)](1.15) = 85{,}490 \text{ kg}$$

5.2   KNOCKOUT DRUMS OR SIMPLE PHASE SEPARATORS

The purpose of knockout drums, or more correctly, simple phase separators, is to provide a location for the disengagement of one phase from another. The most common example is the disengagement of a liquid from a gas or vapor. However, the separation of two liquid phases and indeed the separation of a gas or vapor and two liquid phases are commonly practiced. Equipment in which the separation of solids from either a gas or liquid phase are also common but are not covered here.

5.2.1   Vapor–Liquid (V–L) Separation

The separation of a liquid from a mixture of gas or vapor and liquid occurs in many instances in a chemical plant. For example, during the pressurization of a gas through multiple stages of compression, it is usual for interstage coolers to be employed between successive compression stages. The cooling of the gas reduces the volumetric flow into the subsequent compression stage and hence reduces the overall power requirement for compression. As the gas is cooled, it is possible for condensable components to liquefy, and these must be removed prior to recompression to avoid damage to the compressor. Likewise, it is common for the effluent (exit) stream from a gas-phase reactor to contain condensable components that will liquefy when this stream is cooled prior to separation in a distillation column/tower or other unit operation. Often, gases and liquids are separated prior to being fed to separation equipment to avoid slugs of liquid entering the tower and dislodging trays or temporarily flooding the tower. The reflux drum placed after the condenser in a distillation column (often) may also act as a liquid-vapor separator if noncondensable species are present in the overhead vapor or if a partial condenser is used. In these cases, the liquid is sent back to the column as reflux and may also be the overhead product. Any noncondensed vapor must be separated from the liquid to avoid cavitation in the reflux and/or product pump. Because mechanical damage to a compressor can be very expensive and potentially dangerous, the criteria for particle or droplet removal for compressor knockout drums are generally more stringent than for other applications.

Only the physics of separating the two phases (liquid and vapor) are considered in the design procedure outlined here. The liquid and vapor will generally be in or close to equilibrium with each other, and the amounts of liquid and vapor will be determined by conditions in the V-L separator. A typical vertically oriented V-L separator is illustrated in Figure 5.3. Referring to Figure 5.3, the two-phase V-L mixture enters the side of the drum through a feed nozzle and flows through an inlet device. This device could be an angled piece of metal, as shown in the diagram (called a *splash plate*), or it could be a variety of other devices, such as a welded metal box with open sides,

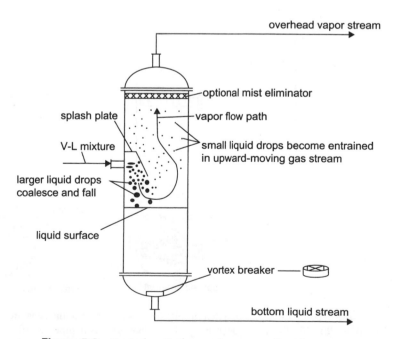

Figure 5.3   Typical vertical, two-phase vapor-liquid separator

a slotted tee distributor, a half-open pipe, a simple elbow, a tangential inlet with an annular ring, or a vane-type device sometimes called a *Schoepentoeter*™. These devices are illustrated in Figure 5.4. The purpose of the inlet device is twofold. First, it protects the opposite side of the vessel from erosion that would be caused by the high velocity V-L mixture impinging on the vessel wall. Second, it dissipates the momentum from the incoming fluid and diverts the liquid stream downward toward the surface of the accumulated liquid. The other inlet devices shown in Figure 5.4 are used

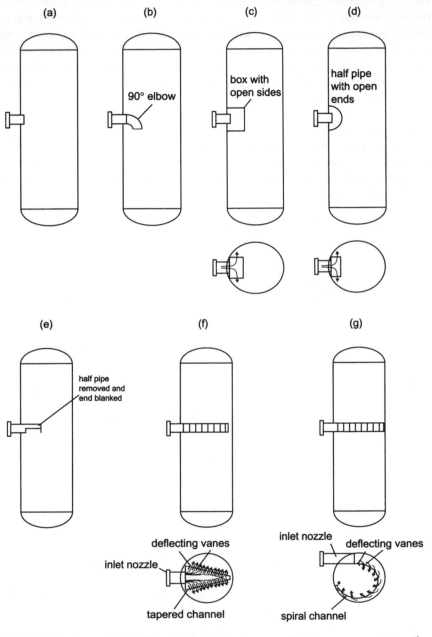

Figure 5.4   Examples of different inlet devices for vapor-liquid separators: (a) straight pipe, (b) 90° elbow, (c) open box/splash plate, (d) open pipe crossflow orientation, (e) open pipe inlet, (f) tapered channel (Schoepentoeter-type), and (g) cyclonic distributor

to provide the same effect as the simple splash plate in Figure 5.3 but are generally more efficient and can be used for higher inlet momentum. They also reduce the production of fine droplets and may be better suited for certain applications. As the vapor disengages from the liquid, it eventually moves upward toward the vapor exit at the top of the vessel. Inevitably, small liquid droplets will be entrained in the upward-moving gas stream and usually must be removed from the vapor leaving the top of the vessel. Liquid droplet removal from the gas is achieved by reducing the gas velocity, which allows the larger drops to fall under the influence of gravity back to the liquid surface via gravity settling. The addition of a mist eliminator is also often employed to capture and return very small droplets that cannot be practically removed via gravity settling. The liquid leaves from the bottom nozzle that normally has a vortex breaker attached. The vortex breaker may be a simple metal cross-welded to the top of the nozzle inside the vessel, or it could be one of several other designs. The purpose of the vortex breaker is to disrupt the vortex that is created when liquid leaves the bottom of the vessel and reduces the possibility that vapor is sucked down into the liquid stream, which becomes more likely when the liquid level drops toward the bottom of the vessel.

The criteria for using vertical or horizontal separators are given in *BN-EG-UE109 Guide for Vessel Sizing* (Red-Bag, n.d.) and are summarized in Table 5.6.

The choice of an inlet device is a balance between cost and a variety of other factors, such as reduction of droplet shattering upon entry, good distribution of gas and liquid, the ability to reduce bubble entrainment in the liquid, pressure drop, and reducing the momentum of the two-phase mixture entering the V-L separator. Common inlet devices used in V-L separators were

Table 5.6 Criteria for Choosing Vertical or Horizontal V-L Separators

Vertical Vessel
Has a small footprint and should be used when plot/site space is at a premium.
For a given flowrate and inlet condition, the separation efficiency is not a function of liquid level, unlike for horizontal vessels, where the cross-sectional area for gas flow decreases with increasing liquid level.
Generally it is easier to remove solids than in horizontal vessels where solids tend to settle out on the horizontal floor of the vessel.
Generally, the calculated required volume is lower.
Preferred orientation for reactor effluent (V-L), compressor knock out drums, gas knockout drums, and condensate flash drums.
Horizontal Vessel
Easier to accommodate large liquid slugs that can be mitigated by including a vertical baffle or liquid distributor to calm the liquid flow.
Less head room (overhead space) is required.
The liquid velocity in the downward direction is lower which improves the removal of small gas bubbles (de-gassing) and improves the breakdown of foam.
A boot can easily be added to allow the removal of small amounts of a heavier liquid phase (water).
Preferred orientation for three-phase reactor effluent separators (V-L/L), reflux drums, flare knockout drums, recycle mixing drums, and steam disengaging drums.

Source: Excerpted from *BN-EG-UE109 Guide for Vessel Sizing* (Red-Bag, n.d.).

illustrated in Figure 5.4. Table 5.7 shows the attributes for these devices. It should be noted that the higher the velocity and/or the amount of liquid in the feed, the more important it is for the distribution of the feed to be uniform. Some typical limits for inlet devices are given in Table 5.8.

The main functions/criteria for the separation of the phases in the vessel are to

- Provide a disengaging space for the two phases.
- Provide sufficient holdup of liquid to maintain stable operations in downstream equipment.
- Reduce the gas velocity sufficiently to allow any entrained liquid droplets to disengage from the gas and be collected.
- Reduce the entrainment of gas/vapor into the liquid.
- Allow sufficient time and volume for two (immiscible) liquid phases to separate based on density difference.

Table 5.7   Relative Merits of Different Inlet Devices

Inlet Device	Provides Low Inlet Pressure Drop	Prevents Entrain-ment	Minimizes Droplet Size and Breakup	Provides Good Bulk Separation of Liquid	Provides Good Gas Distribution	Reduces Bubble Entrainment in Liquid and Defoaming
Straight Pipe	+	−	−	−	−	−
Splash Plate/ Diverter Box	+	−	−	−	−	−
Half Pipe	+	o	o	o	−	−
Vane Type	+	+	+	+	+	o
Cyclonic Type	o	o/+	+	+	o	+

− = poor, o = average, + = good.

Source: From Campbell (2014).

Table 5.8   Criteria for Choosing the Appropriate Inlet Device for Vapor-Liquid Separators Based on the Average Momentum of the Inlet Flow, $<\rho u^2>_{inlet}$

Type of Inlet Device	$<\rho u^2>_{inlet}$ lb/ft/sec²	$<\rho u^2>_{inlet}$ kg/m/s²
No device (straight pipe)	<700	<1040
Diverter/splash plate	<950	<1410
Half pipe	<1400	<2080
Vane type	<5400	<8040
Cyclonic type	<10,000	<14,890

Source: Bothamley (2013).

5.2.2 Design of Vertical V–L Separators

5.2.2.1 Basic Vessel Geometry

One of the basic concerns in designing a V-L separator is the location of the liquid level alarms. Figure 5.5 illustrates the typical locations and spacing between alarm points and the relative location of nozzles and mist eliminator pads for vertically oriented vessels. The terms LLAL and HHAL used in Figure 5.5, stand for low-low alarm level and high-high alarm level, respectively. The difference between these two values represents the span of the level instrument. The normal high and low alarms (HAL and LAL) are usually set to be 90% and 10% of the span, respectively. The height of the liquid, z, and the vessel diameter, D_{ves}, are related to the desired holdup time of liquid, t_{l-h}, and the volumetric flowrate of liquid into the vessel, $\dot{v}_l$.

$$t_{l-h} = \frac{4\dot{v}_l}{\pi D_{ves}^2 z} \tag{5.8}$$

Typical values for liquid holdup in vessels are given in Table 5.9. Once the value of z has been determined using Equation (5.8) and the holdup time is known, the locations of the HHAL and LLAL can be set. It should be noted that the actual holdup of liquid in the vessel is somewhat greater than the value calculated from Equation (5.8), because the volume of liquid contained between the LLAL and lower tangent line and in the curved head at the bottom of the vessel (end section) add to the true liquid holdup. Since operation below the LLAL occurs under undesirable (emergency) conditions, the use of Equation (5.8) gives a reasonable and somewhat conservative estimate.

In determining the aspect ratio of a vessel (length-to-diameter ratio), the recommended minimum values are given for process vessels in Table 5.10. The values given in Table 5.10 should be used only after the vessel has been designed by considering the liquid holdup and gas velocity criteria, which is covered in this and subsequent sections. Often, these design procedures will determine the L/D_{ves} ratio, but consultation with Table 5.10 is recommended after the vessel design has been completed.

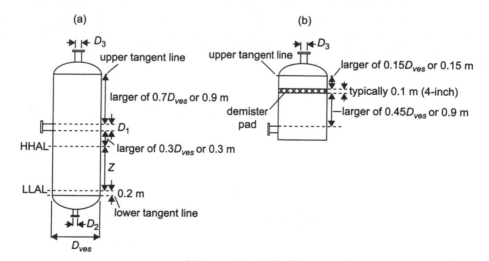

Figure 5.5 Basic dimensions of a vertical vapor-liquid separator: (a) without mist eliminator pad, (b) with mist eliminator pad. Recommended values from the *BN-EG-UE109 Guide for Vessel Sizing* (Red-Bag, n.d.) are shown in terms of a fraction of the vessel diameter or an actual minimum length.

Table 5.9   Recommended Liquid Holdup Times for Vapor-Liquid Separators for Different Process Functions

Downstream Equipment/Process Function	Holdup Time, t_{l-h} (min)
Distillation tower/column	10
Reactor	15
Heater/furnace	30
Compressor knockout drum	10
Condensate flash drum	5
Reflux drum	5
Source: From Red-Bag (n.d.) and Couper et al. (2012).	

Table 5.10   Recommended L/D_{ves} as a Function of Vessel Design Pressure

Vessel Design Pressure (bar)	L/D_{ves}
<18	≥2.5
18–36	3.0–4.0
>36	4.0–6.0
Source: From Red-Bag (n.d.).	

5.2.2.2   Estimation of Gas or Vapor Velocity—Determination of Vessel Diameter

The determination of the vessel diameter is, in part, based on reducing the gas velocity to a level whereby sufficient deentrainment of liquid drops occurs. Referring to Section 1.4.1 in Chapter 1, the terminal velocity for a (spherical) liquid droplet is given by Equation (1.45):

$$u_t^* = \left[\frac{18}{(D^*)^2} + \frac{0.591}{(D^*)^{0.5}} \right]^{-1} \tag{1.45}$$

where

$$u_t^* = u_t \left[\frac{\rho_g^2}{\mu_g(\rho_l - \rho_g)g} \right]^{1/3} \quad \text{and} \quad D^* = D_{sph} \left[\frac{\rho_g(\rho_l - \rho_g)g}{\mu_g^2} \right]^{1/3}$$

and subscripts l and g refer to liquid and gas, respectively.

Table 5.11   Terminal Velocity of Water Droplets in Air at Ambient Conditions, Using Equation (1.45)

Droplet Size (microns, μm)	Terminal Velocity (m/s)
1000	4.168
500	2.354
200	0.776
100	0.252
50	0.071
20	0.012
10	0.003

Typical values of the terminal velocity of water drops in air (at ambient conditions) calculated using Equation (1.45) are given in Table 5.11.

A reasonable criterion for choosing a vessel diameter would be to aim to remove droplets larger than some value from the gas via gravity settling using the appropriate equation for terminal velocity in Chapter 1. The choice of droplet size should lead to reasonable gas velocities (not too low) and reasonable vessel diameters (not too large). The typical drop size used for vessel sizing is 100 μm (100 $\times$ 10^{-6} m or ~0.004-in), and for droplets of this size, the Reynolds number is in the Stokes regime and the drag coefficient, C_D, is given by 24/Re, which gives rise to an analytical expression for terminal velocity.

$$u_t = \frac{D_{sph}^2 (\rho_l - \rho_g) g}{18 \mu_g} \qquad (5.9)$$

By setting D_{sph} equal to 100 μm, Equation (5.9) can be used to determine the appropriate vessel diameter by equating the velocity in the vessel with the terminal velocity of 100 μm droplets.

$$u_g = \frac{4 \dot{v}_g}{\pi D_{ves}^2} = u_t = \frac{D_{sph}^2 (\rho_l - \rho_g) g}{18 \mu_g}$$

$$\therefore D_{ves} = \sqrt{\frac{72 \dot{v}_g \mu_g}{\pi D_{sph}^2 (\rho_l - \rho_g) g}} = 15{,}300 \sqrt{\frac{\dot{v}_g [\text{m}^3/\text{s}] \mu_g [\text{kg/m/s}]}{(\rho_l - \rho_g) [\text{kg/m}^3]}} \qquad (5.10)$$

where $\dot{v}_g$ is the volumetric flowrate of gas in the vessel and D_{sph} is set to 100 μm.

Example 5.2

A vertical vessel is to be designed to remove water drops from air prior to a blower. The vessel operates at 2 bar pressure and 50°C. Determine the appropriate vessel diameter to remove 100 μm water droplets and the corresponding velocity of air in the vessel.

Data: ρ_l = 1000 kg/m³, ρ_g = 2.156 kg/m³, μ_g = 19.62 × 10⁻⁶ kg/m/s, mass flowrate of vapor = 30,000 kg/h → $\dot{v}_g$ = (30,000)/(2.156)/(3600) = 3.865 m³/s.

Solution

From Equation (5.10),

$$D_{ves} = 15,300\sqrt{\frac{\dot{v}_g \mu_g}{(\rho_l - \rho_g)}} = 15,300\sqrt{\frac{(3.865)(19.62\times10^{-6})}{(1000-2.156)}} = 4.218 \text{ m}$$

$$u_g = \frac{4\dot{v}_g}{\pi D_{ves}^2} = \frac{(4)(3.865)}{\pi(4.218)^2} = 0.2766 \text{ m/s}$$

Example 5.3

A vertical V-L separator is used to disengage benzene and toluene from a fuel gas stream at 24 bar pressure. Determine the diameter of the vessel required to ensure that no liquid droplets greater than 100 μm are entrained in the vapor stream and the corresponding gas velocity in the vessel.

Data: ρ_l = 848.5 kg/m³, ρ_g = 7.8839 kg/m³, μ_g = 5.05 × 10⁻⁶ kg/m/s, mass flowrate of vapor = 19,773 kg/h → $\dot{v}_g$ = (19,773)/(7.8839)/(3600) = 0.6967 m³/s.

Solution

From Equation (5.10),

$$D_{ves} = 15,300\sqrt{\frac{\dot{v}_g \mu_g}{(\rho_l - \rho_g)}} = 15,300\sqrt{\frac{(0.6967)(5.05\times10^{-6})}{(848.5-7.8839)}} = 0.990 \text{ m}$$

$$u_g = \frac{4\dot{v}_g}{\pi D_{ves}^2} = \frac{(4)(0.6967)}{\pi(0.990)^2} = 0.9051 \text{ m/s}$$

Use of Equation (5.10) is a reasonable approach to vessel sizing and is recommended for vertical vessels. It should be noted that the flow patterns for both liquid and vapor in V-L separators may be quite complex and that a size distribution of liquid drops with a distribution of velocities will exist in the vessel above the feed inlet. The actual distribution of liquid drops entrained in and leaving with the vapor or gas stream most likely will contain drops both smaller and larger than 100 μm even if Equation (5.10) has been used. Computational fluid dynamics (CFD) software, when used correctly, may be able to give better predictions of the size distribution of liquid drops leaving in the gas phase, and such programs are used by equipment manufacturers.

5.2.3   Design of Horizontal V-L Separators

When large liquid slugs may be present in the feed or for other reasons given in Table 5.6, the preferred orientation of the V-L separator is horizontal.

5.2.3.1   Basic Vessel Geometry

Figure 5.6 shows the basic configuration and recommended dimensions for a horizontal V-L separator.

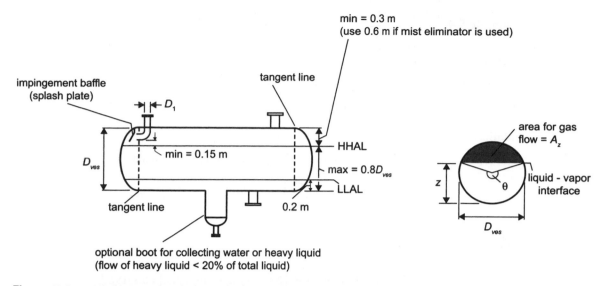

Figure 5.6   Basic dimensions of a horizontal vapor-liquid separator. Recommended values from the *BN-EG-UE109 Guide for Vessel Sizing* (Red-Bag, n.d.) are shown in terms of a fraction of the vessel diameter or an actual length.

5.2.3.2   Estimation of Gas or Vapor Velocity—Determination of Vessel Diameter

It should be noted that for horizontal vessels, the cross-sectional area for gas flow is a function of the liquid level. The cross-sectional area (A_z) above the liquid level (z) is given by

$$A_z = \frac{D_{ves}^2}{8}\left[2\pi - \theta - \sin\theta\right] \tag{5.11}$$

and

$$z = \frac{D_{ves}}{2}\left(1 - \cos\left[\frac{\theta}{2}\right]\right) \tag{5.12}$$

where θ is shown in Figure 5.6 and is measured in radians.

The area through which the disengaged gas flows is A_z; thus, the velocity of the gas above the liquid is $u_g = \dot{v}_g/A_z$. The determination of the maximum size of droplets that will be entrained in the gas is more difficult to determine than for vertically oriented vessels, and one criterion that can be used is to limit the maximum vapor velocity, $u_{g,max}$ using the following relationship:

$$u_{g,\,max} = k_{SB}\sqrt{\frac{\rho_l - \rho_g}{\rho_g}} \tag{5.13}$$

Equation (5.13) is known as the Souders-Brown equation, and the Souders-Brown parameter k_{SB} for a horizontal drum without a mist eliminator can be taken as 0.08 m/s.

Example 5.4

Consider a horizontal drum used to separate a V-L mixture. The HHAL of the liquid is set at 0.8 D_{ves} and the vapor and liquid properties are:

$\rho_l = 1000$ kg/m³, $\rho_g = 2.156$ kg/m³, $\mu_g = 19.62 \times 10^{-6}$ kg/m/s, $\dot{v}_g = 3.865$ m³/s

Determine the diameter of the drum, D_{ves}.

Solution

From Equation (5.12) and using operation at HHAL as the limiting design condition,

$$z = 0.8D_{ves} = \frac{D_{ves}}{2}\left(1 - \cos\left[\frac{\theta}{2}\right]\right) \Rightarrow \cos\left[\frac{\theta}{2}\right] = -0.6 \Rightarrow \theta = 4.4286 \text{ rad}$$

From Equation (5.11),

$$A_z = \frac{D_{ves}^2}{8}\left[2\pi - \theta - \sin\theta\right] = \frac{D_{ves}^2}{8}\left[2\pi - 4.4286 - \sin(4.4286)\right] = 0.3518D_{ves}^2$$

and Equation (5.13) yields

$$u_{g,max} = k_{SB}\sqrt{\frac{\rho_l - \rho_g}{\rho_g}} = 0.08\sqrt{\frac{1000 - 2.156}{2.156}} = 1.721 \text{m/s}$$

Setting $u_{g,max} = u_g$ gives

$$u_{g,max} = 1.721 \text{m/s} = \frac{\dot{v}_g}{A_z} = \frac{3.865}{0.3518D_{ves}^2}$$

$$\Rightarrow D_{ves} = \sqrt{\frac{3.865}{(0.3518)(1.721)}} = 2.527 \text{ m}$$

The distance between liquid surface and top of vessel = $0.2D_{ves}$ = 0.505 m > 0.3 m minimum from Figure 5.6, so design is acceptable.

An estimate of the maximum size of droplets to escape with the gas in Example 5.4 can be made if the length of the vessel is known. This technique is illustrated in the following two examples.

Example 5.5

Using the result from Example 5.4 and assuming a 10-minute holdup time for the liquid and a liquid flowrate of 35 m³/h, determine the length of the vessel.

Solution

Assume that the LLAH is located at 0.2 m above the bottom of the vessel and that the normal operating level (NOL) is halfway between the HHAL and LLAL. The NOL and LLAH are illustrated in Figure E5.5 and the hold-up time is based on the NOL.

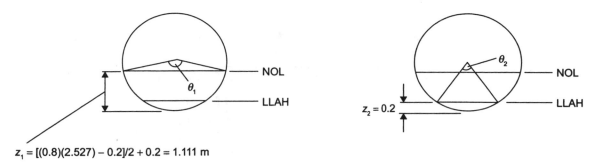

$$z_1 = [(0.8)(2.527) - 0.2]/2 + 0.2 = 1.111 \text{ m}$$

Figure E5.5   Cross-sectional view of horizontal vessel

The area contained between z_1 and z_2 in Figure E5.5 is

$$A_{liq} = \frac{D_{ves}^2}{8}\left(\theta_1 - \sin\theta_1 - \theta_2 + \sin\theta_2\right)$$

where

$$z_1 = 1.111 = \frac{D_{ves}}{2}\left(1 - \cos\left[\frac{\theta_1}{2}\right]\right) \Rightarrow \cos\left[\frac{\theta_1}{2}\right] = 1 - \frac{(2)(1.111)}{(2.527)} \Rightarrow \theta_1 = 2.900 \text{ rad}$$

and

$$z_2 = 0.2 = \frac{D_{ves}}{2}\left(1 - \cos\left[\frac{\theta_2}{2}\right]\right) \Rightarrow \cos\left[\frac{\theta_2}{2}\right] = 1 - \frac{(2)(0.2)}{(2.527)} \Rightarrow \theta_2 = 1.141 \text{ rad}$$

$$A_{liq} = \frac{(2.527)^2}{8}[2.9 - \sin(2.9) - 1.141 + \sin(1.141)] = 1.939 \text{ m}^2$$

If L is the distance between the tangent lines (the length of the vessel ignoring the end sections), then

$$LA_{liq} = t_{l-h}\dot{v}_l \Rightarrow L = \frac{t_{l-h}\dot{v}_l}{A_{liq}} = \frac{(10)(35/60)}{(1.939)} = 3.009 \text{ m}$$

and

$$L/D_{ves} = 3.009/2.527 = 1.191$$

From Table 5.10, the minimum L/D is 2.5 for this low-pressure service, so use $L = (2.5)(2.527) = 6.32$ m.

Example 5.6

Using the results from Examples 5.4 and 5.5, estimate the maximum size of droplet that will be entrained with the gas stream assuming that the liquid level is at the NOL. A diagram of the vessel is shown in Figure E5.6.

Solution

As droplets of liquid fall under the influence of gravity, they are swept along the length of the vessel by the gas. Assume that the distance between the inlet and outlet nozzles is ~L = 6.32 m, and for this liquid level, u_g = 1.256 m/s. The maximum time that the droplet can remain in the vapor before it is swept out of the vapor exit is L/u_g = 6.32/1.256 = 5.03 s. In the worst case, for a droplet not to be swept out with the gas, it must travel a vertical distance of 1.416 m in 5.03 s. Thus, the critical

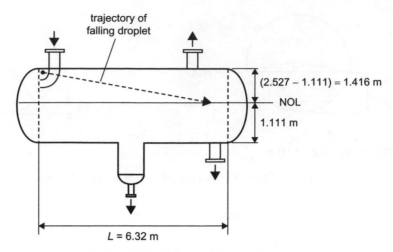

Figure E5.6   Vessel sketch with dimensions

droplet size, $D_{sph-crit}$, is one that has a terminal velocity of $1.416/5.03 = 0.2815$ m/s. Using Equation (5.9), with $u_t = 0.2815$ m/s and the data in Examples 5.4 and 5.5,

$$D_{sph-crit} = \sqrt{\frac{18 u_t \, \mu_g}{(\rho_l - \rho_g) g}} = \sqrt{\frac{(18)(0.2815)(19.62 \times 10^6)}{(1000 - 2.156)(9.81)}} = 100.8 \, \mu m$$

5.2.4   Mist Eliminators and Other Internals

Mist eliminators serve the purpose of removing droplets entrained or carried up with the vapor before leaving the top of the V-L separator. These devices are incorporated in the design for two main reasons. The first reason is to reduce the size of droplets carried out with the gas. This is especially important when the elimination of large droplets entrained in the gas stream are a critical process function, for example, with compressor knockout drums. The second reason is to reduce the overall size of the vessel by allowing higher gas velocities than those calculated from Equation (5.9) and yet still removing droplets smaller than 100 μm.

Two basic designs for mist eliminators are shown in Figure 5.7. The basic principle behind both types of device is that liquid droplets have higher momentum than the entraining gas and tend to deposit on surfaces placed in their path, whereas the gas simply moves around the surface or obstacle. The smaller the liquid droplet, the more it can maneuver around obstacles and the more it behaves like the carrier gas. Thus, the smaller the droplets that must be removed, the thicker and/or more surface area (smaller wire diameter) the mist eliminator pad area must have. In the top sketch of Figure 5.7, a mist eliminator pad is shown that is comprised of a woven/knitted mesh of material (metal or fabric) that has a high surface area but relatively low bulk density. This pad is held in position (at the top of the vessel) by two grid supports that both have a very high open area and hence have very low frictional resistance to the fluid flow. The fine liquid drops impinge on the mesh material and coalesce to form larger drops that eventually detach and fall downward via gravity through the upward-moving fluid. Mist eliminator pads are very effective in removing small liquid drops down to 10 μm in diameter and smaller but do not handle sticky liquids or solids that

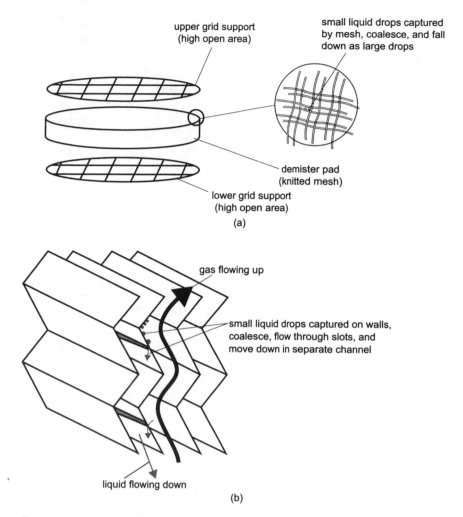

upper grid support
(high open area)

small liquid drops captured
by mesh, coalesce, and fall
down as large drops

demister pad
(knitted mesh)

lower grid support
(high open area)

(a)

gas flowing up

small liquid drops captured on walls,
coalesce, flow through slots, and
move down in separate channel

liquid flowing down

(b)

Figure 5.7 Example of (a) a mist eliminator pad and (b) a vane-type mist eliminator

may plug the wire mesh. In addition, when the gas has a high liquid content, reentrainment of coalesced drops may be problematic for mist eliminators.

In Figure 5.7(b), a typical vane-type mist eliminator is shown in which the V-L flow is sent through an array of zig-zag channels. Again, the small drops tend to impact the channel walls and form beads on the surface. These beads will coalesce and form bigger drops and start to run down the sides of the channels. Some designs incorporate openings that isolate the liquid from the gas and allow the liquid to flow downward in separate channels (as shown in Figure 5.7). Various configurations are available, and each manufacturer has designs with specific features to promote better separation, reduce the chances of liquid reentrainment, and reduce overall pressure loss in the device. Both types of device (wire mesh and vane type) can be located vertically or horizontally depending on the flow situation. Some typical configurations are shown in Figure 5.8.

5.2.4.1 Flooding in Mist Eliminators

Just as in packed and tray towers, the maximum velocity through a V-L separator containing a mist eliminator is limited by flooding. There are two common approaches to determine the design velocity of gas through a mist eliminator, which are discussed in the following sections.

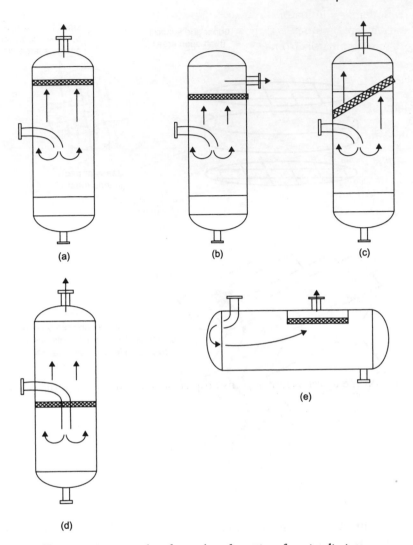

(a)   (b)   (c)

(d)   (e)

Figure 5.8   Examples of typical configurations for mist eliminators

Souders-Brown Equation

To determine the maximum velocity through a mist eliminator, the Souders-Brown (S-B) equation, Equation (5.13), with appropriate values of k_{SB} for a specific mist eliminator type and conditions of the gas and liquid, may be used. The maximum velocity is set at the flooding limit of the mist eliminator, and the design velocity is usually set at between 70% and 80% of the maximum value. Table 5.12 gives typical values for k_{SB} taken from manufacturers' literature. In using the S-B approach, there is no direct consideration of the amount of liquid being removed. Instead, by using a value of 70% to 80% of the maximum vapor velocity, it is assumed that flooding is avoided. A more rigorous approach is to check the flooding condition and then adjust the diameter to be at 70% to 80% of flooding. This approach is discussed next.

Flooding Curve

The second approach to finding the design condition for a mist eliminator is to determine the flooding point for a given set of process conditions and then to design the throughput for between 70% and 80% of flooding. This is similar to the method presented in Chapter 3 for determining the diameter of a tray or packed tower. A standardized flooding curve for two types of knitted wire mesh mist eliminators using data for water and ethylene glycol are given by Bürkholz (1989). These data are replotted and shown as Figure 5.9. Table 5.13 gives the physical properties of wire mesh mist eliminators from technical literature from Rhodius (n.d.).

Table 5.12   Values for k_{SB} for Mist Eliminators to Be Used in Equation (5.13)

Manufacturer/Product	k_{SB}	
Koch-Glitsch	**m/s**	**ft/s**
Demister®, style 421	0.11	0.361
Demister, style 709	0.13	0.427
Demister, style 931	0.102	0.335
Demister, style 708	0.12	0.394
Horizontal flow vane type, Style 250	0.35	1.15
Vertical flow vane type, style 350	0.35	1.15
ACS Separations		
Mist eliminator, horizontal, style 4CA pad	0.107	0.35
Mist eliminator, style 4CA MisterMesh® pad	0.128	0.42
Vane type, Horizontal Plate-Pak™	0.152	0.50
Vane type, Vertical Plate-Pak™	0.198	0.65

Source: From Koch-Glitsch (2015) and ACS Industries (2004) product catalogs.

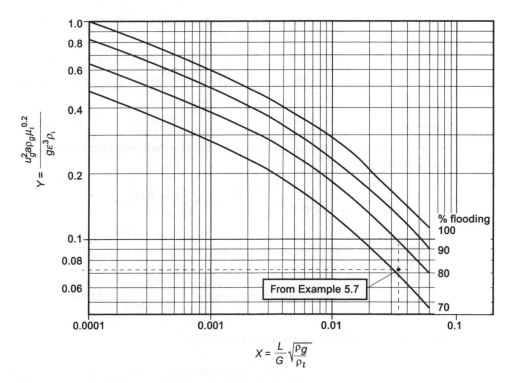

Figure 5.9   Flooding curve for knitted wire mesh separators, from Bürkholz (1989). G and L are the mass flow rates of gas and liquid, respectively.

Table 5.13   Physical Properties of Mesh Droplet Separators from Rhodius*

Material	Density (kg/m³)	Wire Diameter (μm)	Specific Surface Area (a) (m²/m³)	Porosity (Voidage, ε) (-)
Stainless steel	80	280	145	0.990
Stainless steel	110	280	200	0.986
Stainless steel	130	280	236	0.983
Stainless steel	145	280	265	0.981
Stainless steel	175	280	320	0.978
Stainless steel	192	280	350	0.975
Stainless steel	240	280	435	0.970
Stainless steel	240	140	868	0.970
Stainless steel	432	120	1835	0.945
Polypropylene	50	400	170	0.945
Polypropylene	70	400	235	0.923
Polypropylene	100	400	335	0.890
Polypropylene	100	220	610	0.890

*Standard thicknesses are 25, 50, 75, 100, 150, and 200 mm.

Example 5.7

Estimate the diameter of a vertical V-L separator containing a 4-in (100 mm) thick wire mesh mist eliminator. The properties of the gas and liquid are:

ρ_l = 870 kg/m³, ρ_g = 9.41 kg/m³, $\dot{v}_g$ = 7.32 m³/s, $\dot{v}_l$ = 1.553 m³/min = 0.02588 m³/s, μ_l = 0.0019 kg/m/s.

Consider a mist eliminator with a k_{SB} = 0.11 m/s that has a specific surface area, a = 300 m²/m³, and a porosity, ε = 0.98.

Solution

First use the Souders-Brown approach, and then check for the flooding condition.
From Equation (5.13),

$$u_{g,max} = k_{SB}\sqrt{\frac{\rho_l - \rho_g}{\rho_g}} = 0.11\sqrt{\frac{(870-9.41)}{(9.41)}} = 1.052 \text{ m/s}$$

So using a gas velocity of 80% $u_{g,max}$ = 0.842 m/s,

$$\frac{\pi}{4}D_{ves}^2 u_g = \dot{v}_g \Rightarrow D_{ves} = \sqrt{\frac{4\dot{v}_g}{\pi u_g}} = \sqrt{\frac{(4)(7.32)}{\pi(0.842)}} = 3.327 \text{ m}$$

Determine the X and Y parameters used in Figure 5.9:

$$X = \frac{L}{G}\sqrt{\frac{\rho_g}{\rho_l}} = \frac{(0.02588)(870)}{(7.32)(9.41)}\sqrt{\frac{(9.41)}{(870)}} = 0.034$$

$$Y = \frac{u_g^2 a \rho_g \, \mu_l^{0.2}}{g \varepsilon^3 \rho_l} = \frac{(0.842)^2(300)(9.41)(0.0019)^{0.2}}{(9.81)(0.98)^3(870)} = 0.0712$$

The X and Y values on Figure 5.9 indicate operation at about 72% of flooding, which is acceptable. The diameter could be reduced a little to bring operation closer to 80% flooding. For 80% flooding, the diameter would be reduced to ~3.06 m ($Y = 0.1$, $u_g = 0.998$ m/s, and $D_{ves} = 3.06$ m).

5.2.4.2 Droplet Distribution and Separation Efficiency from Mist Eliminators

As stated previously, the purpose of a mist eliminator is to reduce the amount and size of liquid drops leaving with the gas stream from the top of the V-L separator. In the case of knockout drums placed upstream of a compressor, the size distribution of the liquid drops is critical, since large drops entering a compressor may lead to excessive vibration of moving parts and premature failure. According to Barringer (2013), a knockout drum placed upstream of a compressor must be capable of removing 99% of all droplets (and solid particles) greater than 3 to 5 μm. Barringer also presents a figure that shows the compressor life (presumably the time required between extensive maintenance/overhauls) as a function of the size of drops/particles in the inlet. Data shown in Table 5.14 were taken from this curve.

The determination of the droplet size distribution leaving a mist eliminator is a complex calculation that depends on the approach gas velocity, physical characteristics of the device, and the distribution of droplets in the approaching gas. The term η_f is used to determine the efficiency of separation of a device and is defined as

$$\eta_f = \frac{\text{droplets retained in device of size } D_p}{\text{droplets fed to the device of size } D_p} \tag{5.14}$$

To characterize the flow and separation within the device, the following three dimensionless variables are used

$$\text{Euler number} \qquad Eu = \frac{\Delta P}{\rho_g u_g^2}$$

$$\text{Reynolds number} \qquad Re = \frac{\rho_g D_{wire} u_g}{\mu_g}$$

$$\text{Stokes number} \qquad Stk = \frac{\rho_l u_g D_p^2}{\mu_g D_{wire}}$$

Table 5.14   Compressor Life as a Function of Droplet/Particle Size in Inlet

Particle/Droplet Size (μm)	Approximate Compressor Life (days)
1000	14
50	28
10	1×10^2
4	1×10^3
2	1×10^4

Source: Data from Barringer (2013).

where D_{wire} is the diameter of the wire (for a mesh type mist eliminator or the characteristic length for the device), D_p is the droplet diameter, and ΔP is the pressure drop across the device.

In representing the efficiency of separation, it is convenient to combine these three parameters into a single one, namely, an inertial separation parameter, Ψ, where

$$\Psi = 0.25 \mathrm{Re}^{1/3} \mathrm{Eu}^{2/3} \mathrm{Stk} \tag{5.15}$$

For knitted wire meshes, Saemundsson (1968) gives the following correlation for determining the pressure drop through a length of mesh, L_{mesh}:

$$\Delta P = 2 f \rho_g \frac{L_{mesh}}{D_H} \left(\frac{u_g}{\varepsilon} \right)^2 \tag{5.16}$$

where D_H is the hydraulic diameter $D_H = \dfrac{\varepsilon}{(1-\varepsilon)} D_{wire}$, ε is the porosity of the mesh (see Table 5.13 for some typical values), and f is an equivalent friction factor given by Equation (5.17):

$$f = \frac{771.2 - 1.56P}{\mathrm{Re}_{wire}} + \frac{2.72 - 0.0038P}{\mathrm{Re}_{wire}^{0.2}} \tag{5.17}$$

where P is the pressure in bar and $\mathrm{Re}_{wire} = \dfrac{u_g D_{wire} \rho_g}{\varepsilon \mu_g}$

Finally, the information in Equations (5.14) through (5.17) can be combined to give the following correlation attributed to Bürkholz (1989):

$$\eta_f = x \left(1 - e^{-1/x} \right) \tag{5.18}$$

where $x = 0.003 \Psi^2$.

This relationship is shown in Figure 5.10.

Example 5.8 illustrates the use of Equations (5.14) through (5.18).

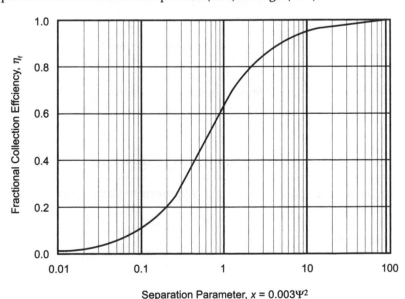

Figure 5.10   Fractional separation efficiency as a function of separation parameter, x (from Bürkholz [1989])

Example 5.8

Determine the pressure drop in the mist eliminator designed in Example 5.7. Using this result,

a. Determine the fraction of 4 μm particles that are captured in the device at the design flow-rate. Would this design be suitable for a knockout drum upstream of a compressor?

b. Repeat Part (a) for an 8-in (200 mm) thick mesh pad. Would this design be suitable for a knockout drum upstream of a compressor?

Solution

a. From Example 5.7,

$\rho_l = 870 \text{ kg/m}^3$, $\rho_g = 9.41 \text{ kg/m}^3$, $\dot{v}_g = 7.32 \text{ m}^3/\text{s}$, $\dot{v}_l = 0.002588 \text{ m}^3/\text{s}$, $\mu_l = 0.0019 \text{ kg/m/s}$, $\mu_g = 23 \times 10^{-6}$ kg/m/s, $a = 300 \text{ m}^2/\text{m}^3$, and porosity, $\varepsilon = 0.98$, $u_g = 0.842 \text{ m/s}$, $D_{wire} = 280 \text{ }\mu\text{m}$, $D_p = 4 \text{ }\mu\text{m}$, $L = 100 \text{ mm}$, $P = 7 \text{ bar}$

$$D_H = \frac{\varepsilon}{(1-\varepsilon)} D_{wire} = \frac{0.98}{(1-0.98)}(280 \times 10^{-6}) = 13{,}720 \times 10^{-6} \text{ m}$$

$$\text{Re}_{wire} = \frac{u_g D_{wire} \rho_g}{\varepsilon \mu_g} = \frac{(0.842)(280 \times 10^{-6})(9.41)}{(0.98)(23 \times 10^{-6})} = 98.43$$

From Equation (5.17),

$$f = \frac{771.2 - 1.56P}{\text{Re}_{wire}} + \frac{2.72 - 0.0038P}{\text{Re}_{wire}^{0.2}} = \frac{771.2 - 1.56(7)}{98.43} + \frac{2.72 - 0.0038(7)}{98.43^{0.2}} = 7.72 + 1.08 = 8.80$$

From Equation (5.16),

$$\Delta P = 2 f \rho_g \frac{L_{mesh}}{D_H} \left(\frac{u_g}{\varepsilon}\right)^2 = (2)(8.80)(9.41)\frac{(0.10)}{(0.01372)}\left(\frac{0.842}{0.98}\right)^2 = 891.1 \text{ Pa}$$

Using the definitions

$$\text{Eu} = \frac{\Delta P}{\rho_g u_g^2} = \frac{(891.1)}{(9.41)(0.842)^2} = 133.6$$

$$\text{Re} = \frac{\rho_g D_{wire} u_g}{\mu_g} = \frac{(9.41)(280 \times 10^{-6})(0.842)}{(23 \times 10^{-6})} = 96.46$$

$$\text{Stk} = \frac{\rho_l u_g D_p^2}{\mu_g D_{wire}} = \frac{(870)(0.842)(4 \times 10^{-6})^2}{(23 \times 10^{-6})(280 \times 10^{-6})} = 1.82$$

From Equation (5.15),

$$\Psi = 0.25 \text{Re}^{1/3} \text{Eu}^{2/3} \text{Stk} = (0.25)(96.46)^{1/3}(133.6)^{2/3}(1.82) = 54.5$$

From Equation (5.18),

$$x = 0.003 \Psi^2 = (0.003)(54.5)^2 = 8.92$$

$$\eta_f = x\left(1 - e^{-1/x}\right) = (8.92)\left(1 - e^{-1/(8.92)}\right) = 0.946$$

The criterion for compressor knockout drums is $\eta_f > 0.99$, so the current design is unsuitable.

b. If L_{mesh} increases to 200 mm, then $\Delta P = (2)(891.1) = 1782.2$ Pa and Eu = 267.1; all other parameters remain unchanged.

Thus, from Equation (5.15),

$$\Psi = 0.25\text{Re}^{1/3}\text{Eu}^{2/3}\text{Stk} = (0.25)(96.46)^{1/3}(267.1)^{2/3}(1.82) = 86.54 \text{ and } x = 22.47$$

$$\eta_f = x\left(1 - e^{-1/x}\right) = (22.47)\left(1 - e^{-1/(22.47)}\right) = 0.978$$

So again, this design is not feasible for a compressor knockout drum. In fact, Barringer (2013) recommends a multilevel approach that may include a variety of staged separation devices, including coalescing filters to condition the gas feed upstream of compressors.

5.2.5   Liquid–Liquid (L–L) Separation

When two immiscible liquids are mixed together, they form a mixture in which one liquid (often the one with the lower volume) forms droplets (dispersed phase) in the other liquid (continuous phase). The effective separation of two immiscible liquids (phases) is strongly dependent on the following three factors:

- Size distribution of the droplets of the dispersed phase
- Relative density difference between the two phases
- Viscosity of the continuous fluid

In certain cases, the dispersed phase can be quite stable, due to surface forces, and the resulting emulsion will not settle substantially over prolonged periods due to the effects of gravity, such as in latex paints. In such cases, the separation or "breaking" of this emulsion can be very difficult. In this chapter, techniques for separating the two liquid phases based on gravity settling are considered. Centrifugal, electrostatic, and chemical methods are not considered here but are covered in standard handbooks of chemical engineering.

5.2.5.1   Size Distribution of Dispersed Phase

The range of expected droplet sizes for different mechanical, heat transfer, and reaction processes are given by Cusack (2009). These data are shown in Figure 5.11.

In general, the wider the size distribution of drops, the more difficult is the separation. In addition, the smaller the droplets, the more difficult they are to coalesce and separate. Thus, looking at Figure 5.11, a two-liquid-phase stream having passed through a static mixer will be easier to separate than one having passed through a centrifugal pump or a two-liquid-phase stream generated by homogenous condensation from a bulk phase.

5.2.5.2   Relative Density Difference between the Two Phases

For small particles, Stokes law settling will govern the gravity separation of the dispersed drops of liquid. Equation (5.9) can be written for L-L systems as

$$u_t = \frac{D_{sph}^2(\rho_{l_1} - \rho_{l_2})g}{18\mu_{l-c}} \tag{5.19}$$

where subscripts l_1 and l_2 refer to the two liquids and $l\text{-}c$ refers to the continuous phase. For comparison purposes, Table 5.15 shows the terminal velocities for dispersed water drops in a continuous phase of oil (μ_{l-c} = 0.001 kg/m/s and ρ_l = 850 kg/m³) compared to the results given previously in Table 5.11 for water drops in air. The results in Table 5.15 are generated using Equation (1.45), although the results using Equation (5.19) for drops 100 μm and smaller are virtually identical.

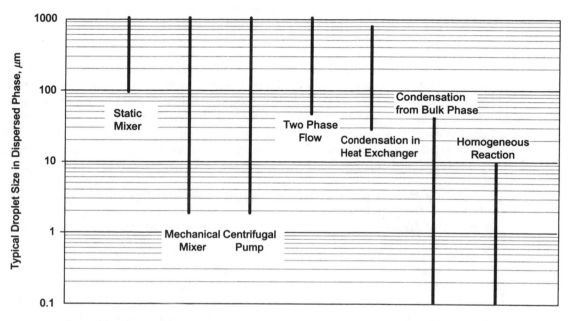

Figure 5.11 Typical drop size distributions (Cusack, 2009) from operations producing or processing liquid-liquid mixtures

Table 5.15 Terminal Velocities for Water Drops in Air and Oil (ρ_l = 850 kg/m³)

Droplet Size (microns, μm)	Terminal Velocity in Air (m/s)	Terminal Velocity in Oil (m/s)
1000	4.168	0.0818
500	2.354	0.0204
200	0.776	0.0033
100	0.252	8.18×10^{-4}
50	0.071	2.04×10^{-4}
20	0.012	3.27×10^{-5}
10	0.003	8.18×10^{-6}

It can be seen that the settling or terminal velocities of water drops in oil are several orders of magnitude smaller than those in air. This result poses additional constraints in the design of L-L settling equipment and suggests that, except for cases when the dispersed phase exists only as large drops, the distance that droplets must travel to become separated must be decreased significantly compared with L-V separators, and the addition of coalescing steps may be required.

5.2.5.3   Viscosity of the Continuous Fluid

The effect of the viscosity of the continuous phase is illustrated in Table 5.16 where the results for terminal velocities of falling water drops in oils with different viscosities are given.

Clearly, the higher the viscosity of the continuous phase, the lower is the terminal velocity of a given drop size and the more difficult is the separation. The simplest way to decrease the viscosity of the continuous phase is to increase the temperature, but this may lead to increased mutual solubility of the phases, which is counterproductive to the separation.

5.2.5.4   Design of L-L Gravity Settling and Coalescence Equipment

L-L separation is usually achieved in a horizontal vessel. For L-L-V separators, vertically oriented vessels may be employed, but horizontal vessels are usually preferred when significant degassing of the liquid is required due to the typically longer liquid residence times in horizontal vessels. Because the terminal velocities of even moderately sized droplets are small, some form of separating device is usually incorporated into the design. A diagram of a parallel plate separator is shown in Figure 5.12. In such a device, the two-phase liquid is forced to flow through a series of narrowly spaced, parallel channels, which according to Cusack (2009) leads to two advantages:

- Decreases the effective diameter for flow and reduces the Reynolds number and associated turbulence in the continuous phase
- Decreases the distance drops need to travel (z) in order to settle on to a surface and coalesce

When the effects of gravity settling are not sufficient to give an effective separation, then it is necessary to place surfaces normal to the flow that will allow drops to impact these surfaces and enhance separation via coalescence. These surfaces can take many forms, from corrugated metal channels running normal to the flow to wire meshes and other types of high surface area media placed in the flow path. In general, the following criteria, from Cusack (2009), enhance separation via coalescence:

- Use a higher droplet (continuous phase) velocity, which promotes higher drop impact efficiency.

Table 5.16   Comparison of Terminal Velocities of Water Droplets in Oils with Different Viscosities

Droplet Size (microns, μm)	Terminal Velocity in Oil ($\mu_{l\text{-}c}$ = 0.001 kg/m/s) (m/s)	Terminal Velocity in Oil ($\mu_{l\text{-}c}$ = 0.01 kg/m/s) (m/s)
1000	0.0818	7.26×10^{-3}
500	0.0204	1.96×10^{-3}
200	0.0033	3.23×10^{-4}
100	8.18×10^{-4}	8.14×10^{-5}
50	2.04×10^{-4}	2.04×10^{-5}
20	3.27×10^{-5}	3.30×10^{-6}
10	8.18×10^{-6}	8.18×10^{-7}

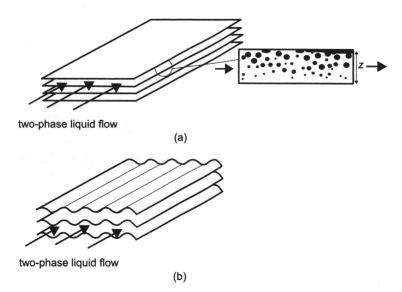

two-phase liquid flow

(a)

two-phase liquid flow

(b)

Figure 5.12 Principles of gravity settling equipment: (a) parallel plate settlers (b) corrugated plate settlers

- Reduce the drag force on droplets of the discontinuous phase by lowering the continuous phase viscosity. This is usually done by increasing temperature, but, as mentioned previously, this may increase mutual solubility of dispersed and continuous phases.
- Decrease the target diameter (diameter of wires in mesh), which allows smaller drops to be separated.
- Increase length or depth of the target device (depth in the direction of continuous phase flow), which provides greater surface area for collection and increases the number of potential impacts with dispersed phase. However, the pressure drop will increase as depth increases.
- Use appropriate materials for target device that wet the dispersed phase. For example, use hydrophilic (fiberglass or stainless steel) material for aqueous drops in an oil phase and oleophilic or oil-loving materials (many polymers) for separating oil drops from an aqueous phase.

5.2.5.5 Typical Layouts for L-L Separators

Some typical arrangements for L-L separators are shown in Figure 5.13. Referring to this figure, feed enters on the left-hand side of the vessel through a slotted pipe or similar distributor that is pointed toward the upstream elliptical head. A liquid distributor plate consisting of a disk with holes, similar to a sieve tray, is placed near the feed end of the vessel, and this plate acts to distribute the liquid and suppresses any wavelike motions that may result from a surge of liquid in the feed. The two-phase liquid mixture passes through the distributor plate and, depending on the size of the droplets in the dispersed phase, either passes directly through a gravity separator, such as a parallel or corrugated plate device, or passes through a series of coalescers and gravity separators. In Figure 5.13(a), the situation shown is for a relatively narrow size distribution of large droplet sizes of dispersed phase where only a single gravity settling stage is required. Upon leaving the parallel plate device, the heavy (continuous phase) and light (dispersed phase) separate easily and an L-L interface forms. The light phase rises to the top of the vessel and exits from the upper nozzle, while the heavy phase leaves from the lower nozzle. The location of the L-L interface is controlled

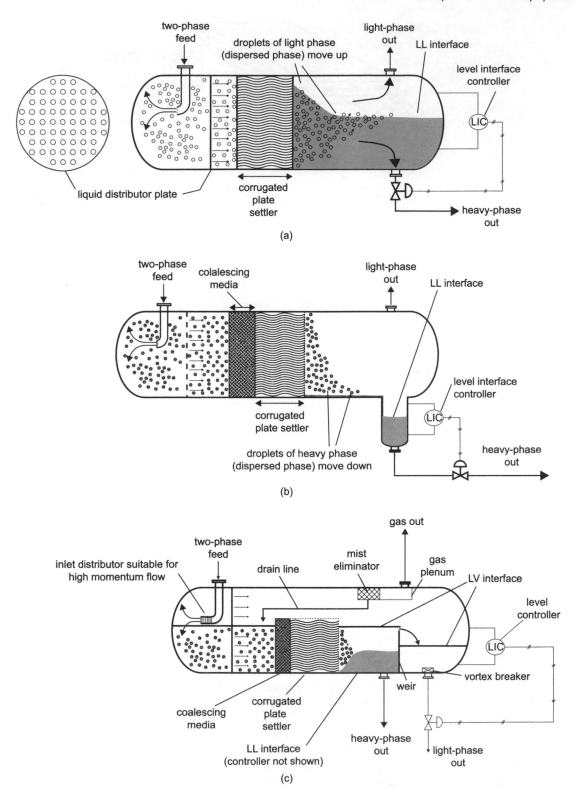

Figure 5.13 Typical arrangements for horizontal liquid-liquid (L-L) and liquid-liquid-vapor (L-L-V) separators: (a) L-L separator for large drops, (b) L-L separator for smaller drops requiring coalescence, (c) L-L-V separator (Adapted with permission from Cusack [2009])

by manipulating the flows of continuous phase from the vessel. The optimal location of the interface is found through trial and error.

In Figure 5.13(b), a situation is shown in which the droplets of dispersed phase are small and have a relatively wide size distribution. In such cases, an additional coalescence step is required. Thus, after the distributor plate, the two-phase mixture passes first through a coalescing medium (perhaps one form of woven mesh) that acts to coalesce the small particles into larger ones. The stream leaving the coalescer is fed directly into a stage of gravity separation, and then the mixture is allowed to separate in the region on the right-hand side of the vessel. The figure illustrates the case when the dispersed phase is the heavy liquid and the volume fraction of dispersed phase is small (a few percent) of the total. In this case, the heavy phase is collected in a boot at the right end of the vessel, and the level is controlled by regulating the flow of heavy liquid product from the vessel. If the size distribution is very wide, the dispersed phase droplets are very small (<10 μm), and/or the purity specification is very stringent, then an additional high-efficiency or "polishing" media stage must be added and accompanied by an additional gravity separation stage. These additional separation stages are not shown in Figure 5.13(b).

In Figure 5.13(c), a three-phase, L-L-V separator is shown. The same principles of operation are illustrated with a few additional features. For example, the three-phase mixture entering the vessel would typically flow through a vane-type distributor suitable for high-momentum flows (see Table 5.8). The level of liquid in the vessel is maintained at the top of the coalescing media and settler by the use of an overflow weir on the right-hand side of the vessel. The flow of the light-phase liquid, which flows over the weir, is regulated by a level controller that maintains the level of light liquid to the right of the weir. The interface between the heavy and light liquids is also sensed and is used to control the flow of heavy liquid that leaves from the bottom of the vessel through a nozzle placed to the left of the weir. However, this liquid interface control loop is not shown in Figure 5.13. The vapor phase leaves the vessel through a nozzle at the top right of the vessel. A gas plenum or box is placed around the nozzle, and a mist eliminator is placed at the entrance of the plenum to ensure that all the gas flows through it. A drain line is added to the mist eliminator, and it routes any collected liquid back to the front end of the vessel. Finally, a vortex breaker is installed at the exit of the light phase to eliminate entrained vapor leaving with the light liquid.

5.3   STEAM EJECTORS

To operate chemical processes at pressures below ambient, it is necessary to remove noncondensable vapors at subatmospheric pressure and deliver them to equipment operating at ambient or above ambient pressure. For example, consider a feed stream of acrylic and acetic acid that must be purified to produce separate high-purity acrylic and acetic acid streams. Acrylic acid is known to polymerize spontaneously at temperatures of 90°C and above. This polymerization is highly exothermic and can lead to an explosion (Kurland and Bryant, 1987). To avoid the disastrous effect of polymerizing acrylic acid in the distillation column, the bottom of the column must be operated at approximately 15.7 kPa, at which pure acrylic acid boils at about 89°C. The pressure drop across the column is somewhere around 8 to 9 kPa, which means that the top vapor (nearly pure acetic acid) leaves the column at about 7 kPa and must be condensed at approximately 48°C in the overhead condenser using cooling water (in at 30°C and out at 40°C). This situation is illustrated in Figure 5.14.

For this system, if the feed contained only acrylic and acetic acids and there was no leakage, then the heat removal by the cooling water in the overhead condenser would be sufficient to maintain the pressure at the top of the column at 7 kPa. This is because vapor-liquid equilibrium exists everywhere in the column and nearly pure acetic acid (>99 mol%) vapor at 48°C condenses at only one pressure, its vapor pressure at 48°C, which is 7 kPa. In reality, this situation never occurs for two reasons. First, the feed stream will almost certainly contain trace amounts

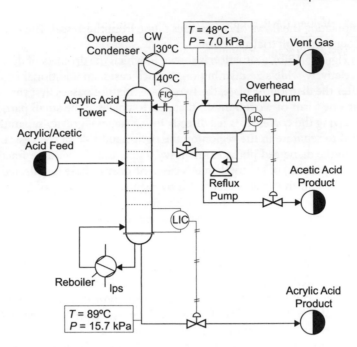

Figure 5.14   Process flow diagram for acrylic acid separation tower

of dissolved gases (N_2, O_2, CO_2, etc.) that will come out of solution and travel to the top of the column. These gases may not fully redissolve at the temperatures in the condenser and will subsequently accumulate in the condenser or in the reflux drum. If they are not removed, then the pressure will slowly rise and the pressure and temperature throughout the column will increase, leading to the polymerization of acrylic acid at the bottom of the column with disastrous consequences. Second, because the system in Figure 5.14 operates below atmospheric pressure, there is a tendency for air to leak into the system through flanged connections. This air will again accumulate at the top of the column with the same consequences as described previously. Therefore, it is imperative that these noncondensable gases be removed from the overhead system of the distillation column. These gases (shown as Vent Gas in Figure 5.14) could be removed using a mechanical vacuum pump, but it is usually more economical to remove them by means of a single-stage or multistage steam ejector.

5.3.1   Estimating Air Leaks into Vacuum Systems and the Load for Steam Ejectors

The amount of dissolved gas in the acrylic acid feed stream in Figure 5.14 can be estimated by using an appropriate vapor-liquid equilibrium (VLE) thermodynamics package in the process simulator, such as Henry's law or a similar package that predicts the amount of dissolved gases at the feed conditions. Likewise, if the acrylic acid feed were to contain dissolved lighter hydrocarbons, then the amount of these compounds that would need to be removed from the overhead condenser should be estimated in the process simulator using the appropriate VLE model and accurately simulating the overhead condensation process. However, the process simulator will not be able to predict the amount of air leaking into the system.

   If a detailed mechanical flow or piping and instrumentation diagram (P&ID) were available for the acrylic acid separations process, then a count of the number, type, and size of all flanged

Table 5.17   Values for Parameter *C* for Use in Equation (5.20)

System Pressure Range		*C*	*C*
mmHg	*kPa*	*(kg/m²/h)*	*(lb/ft²/h)*
90–760	12.0–101	0.9466	0.1939
21–89	2.8–11.9	0.7152	0.1465
3.1–20	0.41–2.67	0.4838	0.0991
1.0–3	0.13–0.40	0.2314	0.0474
<1	<0.13	0.1157	0.0237

connections along with the fluid pressure could be made and the air leak into the system could be predicted through analytical methods. Details and estimates for different types of connections are given by Jackson (1948). However, such an approach is seldom used for preliminary estimates of steam consumption rates when the P&ID may not be available. Instead, we use the method recommended by the Heat Exchange Institute in their *Standards for Steam Jet Vacuum Systems* (2000) for "tight" systems that would be common for chemical processes. The rate ($\dot{m}_{air-leak}$) at which air leaks into a system of volume V is given by Equation (5.20):

$$\dot{m}_{air-leak} = CV^{2/3} \tag{5.20}$$

where *C* is a parameter that depends on the pressure in the system and is given in Table 5.17.

It should be noted that the values of *C* are higher with higher system pressure, which seems counterintuitive, as it would be expected that the air leakage would increase with lower system pressure, that is, higher vacuum. However, the values of *C* in Table 5.17 also reflect the improved maintenance and system "tightness" that typically exists for higher vacuum systems.

Examples of how to estimate leaks for "loose" vacuum systems using a hybrid approach are given in McKetta (1997).

Example 5.9

Estimate the leak of air into a system comprising the acrylic acid tower and associated equipment and piping for the process flow diagram shown in Figure 5.14.

Solution

From Turton et al. (2012), Appendix B, the tower, and overhead condenser sizes are given as

Acrylic acid tower: Diameter = 2.3 m, height = 25 m

Reflux drum: Diameter = 1.0 m, length = 2.5 m

Ignoring the volume of the vessel heads, the volume of these two pieces of equipment is

$$V \approx \frac{\pi}{4} D_{tower}^2 L_{tower} + \frac{\pi}{4} D_{drum}^2 L_{drum} = \frac{\pi}{4}[(2.3^2)(25)+(1^2)(2.5)] = 105.8 \text{ m}^3$$

Adding a 25% factor to include the volume of the heat exchangers and connecting pipe gives:
$V_{system} = (1.25)(105.8) \approx 132 \text{ m}^3$

Using Equation (5.20) with a system pressure of 7 kPa and using a value of C from Table 5.17 of 0.7152 (kg/m²/h) gives

$$\dot{m}_{air-leak} = CV^{2/3} = (0.7152)(132)^{2/3} = 18.54 \text{ kg/h}$$

5.3.2   Single-Stage Steam Ejectors

A schematic diagram of a single-stage steam ejector is given in Figure 5.15. The ejector consists of a converging-diverging supersonic steam nozzle. As the steam expands through the nozzle, it becomes supersonic (between Mach 3 and 4), and its pressure drops below that of the process gas. In this way, the movement of steam through the nozzle creates a vacuum that educts the low-pressure process stream, and then the two streams mix in a mixing zone. The mixed gas stream is then expanded in the diffuser section of the ejector, which reduces the gas velocity and increases the stream pressure to well above the process gas pressure. When a single steam ejector is required, the exit gas pressure (P_{out}) must be maintained at above atmospheric pressure. With multistage ejectors, only the exit pressure from the final stage must be greater than atmospheric.

The optimal design of single-stage ejectors was studied by DeFrate and Hoerl (1959). The results of their analysis are quite involved, and the results are presented graphically and are reproduced in Figures 5.16 and 5.17. The notation for the different terms in the axes of these figures corresponds to that given in Figure 5.15. These figures apply to the case when the motive gas and the educted gas have the same molecular weight and temperature. For the normal case when these two gases have different compositions and temperatures, DeFrate and Hoerl recommend the following correction factor:

$$\left[\frac{\dot{m}_b}{\dot{m}_a}\right]_2 = \left[\frac{\dot{m}_b}{\dot{m}_a}\right]_1 \sqrt{\left(\frac{M_b}{M_a}\right)_2 \left(\frac{T_a}{T_b}\right)_2} \qquad (5.21)$$

where the subscripts 2 and 1 refer to the case for unequal molecular weights and/or temperatures and the base case condition given in Figures 5.16 and 5.17, respectively. The gas velocities into and out of the nozzle may be considered small, and thus the kinetic energy terms for the inlet and

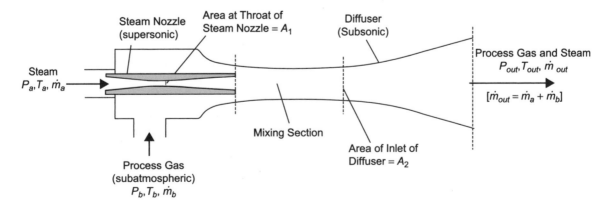

Figure 5.15   Diagram of the processes and notation used in the design of a single-stage steam ejector

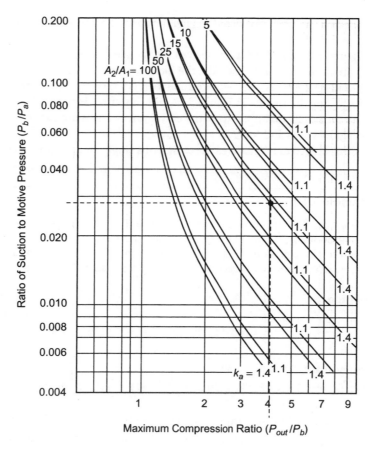

Figure 5.16 Maximum compression ratio for constant-area, single-stage steam ejectors ($M_a = M_b$ and $T_a = T_b$) (Figure redrawn from DeFrate and Hoerl [1959])

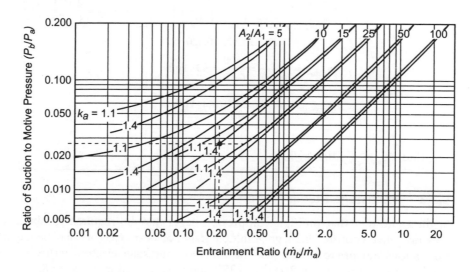

Figure 5.17 Entrainment ratio for constant-area, single-stage steam ejectors ($M_a = M_b$ and $T_a = T_b$) (Figure redrawn from DeFrate and Hoerl [1959])

outlet streams may be considered negligible; although, the kinetic energy transformations in the ejector are very significant.

Figures 5.16 and 5.17 were derived for the ideal case, when the flow processes (apart from mixing) are reversible and adiabatic, the ideal gas law holds, and the heat capacity ratios of both gases (k_a and k_b) are constant. DeFrate and Hoerl (1959) also show that there is a small effect on the ratio of k_a/k_b for the results in Figure 5.17, but this effect is not included here. From comparisons with tests on actual steam ejectors, the entrainment ratios are expected to be about 90% of those given in Figures 5.16 and 5.17.

Example 5.10 demonstrates the design of a single-stage steam ejector.

Example 5.10

Estimate the flowrate of medium pressure steam ($T_a = 180°C$, $P_a = 1000$ kPa) required to remove an air leak of 300 kg/h into an overhead condenser system operating at 27.5 kPa pressure and 50°C using a single-stage steam ejector. You should assume that the steam and air stream will be partially condensed in a heat exchanger operated at an inlet pressure of 110 kPa (P_{out}).

Solution

$$P_a = 1000 \text{ kPa}, T_a = 180°C, P_b = 27.5 \text{ kPa}, T_b = 50°C, P_{out} = 110 \text{ kPa}$$

Assume that $k_a = 1.38$ ($k_b = 1.40$).

From Figure 5.16, $X = P_{out}/P_b = 110/27.5 = 4.0$ and $Y = P_b/P_a = 27.5/1000 = 0.0275$, giving an area ratio $A_2/A_1 \approx 15$ (shown on Figure 5.16).

From Figure 5.17, using $A_2/A_1 = 15$ from Figure 5.16, with $P_b/P_a = 0.0275$, gives $\dot{m}_b/\dot{m}_a = 0.21$.

Correcting for temperature and molecular weight using Equation (5.21) and accounting for an efficiency of 90% gives

$$\left[\frac{\dot{m}_b}{\dot{m}_a}\right]_{2,theor} = \left[\frac{\dot{m}_b}{\dot{m}_a}\right]_{1,theor}\sqrt{\left(\frac{M_b}{M_a}\right)_2\left(\frac{T_a}{T_b}\right)_2} = (0.21)\sqrt{\frac{29.8}{18}\frac{(180+273.2)}{(50+273.2)}}$$

$$= (0.21)(1.52) = 0.320$$

$$\dot{m}_{a,theor} = \frac{\dot{m}_b}{(0.320)} = \frac{300}{0.320} = 938 \text{ kg/h}$$

$$\dot{m}_{a,actual} = \frac{\dot{m}_{a,theor}}{\varepsilon} = \frac{938}{0.9} = 1042 \text{ kg/h}$$

Thus, $\dot{m}_{steam} = \dot{m}_a = 1042$ kg/h.

5.3.3   Multistage Steam Ejectors

In the same way that multiple compressor stages may be required to obtain a desired outlet pressure when compressing a gas, multiple stages of steam ejectors may be required to educt gas from a low pressure to a high one. The limits for multistage ejectors with and without interstage condensation are given by Berkeley (1957) and are repeated in Table 5.18. For multistage steam ejectors, it is possible to operate ejectors in series without interstage condensation, because at the low intermediate pressures in these systems, condensation of steam will not occur at the

Table 5.18 Limits for the Operation of Multistage Steam Ejectors for Steam at 790 kPa and Cooling Water at 30°C

Number of Stages	Type	Range of Operation			
		From (mm Hg)	To (mm Hg)	From (kPa)	To (kPa)
One (1)	A	760.0	50.0	101.00	6.70
Two (2)	AB	100.0	4.2	13.30	0.56
Two (2)	A-X-B	80.0	3.0	10.70	0.40
Three (3)	AB-X-C	6.0	1.5	2.00	0.80
Three (3)	A-X-B-C	80.0	1.8	10.70	2.40
Four (4)	AB-X-C-X-D	8.0	0.15	1.07	0.20
Five (5)	ABC-X-D-X-E	0.7	0.035	0.09	4.7×10^{-3}

Ejector stages designated as A, B, C, D, and E

Location of condensing heat exchangers designated by X.

Source: Data from Berkeley (1957).

temperature of cooling water. The purpose of interstage condensation is to reduce the flow of gas between stages and hence reduce the amount of motive steam required by the later ejector stages and help reduce the operating costs.

Returning to the case of the acrylic acid separation process given in Figure 5.14 and Example 5.9, it is noted that the top operating pressure of the tower is 7 kPa, which is close to the minimum pressure for single-stage ejectors shown in Table 5.18. This suggests that a multiple-stage ejector may be required for this design. Example 5.11 investigates the acrylic acid tower and illustrates the use of a two-stage steam ejector.

Example 5.11

Design a steam ejector system for the acrylic acid distillation tower shown in Figure 5.14 and discussed in Example 5.9. Steam is available at 180°C and 1000 kPa, and the air leaking into the system must be delivered to the battery limit at a pressure of 120 kPa. For this system, determine the flow of steam needed to educt the air flow determined from Example 5.9.

a. Using a single-stage ejector.

b. Using a two-stage, noncondensing ejector.

Solution

The system is shown in the Figure E5.11A. For this system, there will be some uncondensed acetic acid in the stream that leaves the overhead condenser due to the presence of the air. For this problem, the amount of acetic acid in this vapor will be considered negligible; however, this should be checked by simulating the overhead condenser and reflux drum for the condition stated in the problem. In addition, the pressure drop across the reflux drum is assumed here to be zero.

$$P_a = 1000 \text{ kPa}, T_a = 180°C, P_b = 7.0 \text{ kPa}, T_b = 48°C, P_{out} = 120 \text{ kPa}, k_a = 1.40$$

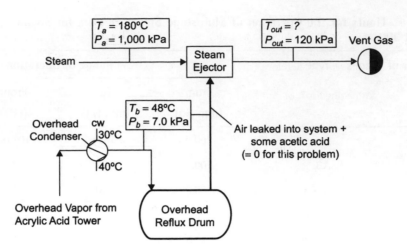

Figure E5.11A   Overhead system of acrylic acid column

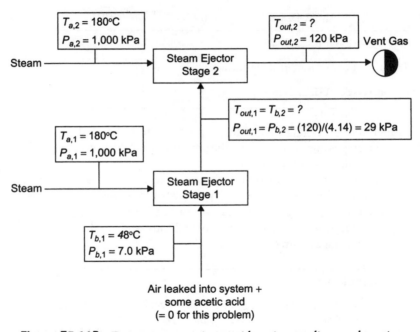

Figure E5.11B   Two-stage steam ejector without intermediate condensation

a. $X = P_{out}/P_b = 120/7.0 = 17.1$ and $Y = P_b/P_a = 7/1000 = 0.007$; this point (X, Y) is off the right-hand side of Figure 5.16. Extrapolating gives an area ratio $A_2/A_1 \approx 20$. For this area ratio and $P_b/P_a = 0.007$, the point is located way to the left of Figure 5.17—even if this point could be estimated, it would give a value of $\dot{m}_b/\dot{m}_a$ close to zero, leading to an extremely large steam flowrate. This suggests that this design cannot be accomplished using a single-stage ejector, confirming the range of suction pressures shown in Table 5.18.

b. The steam ejector will consist of two ejectors placed in series, as shown in Figure E5.11B. To solve this problem, the relative pressure ratios for both ejector stages must be chosen. As is the case for compressor stages, the optimum should be when the pressure ratios for each stage are equal. The overall pressure ratio is $P_{out}/P_a = 120/7 = 17.1$; thus, for each stage, the pressure ratio is $\sqrt{17.1} = 4.14$. This pressure ratio and other key data are shown in Figure E5.11B.

Stage 1

$X = P_{out,1}/P_{b,1} = 29/7.0 = 4.14$ and $Y = P_{b,1}/P_{a,1} = 7/1000 = 0.007$, and from Figure 5.16, this point (X, Y) gives a value of $A_1/A_2 \approx 75$. With $A_1/A_2 = 75$ and $P_{b,1}/P_{a,1} = 7/1000 = 0.007$ from Figure 5.17, $\dot{m}_b/\dot{m}_a \approx 0.35$. Therefore, using the result from Example 5.9, $\dot{m}_{b,1} = 18.54$ kg/h.

Adjusting for T, M, and efficiency gives

$$\left[\frac{\dot{m}_b}{\dot{m}_a}\right]_{2,\,theor} = \left[\frac{\dot{m}_b}{\dot{m}_a}\right]_{1,\,theor}\sqrt{\left(\frac{M_b}{M_a}\right)_2\left(\frac{T_a}{T_b}\right)_2} = (0.35)\sqrt{\frac{29.8}{18}\frac{(180+273.2)}{(48+273.2)}}$$

$$= (0.35)(1.53) = 0.535\text{---entrainment ratio at actual conditions}$$

$$(\dot{m}_{a,1})_{theor} = \frac{\dot{m}_{b,1}}{(0.535)} = \frac{(18.54)}{(0.535)} = 34.7\text{ kg/h}$$

$$(\dot{m}_{a,1})_{actual} = \frac{(\dot{m}_{a,1})_{theor}}{\varepsilon} = \frac{(34.7)}{(0.90)} = 38.5\text{ kg/h}$$

A mass and energy balance on Stage 1 of the ejector gives

$$\dot{m}_{out,1} = \dot{m}_{b,2} = \dot{m}_{a,1} + \dot{m}_{b,1} = 18.5 + 38.5 = 57.0\text{ kg/h} \tag{E5.11a}$$

$$\dot{m}_{out,1}h_{out,1} = \dot{m}_{b,2}h_{b,2} = \dot{m}_{a,1}h_{a,1} + \dot{m}_{b,1}h_{b,1} \tag{E5.11b}$$

The only unknown in Equation (E5.11b) is the temperature of stream $m_{out,1}$. Solving for this gives $T_{out,1} = T_{b,2} = 126.9°C$. The energy balance was solved using the CHEMCAD simulator but could also be estimated using $h = c_p\Delta T$ for each stream.

The average molecular weight of the outlet stream is given by

$$\frac{\dot{m}_{out,1}}{M_{out,1}} = \frac{\dot{m}_{b,2}}{M_{b,2}} = \frac{\dot{m}_{a,1}}{M_{a,1}} + \frac{\dot{m}_{b,1}}{M_{b,1}} \tag{E5.11c}$$

$$\therefore MW_{out,1} = \frac{\dot{m}_{out,1}}{\dfrac{\dot{m}_{a,1}}{M_{a,1}} + \dfrac{\dot{m}_{b,1}}{M_{b,1}}} = \frac{57}{\dfrac{18.5}{29.8} + \dfrac{38.5}{18}} = 20.66$$

Stage 2

$X = P_{out,2}/P_{b,2} = 120/29 = 4.14$ and $Y = P_{b,2}/P_{a,2} = 29/1000 = 0.029$, and from Figure 5.16, this point (X, Y) gives a value of $A_1/A_2 \approx 14$. With $A_1/A_2 = 14$ and $P_{b,1}/P_{a,1} = 29/1000 = 0.029$ from Figure 5.17, $\dot{m}_b/\dot{m}_a \approx 0.19$.

Adjusting for T, M, and efficiency gives

$$\left[\frac{\dot{m}_b}{\dot{m}_a}\right]_{2,\,theor} = \left[\frac{\dot{m}_b}{\dot{m}_a}\right]_{1,\,theor}\sqrt{\left(\frac{M_b}{M_a}\right)_2\left(\frac{T_a}{T_b}\right)_2} = (0.19)\sqrt{\frac{20.66}{18}\frac{(180+273.2)}{(126.9+273.2)}}$$

$$= (0.19)(1.14) = 0.217\text{---actual entrainment ratio}$$

$$\dot{m}_{b,2,\,theor} = \frac{\dot{m}_{a,2}}{(0.217)} = \frac{(57)}{(0.217)} = 262.7\text{ kg/h}$$

$$\dot{m}_{b,2,\,actual} = \frac{\dot{m}_{b,2,theor}}{\varepsilon} = \frac{(262.7)}{(0.9)} = 291.9\text{ kg/h}$$

This gives the total steam flow to both ejector stages as

$$\dot{m}_{steam} = \dot{m}_{b,1} + \dot{m}_{b,2} = 38.5 + 291.9 = 330.4 \text{ kg/h}$$

Because there is no condensation of the intermediate stream between stages, the mass flow of gas to Stage 2 is much higher than to Stage 1, and hence the amount of steam required in Stage 2 is 7.6 times that required for Stage 1. Based on the conditions of the outlet stream from Stage 1, some condensation could be achieved using typical conditions for cooling water, resulting in a decrease in the steam flow needed for the second stage with a corresponding savings in operating costs. These savings in costs would need to be compared to the additional cost of a heat exchanger, cooling water, and L-V separator required to affect the condensation. This problem is considered further in the end-of-chapter problems.

5.3.4   Performance of Steam Ejectors

As pointed out by DeFrate and Hoerl (1959), the curves in Figures 5.16 and 5.17 for a given value of A_2/A_1 actually represent the locus of the optimum designs for a single-stage ejector at different conditions. However, for practical purposes, a curve for a given value of A_2/A_1 can be used to estimate the performance of that ejector over the range that covers 20% to 130% of the design suction pressure (P_b) and 80% to 120% of the motive steam pressure (P_a). It should be noted that performance (entrainment ratio) drops off rapidly if the maximum compression ratio is exceeded but is not affected significantly if the outlet pressure decreases below the optimum value. Example 5.12 shows a performance problem for a single-stage ejector.

Example 5.12

Consider the problem given in Example 5.10 and assume that the ejector rated for the design conditions given in Example 5.10 has been purchased and is operating. However, during operation it has been noted that the air leak into the system has increased and the pressure at the top of the column is around 36.7 kPa instead of the design value of 27.5 kPa. For this situation, answer the following:

   a. Estimate the rate of air leak into the tower assuming that the flow and pressure of steam is the same as the design condition.

   b. Could increasing the flow of 1000 kPa steam reduce the column pressure to the desired value of 27.5 kPa?

Solution

   a. $P_a = 1000 \text{ kPa}, T_a = 180°\text{C}, P_b = 36.7 \text{ kPa}, T_b = 50°\text{C}, P_{out} = 110 \text{ kPa}$

Assume that $k_a = 1.38$.

From Figure 5.16, with $Y = P_b/P_a = 36.7/1000 = 0.0367$ and $A_2/A_1 = 15$, this gives a value of $X \approx 3.3$, which is less than the design value of $X = P_{out}/P_b = 110/27.5 = 4.0$. So the ejector does not violate the maximum compression ratio, and performance should be as predicted by Figure 5.17.

Using $A_2/A_1 = 15$ from Figure 5.17, with $P_b/P_a = 0.0367$ gives $\dot{m}_b/\dot{m}_a \approx 0.32$.

Correcting for temperature, M, and efficiency gives

$$\left[\frac{\dot{m}_b}{\dot{m}_a}\right]_{2, theor} = \left[\frac{\dot{m}_b}{\dot{m}_a}\right]_{1, theor} \sqrt{\left(\frac{M_b}{M_a}\right)_2 \left(\frac{T_a}{T_b}\right)_2} = (0.32)\sqrt{\frac{29.8}{18}\frac{(180+273.2)}{(50+273.2)}}$$

$$= (0.32)(1.52) = 0.486$$

$$\dot{m}_{a, theor} = \varepsilon \dot{m}_{a, actual} = (0.9)(1044) = 940 \text{ kg/h}$$

$$\dot{m}_b = (0.486)(940) = 457 \text{ kg/h}$$

So the air leak has increased from 300 to 457 kg/h.

b. In principle, increasing the steam flowrate in direct proportion to the airflow rate should solve the problem; thus, if $\dot{m}_{steam}$ = (457)(1044)/300 = 1590 kg/h, then the conditions in the problem are the same as in Example 5.10. However, the ejector was originally designed for the flowrates in Example 5.10, and to accommodate the higher flows, a larger ejector with the same A_1/A_2 value would be required. This information would have to be obtained from a manufacturer.

WHAT YOU SHOULD HAVE LEARNED

- The basic equations governing the wall thickness for pressure vessels, including the shell, heads, and nozzles
- Estimation techniques to determine the mass of pressure vessels made from common materials
- The basic principle of operation of V-L, L-L, and L-L-V separators or knockout drums
- How to determine the length and diameter of knockout drums for separating entrained liquid drops in V-L separators
- How to estimate the pressure drop, flooding condition, and droplet size distribution for mist eliminators
- How to predict the performance of other phase separators
- The principles of operation of steam ejectors
- How to estimate the air leak into a chemical process operating at vacuum
- How to estimate the steam flowrate for single-stage and multistage steam ejectors
- How to predict the performance of steam ejectors

NOTATION

Symbol	Definition	SI Units
a	specific surface area of packing	m^2/m^3
A	area	m^2
C	constant in Equation (5.20)	$kg/m^2/h$
D	diameter	m
D^*	dimensionless diameter	
E	weld efficiency	
f	friction factor	
F	friction factor	
g	acceleration due to gravity	m/s^2
G	mass flowrate of gas	kg/h
h	height of vessel head	m

Symbol	Definition	SI Units
h	specific enthalpy	J/kg
k	ratio of specific heat capacities of gas	
k_{SB}	Souders-Brown constant (Equation [5.13])	m/s
K	geometric factor for elliptical heads	
L	mass flowrate of liquid	kg/h
L	length	m
m	mass	kg
$\dot{m}$	mass flowrate	kg/h, kg/s
M	stress intensity factor for dished heads	
M	molecular weight	g/mol
P	pressure	kPa, mmHg
r_k	knuckle radius for dished heads	m
S	allowable stress for material in vessel design	Pa
t	thickness	m
t	time	s or h
T	temperature	°C or K
u	velocity	m/s
$\dot{v}$	volumetric flowrate	m³/s
V	volume of process system	m³
x	a parameter in Equation (5.18)	
X	abscissa value for flooding curve	
Y	ordinate value for flooding curve	
z	liquid level	m

GREEK SYMBOLS

ε	voidage or porosity of packing	
η_f	efficiency of separation (based on droplet size distribution)	
μ	viscosity	kg/m/s
θ	angle subtended by the liquid surface in a horizontal vessel	rad
ρ	density	kg/m³
Ψ	inertial separation parameter	

SUBSCRIPTS

Symbol	Definition
a, b	motive and process stream designations for steam ejectors
air-leak	air-leak due to vacuum conditions
ca	corrosion allowance
dished	dished head
elliptical	elliptical head
g	gas
H	hydraulic equivalent
i	inner
l	liquid
l-h	liquid-holdup
max	maximum
mesh	mesh
out	outlet condition
p	particle
SB	Souders-Brown
sph	spherical or equivalent spherical
t	terminal
ves	vessel
wire	wire
z	referring to area for flow in a horizontal vessel
1,2	locations in steam ejector or reference to base case and new case conditions

ABBREVIATIONS

Eu	Euler's number
Re	Reynolds number
Stk	Stokes number
NOL	normal operating level
HHAL	high-high alarm level
LLAL	low-low alarm level

REFERENCES

ACS Industries. 2004. *The Engineered Mist Eliminator, ACS Separations and Mass-Transfer Products.* http:// amacs.com/wp-content/uploads/2012/09/Mist-Elimination.pdf.

American Society of Mechanical Engineers. 2015. *Boiler and Pressure Vessel Code, Section VIII, Pressure Vessels.* New York: American Society of Mechanical Engineers.

Barringer, P., 2013. "Compressors and Silent Root Causes for Failure." http://www.barringer1.com/ dec08prb.htm.

Berkeley, F. D. 1957. "Ejectors Give Any Suction Pressure." *Chem Eng* 64(4): 1–7.

Bothamley, M. 2013. "Gas-Liquid Separators—Quantifying Separation Performance, Part 1. *SPE Oil and Gas Facilities* (Aug. 22–29).

Bürkholz, A. 1989. *Droplet Separation.* New York: VCH Publishers.

Campbell, J. M. 2014. *Gas Conditioning and Processing, Volume 2: The Equipment Modules,* 9th ed., 2nd printing. Norman, OK: Campbell Petroleum Series.

Chattopadhyay, S. 2005. *Pressure Vessels: Design and Practice.* Boca Raton, FL: CRC Press.

Couper, J. R., W. R. Penney, J. R. Fair, and S. M. Walas. 2012. *Chemical Process Equipment: Selection and Design,* 3rd ed. Waltham, MA: Butterworth-Heinemann.

Cusack, R. 2009. "Rethink Your Liquid-Liquid Separations." *Hydrocarb Process* 88(6): 53–60.

DeFrate, L. A. and A. E. Hoerl. 1959. "Optimum Design of Ejectors Using Digital Computers." *Chem Eng Progr Symp Ser* 21(55): 43–51.

Heat Exchange Institute. 2000. *Standards for Steam Jet Vacuum Systems,* 5th ed. Cleveland, OH: Heat Exchange Institute.

Jackson, D. H. 1948. "Selection and Use of Ejectors." *Chem Eng Prog* 44: 347.

Koch-Glitsch. 2015. Mist Elimination Liquid-Liquid Coalescing [product catalog], Bulletin MELLC-02, Rev. 3. www.koch-glitsch.com/Document%20Library/ME_ProductCatalog.pdf.

Kurland, J. J., and D. B. Bryant. 1987. "Shipboard Polymerization of Acrylic Acid." *Plant Operations Progress* 6(4): 203–207.

McKetta, J. J. 1997. "Vacuum System Leakage." In *Encyclopedia of Chemical Processing and Design,* Vol. 61. Boca Raton, FL: CRC Press.

Perry, R. H., and C. H. Chilton. 1973. *Chemical Engineers' Handbook,* 5th ed. New York: McGraw-Hill.

Red-Bag. n.d. *BN-EG-UE109 Guide for Vessel Sizing.* http://red-bag.com/engineering-guides/249-bn-eg-ue109-guide-for-vessel-sizing.html

Rhodius. n.d. *Droplet Separation—Technical Literature.* www.colasit.be/pdf/rhodius.pdf.

Saemundsson, H. B. 1968. *Absceidung von Öltopfen aus strömender Luft mit Drahtgestrickpaketen Verfahrenstecknik 2,* Nr. 11, 480–486. Cited from Rhodius, *Droplet Separation—Technical Literature,* www.colasit.be/pdf/rhodius.pdf.

Sandler, H. J., and E. T. Luckiewicz. 1987. *Practical Process Engineering, a Working Approach to Plant Design.* Philadelphia: XIMIX, Inc.

Turton, R., R. C. Bailie, W. B. Whiting, J. A. Shaeiwitz, and D. Bhattacharyya. 2012. *Analysis, Synthesis, and Design of Chemical Processes,* 4th ed. Upper Saddle River, NJ: Prentice Hall.

PROBLEMS

Short Answer Problems

1. The term E (weld efficiency) is used in the basic equation for determining the thickness of a pressure vessel. Does this term appear on the top or the bottom of the equation?

2. What is the corrosion allowance, why is it important, and what are typical values?

3. What is the minimum thickness of a pressure vessel? Would you still specify this thickness for a vessel that operated at an absolute pressure of 1.01 bar (1 atm)? Why?

4. What are the two most common types of head (end piece) used in the construction of pressure vessels?

5. What is meant by a design-by-rule philosophy?

6. What is the purpose of a vapor-liquid (V-L) knockout drum?

7. For a simple V-L knockout drum, give three reasons for using a vertically oriented drum.

8. For a simple V-L knockout drum, give three reasons for using a horizontally oriented drum.

9. Sketch three typical inlet devices for a V-L drum.

10. What is the meaning of HHAL and LLAL in the design of a vertical vessel?

11. In the design of a vertical V-L separator, what geometric parameter of the drum (dimension) controls the size of liquid drops leaving with the vapor? Explain your answer.

12. What is the purpose of a mist eliminator?

13. Does a mist eliminator capture equally drops of different sizes in the vapor?

14. For what process function is the use of a V-L separator most critical?

15. For an L-L separator, do liquid drops separate slower or faster than drops of liquid in a V-L separator? Explain your answer.

16. What is the main principle of operation for a parallel plate L-L separator?

17. Why are steam ejectors needed for systems that operate below atmospheric pressure?

18. What is the advantage of partially condensing the intermediate stream in a multistage steam ejector?

19. Do steam ejectors typically operate in the subsonic or supersonic region?

20. Give two sources of noncondensable gas that may contribute to the vent stream leaving the reflux drum on a distillation tower that operates at vacuum conditions.

Problems to Solve

21. Determine the mass of a pressure vessel made of 304 stainless steel (SS), with standard (ASME) dished heads. The dimensions of the vessel are a diameter of 2.4 m and a length of 8.5 m (tangent-to-tangent). The design pressure and temperature of the vessel are 3.7 MPa and 450°C, respectively. You should assume a corrosion allowance of 6.35 mm and that nozzles and manways will contribute an additional 15% to the mass of the vessel. In addition, you should assume that all welds are double-welded butt joints with spot radiographed examination.

22. Repeat Problem 5.21 but using 316 SS as the material of construction. If the relative costs of 316 to 304 SS are 1:1.29, and both materials are suitable for the service, which material do you recommend?

23. If the vessel in Problems 5.21 and 5.22 were to handle 50% nitric acid, what material of construction would you recommend, and why?

24. Estimate the mass of the vessel designed in Problem 5.25 (below), assuming a corrosion allowance of 6.35 mm and a material of construction of SA515-70 carbon steel. Use 2:1 elliptical heads but do not consider any additional mass for nozzles and manways. You should assume that double-welded butt joints that are fully radiographed are used throughout the construction of the vessel.

25. A V-L separator drum is to be designed downstream of a reactor. The stream leaving a partial condenser that is the feed to the knockout drum has the following properties:

    Vapor mass fraction, $x = 0.45$

    Mass flowrate of feed stream = 70.5 kg/s

    M of vapor = 59 g/mol (assume ideal gas behavior)

    M of liquid = 80 g/mol, density of liquid = 1080 kg/m^3

    Gas viscosity = 9.8×10^{-6} kg/m/s

    Temperature of stream = 93°C, pressure of stream = 1075 kPa

    Answer the following questions:

    a. What should the orientation of the vessel be and why?

    b. Determine the diameter of the vessel needed to remove 100 μm diameter drops.

    c. Determine the minimum pipe diameter of schedule 40 pipe required such that a vane-type inlet device can be used. Use the data from Table 5.8 and assume that the flow is homogeneous; that is, the velocity of gas and liquid are the same—no slip between phases.

    d. Determine the appropriate holdup time and volume of liquid holdup.

    e. Determine all the key vessel dimensions such as NOL, HHAL, and LLAL. Sketch a vessel diagram showing the location of the inlets and outlets, and note all key dimensions on the diagram.

26. Estimate the diameter of a vertical V-L separator containing a 6-in (150 mm) thick stainless steel wire mesh mist eliminator. The properties of the gas and liquid are:

$$\rho_l = 940 \text{ kg/m}^3, \rho_g = 7.36 \text{ kg/m}^3,$$

$$\dot{v}_g = 11.45 \text{ m}^3/\text{s}, \dot{v}_l = 17 \text{ m}^3/\text{h}, \mu_l = 0.0019 \text{ kg/m/s}.$$

    For the stainless steel mist eliminator, the following properties should be used: density = 192 kg/m^3; wire diameter = 280 μm; specific surface area, $a = 350$ m^2/m^3; and voidage, $\varepsilon = 0.975$.

    Hint: You should base the design on 80% of flooding.

27. What is the equivalent value of k_{SB} for the mist eliminator in Problem 5.26 that would give the same diameter for the vessel?

28. Determine the pressure drop in the mist eliminator designed in Problem 5.26. Using this result, determine the maximum size of droplet for which the capture rate is 99% in this device at the design flowrate. You may assume that the absolute pressure for the inlet stream is 3 bar and the gas viscosity, $\mu_g = 16 \times 10^{-6}$ kg/m/s.

29. Consider the horizontal V-L drum designed in Examples 5.4, 5.5, and 5.6. Based on the conditions at the operating design point (using NOL), determine the maximum size of gas bubble that would not rise to the liquid surface and be released to the vapor phase. Use a liquid viscosity of 700×10^{-6} kg/m/s.

Hint: You should assume that a gas bubble of size D located at the bottom and left-hand side of the drum will move to the right-hand side at the velocity of the liquid. Simultaneously, this bubble will rise through the liquid at its terminal velocity. If the bubble reaches the surface of the liquid, it will be released; however, if the bubble does not reach the surface, then conservatively, it may be considered to be entrained in the liquid leaving from the bottom of the vessel. This concept illustrates the degassing behavior of horizontal drums.

30. Repeat Problem 5.29 for the case when the liquid residence time is 5 min instead of 10 min for Examples 5.4, 5.5, and 5.6. You may assume that the heights of liquid in the drum are the same.

31. A liquid-liquid separator, with a parallel plate settler installed horizontally, is 1.4 m in length and has a separation distance of 30 mm between the plates. The densities of the two liquids to be separated are $\rho_{l,1} = 840$ kg/m^3 (dispersed phase) and $\rho_{l,2} = 1020$ kg/m^3 (continuous phase). The horizontal velocity of the mixture of two liquids is 1 m/min. Determine the maximum size of droplet of the dispersed phase that will not be collected or coalesced while moving through the settler. You should assume that the droplets of dispersed phase move vertically at their terminal velocity.

32. For Example 5.11, Part (b), rework the problem for the case when the stream leaving the first stage of the ejector is condensed using cooling water (in at 30°C and out at 40°C) with a 5°C approach (leaves the condenser at 45°C). What is the reduction in steam flow to the second stage of the condenser?

    Hint: You will have to set this problem up and solve it using a process simulator; you should use ideal gas and latent heat enthalpy options.

33. For Problem 5.32, determine whether it is more economical to use the two-stage ejector with intermediate condensation than the noncondensing option given in Example 5.11, Part (b). Use the following information:

    Overall heat transfer coefficient for the partial condenser = 1500 W/m^{2}K

    Cost of steam = $0.03/kg

    Cost of cooling water = $15 × 10^{-6}/kg (for $\Delta T = 10$°C)

    Cost of heat exchanger = $30,000A$^{0.43}$ where A = heat exchanger surface area in (m)

    Cost of L-V separator = 20% of cost of heat exchanger

    To account correctly for the time value of money, you should divide the costs of the exchanger and vessel by a factor of 5 to convert the one-time cost ($) to an equivalent yearly cost ($/y).

34. The overhead reflux drum for a vacuum column in an oil refinery operates at 25 mmHg. The overhead temperature is 120°F, and the composition and flowrate of the vapor in the vent gas (not including any air leaks) is

    H_2 = 1.2 lb/h

    CH_4 = 20 lb/h

    C_2H_6 = 45 lb/h

    C_3H_8 = 25 lb/h

    The vacuum tower and associated equipment and pipes have an approximate volume of 21,000 ft^3.

Answer the following:

a. Estimate the air leak into this ("tight") system.

b. Determine how many stages would be required for a steam ejector to remove the vent gas plus the air leak into the system and deliver it to a vent gas–handling system operating at 1.2 bar.

c. Estimate the amount of low-pressure steam (70 psig saturated) required by each stage of ejector assuming there is no intermediate condensation.

d. If cooling water is available at 86°F and must be returned to the cooling tower at 104°F, determine if a partial condenser could be employed and where it should be placed. Also determine if more than one condenser could be employed.

e. Determine the amount of steam required if a single partial condenser is used. You should assume that the condenser is placed between the last two stages of the ejector and a temperature approach of 10°F is used.

35. It is desired to maintain a vacuum of 100 mmHg absolute pressure in a drum by removing accumulated vapors (average M = 40 g/mol, flowrate 200 kg/h, temperature = 50°C) and discharging them to an incinerator operating at 115 kPa. Determine the steam flowrate required for a single-stage ejector system for the following steam pressures:

a. 20 bar saturated

b. 10 bar saturated

c. 5 bar saturated

36. A single-stage steam ejector is designed to remove 50 kg/h of air from a reflux drum operating at 25 kPa and 50°C and discharge it to the atmosphere through a vent using a steam ejector with saturated steam at a pressure of 1 MPa. After the system has been designed, it is found that the pressure in the reflux drum has risen to 30 kPa (still at 50°C). Estimate the additional air flowrate (due to leakage) that is responsible for the increase in reflux drum pressure.

Index